Food Futures

FOOD FUTURES

Growing a Sustainable Food System for Newfoundland and Labrador

Edited by
Catherine Keske

ISER Books

Library and Archives Canada Cataloguing in Publication

Food futures : growing a sustainable food system for Newfoundland and Labrador / edited by Catherine Keske.

(Social and economic papers ; no. 35)
Includes bibliographical references.
Issued in print and electronic formats.
ISBN 978-1-894725-45-3 (softcover).--ISBN 978-1-894725-46-0 (PDF)

1. Food supply--Newfoundland and Labrador. 2. Food security--Newfoundland and Labrador. 3. Sustainable agriculture--Newfoundland and Labrador. I. Keske, Catherine M., editor
II. Series: Social and economic papers ; no. 35

S451.5.N4F66 2018 363.809718 C2018-901629-9
C2018-901630-2

Cover design: Kimberley Devlin
Design and typesetting: Kimberley Devlin
Copy editing: Richard Tallman

Published by ISER Books
Faculty of Humanities and Social Sciences Publications
Institute of Social and Economic Research
Memorial University of Newfoundland
297 Mount Scio Road
St. John's, NL, Canada, A1C 5S7
www.hss.mun.ca/iserbooks/

Contents

List of Figures

List of Tables

Jeremy Charles, chef and co-owner, Raymonds Restaurant and The Merchant Tavern, St. John's, Newfoundland. (Photo by Per-Anders Pettersen)

Foreword

Jeremy Charles

I grew up running around the gardens in Old Perlican. Dodging the hens and roosters, watching the fishermen trench the rows and toss in capelin and cod guts for fertilizer. Playing hide-and-seek in root cellars, amongst the pickles, preserves, and corned fish. It was a way of life. If you wanted to eat in Newfoundland, you had to grow and gather your food. We have survived for generations this way. Everything was organic before it became the loose term used to describe almost anything nowadays. Then came the modern supermarkets and the big-box stores. We lost our way for a bit, at least my generation had. Everything was more convenient, accessible, and "fresh." After living with food that comes from all corners of the globe, food that is modified, treated, and altered, we now find ourselves going back to our roots, to our gardens.

Over the last few years Newfoundland has seen its agricultural heritage start to re-emerge. I get great pleasure in opening my kitchen door at Raymonds and greeting one of the many purveyors. It could be beautiful lamb from the southern shore, raised by the Morrys or the Mooneys. Vegetables from Mike Rabinovitz or Mary Lester, here on the Avalon, or maybe fresh seafood from Gerry Hussey in Bonavista. Every day it seems someone else has started a new venture in agriculture, inspired by previous generations, or maybe they just took a walk through the Pattersons' garden in Upper Amherst Cove and fell in love like I do every time I visit.

This book shows us how great and unique our history of agriculture is in Newfoundland. It's inspiring our cooks, our farmers, and the next generation. Sustainable eating and living is possible in this province, and we have the history to prove it.

Strawberries picked at Lester's Farm U-Pick, St. John's, Newfoundland.
(Photo by Kimberley Devlin)

Acknowledgements

Many of us share the vision that no one should go hungry or experience malnourishment, and believe that academic research provides an important role in achieving these long-term goals. This book was made possible through the hard work and dedication of many fine professionals who have spent their careers advancing food and agricultural research. Their work speaks for itself.

In particular, I would like to thank the following people for their support and contributions:

Lynne Phillips provided invaluable vision, advice, patience, and wisdom throughout the editorial process. Without her presence, this book would not have been possible.

Martha Traverso-Yepez, Paul Foley, and Lynne Phillips served as assistant editors and conducted several chapter reviews, often at short notice. Their guidance improved the cohesiveness of the material and ensured that high scholarly integrity was maintained.

Norman Goodyear provided strategic suggestions, recruited chapter contributions, and conducted several chapter reviews.

Nancy Pedri, interim academic editor, Fiona Polack, academic editor, Alison Carr, managing editor, Kimberley Devlin, acting managing editor, and Randy Drover, Publications Assistant of ISER Books, were a pleasure to work with throughout the publishing process and provided excellent professional guidance.

Lisa Rankin co-ordinated external reviews. We thank Dr. Rankin and the anonymous reviewers for their valuable feedback and suggestions.

Julie Sircom, Greg Goff, Dana Hoag, Sharon Roseman, and Kristie Jameson provided professional, constructive reviews of several chapters that improved the quality of the material.

Thanks to Jeremy Charles of Raymonds for expressing his enduring support for serving locally sourced food, and his heartfelt pride of Newfoundland food systems, past, present, and future.

Vanessa Kavanagh and Sabrina Ellsworth of the Agrifood Development Branch of the Newfoundland and Labrador government contributed material to the Epilogue, as well as insight into several provincial government research and policy initiatives.

Food First NL, with its engaging, practical work in the province, provided rich examples for several chapters and served as a source of inspiration for many authors.

Myron King of the Environmental Policy Institute at Memorial University, Grenfell Campus, created all of the maps that appear in this book, except as otherwise noted.

Collin Campbell and Bob Edmondson delivered quality copyediting to prepare the manuscript for review. Richard Tallman provided final copyediting. Thanks for the polish that made the material shine.

Henry, Euan, and William Devereaux blueberry picking in Blackhead, Newfoundland.
(Photos by Kimberley Devlin)

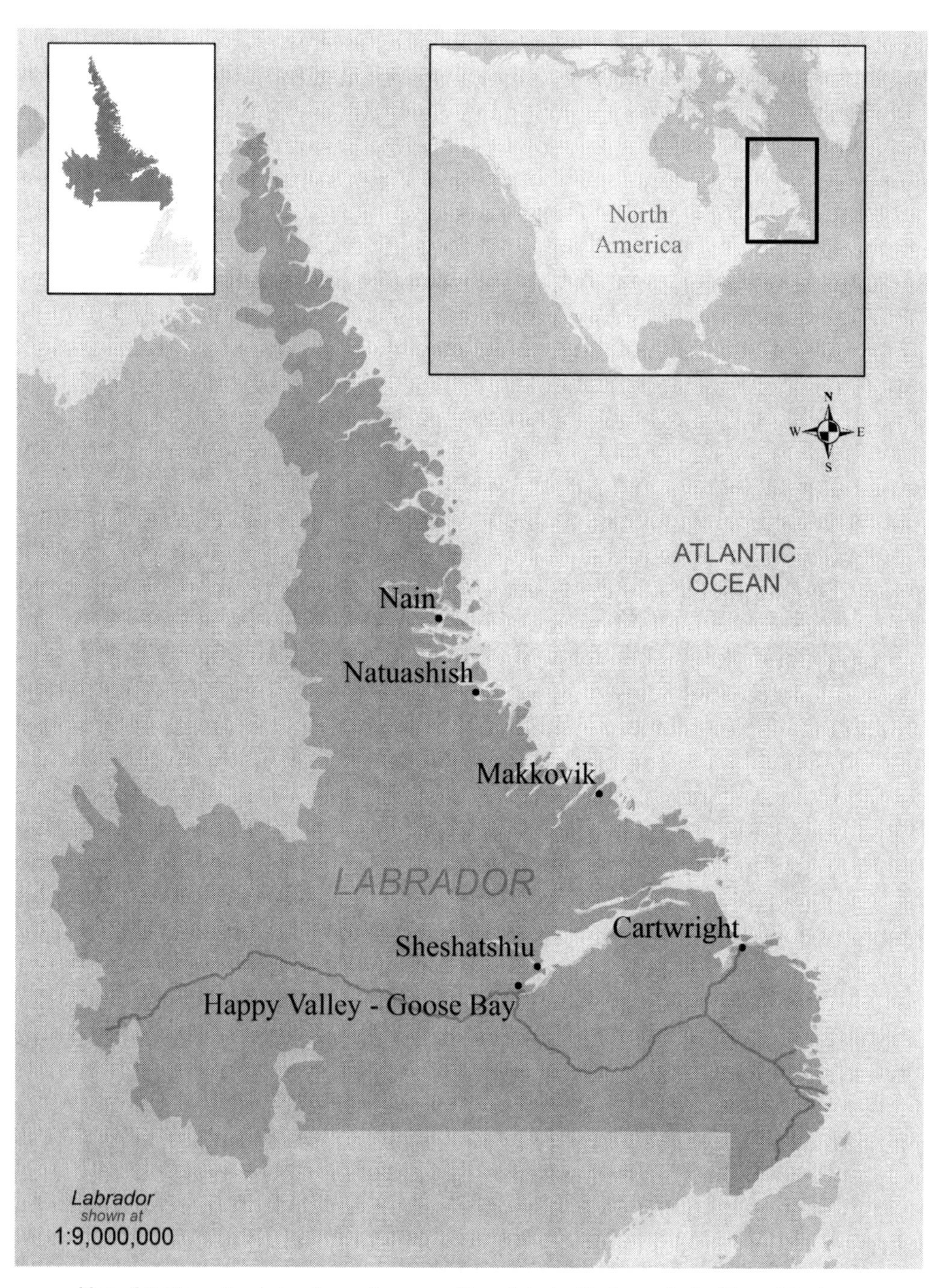

Map of study region: Labrador and communities of study. (Cartography by Myron King)

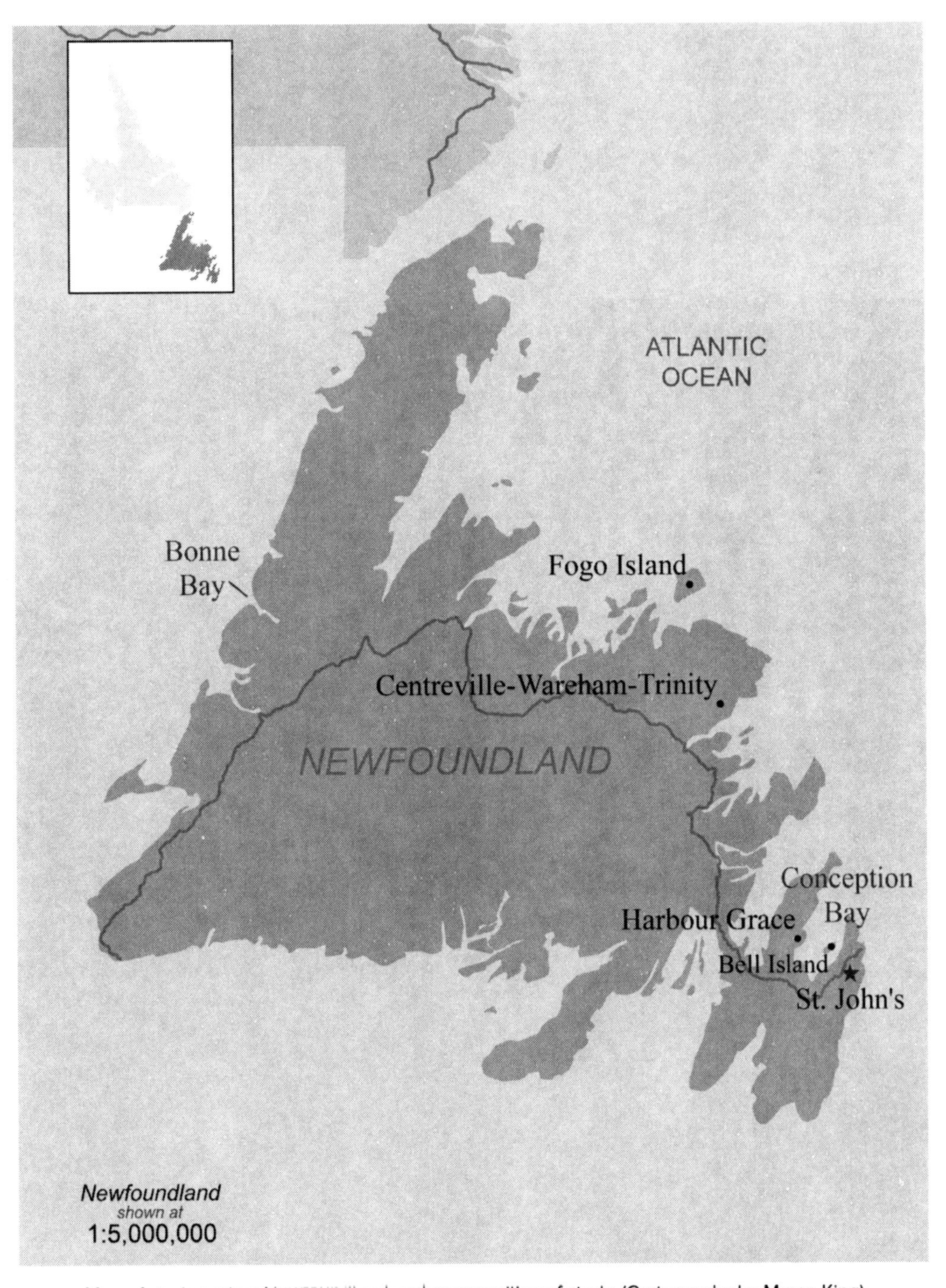

Map of study region: Newfoundland and communities of study. (Cartography by Myron King)

Beets © Dave Howells

Editor's Introduction

The Promise and Precariousness of Newfoundland and Labrador's Food System

Catherine Keske

Picture wintertime in Newfoundland.

Ships carrying imported food and staples are unable to reach the ports. They are barricaded by ice. There is general uneasiness among the public about disruptions to the food supply. Complicated relationships between trade partners and the government compound the situation, rendering the arrival of ships' cargo unpredictable.

Those with means and foresight have already stocked up in anticipation of the long winter, although the end of last year's harvest is in sight. The remaining potatoes and other fresh produce have a withered appearance. Grocery shelves are bare. Households have already incurred considerable debt to make ends meet. With careful planning and a bit of luck, consumers will be able to pay down their debt before it's time to store up for the next winter and repeat the cycle.

Is this vignette set in March 1818 or March 2018? Although specific details may vary, the storyline transcends time. Themes like harsh climate, trade reliance, and precarious resource availability are of the ages in Newfoundland and Labrador. So, too, is the resilience of its people.

Over the centuries, people living in the area that is now the Canadian province of Newfoundland and Labrador have demonstrated remarkable levels of resourcefulness to overcome climatic and governance challenges in order to reap the bounty of both sea and land. Indigenous persons understood the cycles of the seasons and the sustenance of traditional "country foods"[1] gleaned from local stocks, such as mammals, birds, fish, and berries (Wein, Sabry, and Evers, 1991; Van Oostdam et al., 1999; Hanrahan, 2008). Fish (including salmon, Arctic char, and northern cod) were an important source of food and culture among Indigenous peoples (Hanrahan, 2008), with cod being of pre-eminent importance among European settlers, particularly for trade. Sixteenth-century migratory fishing and transient European settlement gave way to unregulated permanent English settlement that increased throughout the eighteenth century (Cadigan, 1992; Cadigan, 2009). Over the course of centuries, Indigenous persons became familiar with foods introduced by settlers such as root vegetables and "market foods" like sugar and coffee (Hanrahan, 2008). Permanent European settlement was sustained, in part, by the exchange of cod for market goods and by subsistence agriculture. As documented by previous authors (Omohundro, 1994; Murray, 2002; Murton, Bavington, and Dokis, 2016) and explored throughout this book, subsistence gardening and local agriculture were prominent from the settlement period through the first half of the twentieth century, at which time a demand for convenience and pre-packaged foods emerged and continued to grow. Packaged and prepared foods reflected a lifestyle based on convenience and increased leisure time that eschewed poverty and manual labour (Pottle, 1979).

These earlier patterns of livelihood gave rise to distinctive food practices and a food culture that are being celebrated (and even rediscovered) by growing numbers of people. However, like most scientific lines of inquiry, Newfoundland and Labrador food system research has been aligned with disciplinary trajectories rather than unfolding as interdisciplinary discourse around a complex issue. For example, several folklore researchers have contextualized the origin of locally accessible Newfoundland and Labrador products like blueberries, root vegetables, and fish; unique dishes like Jiggs dinner[2] and flipper pie; and imported staples like molasses and bologna (Everett, 2012; Tye, 2010). Social scientists such as Adrian Tanner (2014 [1979]) and Sean Cadigan (1992, 2009) conducted their respective research on Indigenous and settler food systems by employing anthropological and historical research methods. As we elaborate upon later, fisheries research has been undertaken by esteemed natural and social scientists including Ian Fleming and Barbara Neis. Medical, health, and nutrition research dates back to the nineteenth century with the Sir Wilfred Grenfell mission to rural and impoverished regions of Newfoundland and Labrador (Paddon, 2002; Phillips, Chapter 1), and the work continues through several Memorial University programs, including the Faculty of Medicine (see Traverso-Yepez, Sarkar, Gadag, and Hunter, Chapter 5) and the Western Regional School of Nursing. Unfortunately, as is the case in academia, many meritorious studies remain isolated in the literature associated with their specific disciplines.

It is with this in mind that we present *Food Futures: Growing a Sustainable Food System for Newfoundland and Labrador*. In contrast to prior food systems research, we organize our multidisciplinary efforts around the working perspective that all Newfoundlanders and Labradorians experience "food insecurity" — as compared to "food security" (defined below) — to a certain extent due to the province's geographic isolation and the harsh climatic conditions that make food production difficult. As discussed throughout the book, many factors challenge household access to food, agricultural production, and other related food sovereignty[3] issues. Hence, organizing our research with mindfulness towards upholding food security and food sovereignty goals provides the potential to improve these vulnerabilities and to foster a sustainable food system. We believe that the present Newfoundland and Labrador food system is at a crossroads where challenges like scarce resources, hurried modern lifestyles, processed foods, and twenty-first-century palates could exacerbate food insecurity. Or, this could be a time of true awakening, a renaissance if you will, and an opportunity to improve food security and food accessibility for all while celebrating Newfoundland and Labrador's unique food heritage.

We believe that this collection of timely research studies about Newfoundland and Labrador food contributes to the wider Canadian and international food studies literature that has expanded considerably during the past decade. Notable books and edited volumes, including *Edible Histories, Cultural Politics: Towards a Canadian Food History* (Iacovetta, Korinek, and Epp, 2012), *Food Sovereignty in Canada: Creating Just and Sustainable Food Systems* (Desmarais, Wiebe, and Wittman, 2011), and *Critical Perspectives in Food Studies* (Koç, Sumner, and Winson, 2012), have explored multicultural and historical influences upon which the Canadian food culture was built, including colonialism, climate, and centuries of Indigenous and settler culture. Respected peer-reviewed journals such as *Cuizine: The Journal of Canadian Food Cultures* and *Canadian Food Studies* have emerged within the past decade to advance the discourse of food as a complex "system" that consists of important, interrelated social dimensions and nuances (Desjardins, 2016).

Definitions vary about what comprises a sustainable food system, but there appears to be consensus that food system research should consider interdisciplinary approaches, social and biophysical aspects, and multiple dimensions over time and space. In an influential reflection about the importance of maintaining community food systems, Kneen (1995: 11) defines a "food system" as "a highly integrated system that includes everything from farm input suppliers to retail outlets, from farmers to consumers." A seminal publication about food and the nutrition system by Sobal, Kettel Khan, and Bisogni (1998) also addresses the importance of integrating social and biophysical research into the food systems discourse. These authors note that food research is commonly addressed through a disciplinary lens. Alternatively, the authors posit that an interdisciplinary approach should be taken to integrate both biophysical and social dimensions. We concur. Hence, we have compiled contributions from both social and biophysical scientists in the province of Newfoundland and Labrador. We recognize the value that different disciplinary perspectives offer for tackling the complexities of food security and food sovereignty. We embrace the value of bringing together diverse experiences, training, and knowledge into an anthology on Newfoundland and Labrador food. There is strength in collective focus.

Food security goals essentially address poverty alleviation through increased food production. One of the most commonly used definitions of food security is "a situation that exists when all people, at all times, have physical, social and economic access to sufficient, safe and nutritious food that meets their dietary needs and food preferences for an active and healthy life" (World Food Summit, 1996, quoted in Shaw, 2007). Several factors contribute to Newfoundland and Labrador's chronic food security vulnerabilities and subsequent side effects.

Newfoundland and Labrador consists of a relatively large and sparsely populated island (Newfoundland) and a northern rural mainland (Labrador). With a 2016 census population of 178,427 people (Statistics Canada, 2017), St. John's, the province's population centre, is located on North America's most eastern geographical boundary. It is one of the oldest colonial settlements in North America. St. John's has served as an international port for generations, although its proximal location from otherwise relatively rural communities presents food accessibility and distribution challenges. Approximately one-third of the province's 519,716 inhabitants reside within the greater St. John's area (Statistics Canada, 2017). As recently as the latter part of the twentieth century, many rural "outport" communities were still only accessible by boat or air (and this still holds true for some communities, particularly in Labrador). However, the Trans-Canada Highway across the island of Newfoundland and other infrastructure still yield inefficient transportation and distribution. Food is often transported from Sydney, Nova Scotia, by ferry to Port aux Basques at the southwestern corner of the island. Semi trailers packed with food typically bring their hauls to St. John's, approximately 10 to 12 hours away, before it is distributed elsewhere in the province. As discussed further by others in this book (see Lowitt and Neis, Chapter 8; Vodden, Keske, and Islam, Chapter 4; Schiff and Bernard, Chapter 7), these practices often leave rural communities in precarious positions, bound by a dysfunctional cycle when it comes to accessibility to nutritional food.

Although the province has a rich cultural history of food subsistence activities, geo-climatic challenges make large-scale, industrialized food production challenging (Keske, Dare, Hancock, and King, 2016). Farms have gone out of business over time due to difficulties producing at cost-effective economies of scale, in part due to high import costs for farm supplies/production inputs like livestock feed, fertilizer, and labour. These costs are compounded by the inefficient transportation and distribution structure (Murray, 2002; Keske, 2014, 2015, 2016; Statistics Canada, 2010). There are success stories about newly emerging farms (Bird, 2015) and multi-generational establishments like Lester Farms (Murray, 2002; CBC News, 2014), but over the past decade the province has lost farms at a higher rate than the rest of Canada, as noted in Census of Agricultural reports (Statistics Canada, 2009, 2011, 2017). Many agricultural production operations, such as the dairy and chicken industries, rely on government supports and quotas to ensure an available supply of food and sufficient profits for farms to remain in business. In general, these economic conditions make cost competitiveness with imported foods (which are typically produced at lower costs) difficult to achieve. As a result of limited regional food production, the province is estimated to import approximately 90 per cent of fresh food requirements, and the supply depends heavily on marine transportation (Food First NL, 2015).

Limited access to healthy food has played a role in the increasing number of health disparities experienced by the people of the province. As has been noted elsewhere, the people of Newfoundland and Labrador have among the highest rates of heart disease and obesity in Canada. In 2014, over 50 per cent of Newfoundland and Labrador inhabitants over the age of 18 were considered either overweight or obese (Statistics Canada, 2016). As of 2014, only 25.7 per cent of the NL population ate five or more servings of fruits and vegetables per day, compared to the national average of 39.5 per cent (Statistics Canada, 2015).

Since limited access to food and dependency on imports are not novel circumstances for the province's inhabitants, as outlined by several authors in this book (e.g., Roseman and Royal, Chapter 2; Lowitt and Neis, Chapter 8; Tanner, Chapter 6; King, Chapter 10) many people actively participate in self-sufficiency activities such as fishing, hunting, and subsistence gardening. However, for the most part, the majority of Newfoundland and Labrador's culturally embedded foods, such as carrots and potatoes, are grown elsewhere in Canada or in other countries. Compounding matters, locally caught seafood that initially served as a driver for European colonialism is typically exported, making it difficult for locals to consume fish as food. As discussed by many authors in this book, since the times of European colonialism Newfoundland and Labrador "has always been built on the export economy" emanating from the fisheries (Song and Cheunpagdee, 2015: 445; Foley and Mather, Chapter 9). There have been long-standing challenges within the province for local food system control, although these issues have arguably been exacerbated since Confederation with Canada in 1949, when Newfoundland and Labrador relinquished its status as a stand-alone entity within the British Commonwealth. This warrants a deeper discussion about food sovereignty, a concept related to food security that encompasses a distinct research thrust and a rapidly developing literature.

Food sovereignty is a dynamic political framework that espouses local-level control of the food system (La Vía Campesina, 2002). As described in greater detail in Desmarais (2012) and McMichael (2014), food sovereignty also is understood as a political movement to empower local peasant farmers to take control of their food sources and for communities to learn from this knowledge. The definition of "food sovereignty" used throughout several chapters in this book was instituted by La Vía Campesina movement, which arose globally in 1993 as a voice for peasants and small-scale farmers at a time when agricultural policies and agribusiness were becoming globalized and disparities grew within the food system. The movement has defined food sovereignty as "the right of peoples to healthy and culturally appropriate food produced through ecological and sustainable methods, and their right to define their own food and agriculture systems" (La Vía Campesina, 2002).[4] La Vía Campesina food sovereignty movement has gained attention in Quebec and in other parts of Canada, as further discussed by Desmarais and Whittman (2014), and this framework holds promise for Newfoundland and Labrador.

The food sovereignty literature evolved from negative ecological, social, and economic consequences of contemporary industrial, globalized food systems, such as loss of biodiversity, health disparities, and food dumping (Binimelis et al., 2014; Windfuhr and Jonsén, 2005; Lavallée-Picard, 2016). In response, activists and scholars have called for a shift away from traditional food security research focused on large-scale agricultural production to a more interdisciplinary and systematic understanding of issues and options for sustainable and just food systems (Wald and Hill, 2016; Lavallée-Picard, 2016). Food sovereignty highlights the importance of improving resource access rights, facilitating equitable trade policies, and transitioning towards sustainable production practices and "right to food" approaches (Windfuhr and Jonsén, 2005: 4). This definition of food sovereignty was further advanced to include five action axes of food sovereignty (Binimelis et al., 2014: 326):

1. *Access to resources:* Food sovereignty aims to support individuals and community processes to ensure equal access to resources (including production resources).
2. *Production model:* Food sovereignty strives to facilitate local production based on a diversified, agro-ecological production model that values local/traditional knowledge.
3. *Transformation and commercialization:* Food sovereignty preserves rights of farmers and workers, fishers, pastoralists, and Indigenous people to be competitive in the domestic market while being protected from global market disruptions.

4. *Food consumption and right to food:* Food sovereignty necessitates the people's rights to safe, nutritious, and culturally appropriate food.
5. *Agricultural policies and civil society organizations:* Food sovereignty asserts public policies relating to the food system should be directly deliberated between producers and consumers.

However, policy implementation problems can arise within wealthy, developed nations such as Canada with elaborate social infrastructures and conflicting incentives that may perpetuate a cycle of less nutritious and industrially produced food. As noted in the prevailing literature, income and wealth can provide demarcation between those who are able to enjoy "good food" and those who struggle to gain access to just "food" (Hochedez and LeGall, 2016). When harsh climate makes agricultural production expensive and unpredictable, as is the case in Newfoundland and Labrador, local food production can be infeasible from a practical perspective (Keske et al., 2016). Recent improvements in food security among the very poor have been attributed to such interventions as increased government income support and affordable housing (Tarasuk, Mitchell, and Dachner, 2016; Loopstra, Dachner, and Tarasuk, 2015), rather than to a focus on increased local agricultural production within the province. In a comprehensive discussion paper, several authors who have contributed to *Food Futures* have reflected upon how societal practices could be reframed in order to cultivate a sustainable food system throughout the province across all stages of life and situations.[4] Well-intended policies and mislaid incentives may inadvertently undercut the economies of scale and scope, as well as the public support necessary to build the momentum needed to facilitate food sovereignty.

Nonetheless, much like Desmarais and Whittman (2014), we believe the food sovereignty movement holds promise for Canada, including Newfoundland and Labrador. These authors point out that the remarkable diversity of national, regional, and cultural identities within larger regions and countries like Canada delivers both challenges and opportunities. On the one hand, "Distinct national, provincial, regional and cultural concerns in terms of community identity and subjectivity, and relationships to political and institutional authority, mean food sovereignty doesn't map tidily onto a national, or even provincial, scale" (Desmarais and Wittman, 2014: 1167).

Yet, Canada exhibits a "unity in diversity" in its food sovereignty movements towards transformative food system change that "has resulted in a reshaping of the political spaces . . . related to how and what food is produced, accessed and consumed" (Desmarais and Wittman, 2014: 1167). If core food sovereignty and "unity in diversity" principles are maintained, there is considerable opportunity to improve connectivity of several successful initiatives across the nation. Fostering a systems-based perspective that allows for local control and multiple targets can be a step towards developing and maintaining sustainable food systems in a more efficient and socially just manner.

The several case studies presented in this book (Roseman and Royal, Chapter 2; Doyle and Traverso-Yepez, Chapter 3; Vodden, Keske, and Islam, Chapter 4; Lowitt and Neis, Chapter 8) reflect success stories of communities implementing food sovereignty strategies that uphold values, identities, and customs of individual communities that developed over years, but that also continue in a modern context that includes provincial and national governance. Other dynamic food sovereignty aspects being addressed within the province include the facilitation of local access to fish as food and the co-integration of fish into other economic sectors, such as recreation and tourism. In October 2015, the provincial government amended the Fish Inspection Act and the Food Premises Act to permit the direct sale of fish from harvesters to consumers and restaurants. This amendment followed a fisheries-tourism pilot project in Bonne Bay that investigated market demand for local fish and fisheries-tourism synergies (Lowitt, 2009, 2014). Previous legislative requirements inhibited local access to fresh seafood because fish harvesters were required to sell their catch to licensed processors. Furthermore, as of 2016, the recreational cod-fishing season was extended by 14 days to "reflect [the] government's commitment to honouring the peoples' deep historical and cultural attachment to the cod fishery in this province" (Wall, 2016). Changes to 2017 food fishery policy seem congruent with efforts to restore access to cultural fishing customs and to fresh-caught fish as food. The Department of Fisheries and Oceans (DFO) announced that it will maintain both the 46-day season expanded in 2016 and the per-person allowable catch, in addition to removing barriers to tour boat operators and updating the cumbersome licensing system. Bartlett (2017) reports that the policies of the federal Minister of Fisheries and Oceans, Dominic LeBlanc, were based on community feedback: "I have heard you loud and clear,"

LeBlanc stated. "There is very little public appetite for the use of tags, and we will therefore not be implementing that regime."

A closer look at the recovering cod fishery reveals the emerging opportunity for expanding community food sovereignty strategies, starting with the Fogo Island Co-operative that embraced these principles back in 1967 (Low, 1967). Owned by inshore fishers and plant workers, it provides the quintessential and perhaps most renowned example of food sovereignty stemming from the residents' awareness of the need to maintain local control over resources. The Fogo Island Co-op was initially established as an alternative to federal and provincial initiatives to relocate residents to the mainland. Rather than relocate to mainland Newfoundland like most communities, local residents instead "rebuilt" the local fishing and harvesting processes into a community-owned enterprise, in a manner that has been dubbed the "Fogo Process" (Low, 1967).

This co-operative was established in defiance of national and provincial policies to relocate residents with the aim of empowering them to have equitable access to production resources throughout the supply chain, including fishing, processing, and distribution. Instead of being uprooted, the empowered residents were able to remain in the region by retaining control over their economic development and food production. To quote Foley and Mather (2016: 968), the co-operative on Fogo Island formed:

> around a community revitalization process that allowed fish harvesters and residents to transcend deeply rooted intercommunity and interreligious tensions on the island, which includes about 11 distinct communities. The Co-op has since provided a mechanism through which to unite the island's residents and communities, and facilitated the forging of an island-wide identity.

The Fogo Island Co-op has garnered worldwide name recognition and fostered the regional momentum towards resource amenity tourism, including attracting high-end tourists. Synergy behind the Fogo Island mystique has inarguably been enhanced by social enterprises such as Shorefast Foundation, which sponsors artist residencies, micro-lending, and historic home preservation (Lionais, 2015). The Fogo Island Inn presents an "off the beaten path" retreat for luxury tourists whose interests also often overlap with interests in "socially responsible" and "locally sourced" food, like the principles instituted by the Co-op (Brinklow, 2015). In addition to providing a local supply of fish/seafood, the Fogo Island Co-op offers direct buying programs across the world and has promulgated brand recognition for its environmental and socially sustainable practices.

The laudable Fogo Island Co-op and Shorefast Foundation serve as prime examples of food sovereignty in action that could be replicated within the province and in communities throughout Canada. However, we would be remiss if we didn't point out some of the potential unintended side effects of the growing demand for local, small-scale, or craft food production among the middle class, often self-described as "foodies" (Johnston and Baumann, 2009; Inglis, 2009). Although technological advances allow some savvy consumers to connect with their seafood and trace the origins of their dinner to the fisher who caught it that same morning, as noted in Chapter 5 by Traverso-Yepez et al., many people are struggling to secure their daily meals. On a positive note, the increasing awareness of local production and traditional food preparation practices has brought more recognition to the value of sustainable local food systems, including the benefits of attaining this goal in Newfoundland and Labrador. The local food movement also provides a forum for discourse that encourages consumers to reconnect with their food sources, and to re-engage in their own home food production with ingredients that may be produced, at least ostensibly, in the province. Foley and Mather, in Chapter 9, use a food regime theory framework to explore provincial examples of the Newfoundland and Labrador seafood industry. Even large grocery and retail chains actively seek and promote their acquisition of locally sourced produce, recognizing the potential for lower costs and higher-quality products, as well as the growing consumer demand for fresh products (Baumann, Engman, and Johnston, 2015; Johnston and Bauman, 2015).

At the same time, the increased demand for locally grown food also heightens the potential for consumers to unknowingly disrupt food sovereignty initiatives and to create unintended consequences for marginalized populations. Specifically, consumers may unwittingly increase their carbon and water footprints by consuming products they believe are being produced in an environmentally or socially sustainable manner. Examples include production encumbrances that ensure compliance with the Canadian Food

Inspection Agency's organic foods criteria, increased transportation, burdens on infrastructure, and increased packaging waste in the preparation of products. As further discussed by Butler, Dabrowska, Neis, and Vincent (2017), some health and safety issues associated with agricultural labour, particularly among international workers, are hidden and difficult to regulate. Other food qualities deemed as having environmental or socially desirable attributes, like "non-GMO" or organic foods, are still being defined through science and policy, and misinformation abounds. Furthermore, strong interest in food products and/or producer cartels may drive the prices for certain products much higher, thus making some nutritious food unattainable for low-income consumers, as recently shown in the almond industry (Daniels, 2016). The availability of natural and antibiotic-free meat at fast-food restaurants has, ironically, been noted as a contributing factor in rising sales for these restaurants where menus offer notoriously unhealthy options (Charlebois et al., 2016).

Irrespective of these potential downsides, we advocate for advancing local food production within Newfoundland and Labrador, in concert with improved local control and access to healthy, culturally appropriate food. By way of example, several authors in this collection present research results on Newfoundland food and agricultural products that include aquaculture, bees, and berries. Couturier and Rideout, in Chapter 11, discuss the evolution of the aquaculture program at Memorial University's Marine Institute from the 1960s to the present. In doing so, they chronicle the ebbs and flows of aquaculture production trends, including ecosystem considerations for farming high-value species like shrimp and salmon. Aquaculture clearly plays an important role for worldwide food security. The authors note that farmed seafood presently accounts for more than 50 per cent of the aquatic protein consumed by humans and that, in 2013, according to the Food and Agriculture Organization of the United Nations, the total global volume of aquatic farmed food was estimated at 97 million tonnes, with a value of US$157 billion. Couturier and Rideout report that this is the second largest economic sector in the province.

The "natural advantage" of Newfoundland's isolation from the mainland is highlighted in Chapter 12 by Walke and Wu, who present results from their socio-ecological study of the province's emerging apiculture industry, which is swarming with opportunity from one of the world's only disease-free bee populations. Geographical isolation and the limited use of pesticides (typically associated with commercial agriculture) likely contribute to this serendipitous honeybee environment. They use a mixed-methods approach to obtain both quantitative information (population size, distribution, and source stock) and qualitative information (opportunities, challenges, and influencing factors) to create an industry profile of apiculture on the island of Newfoundland. Results from the authors' questionnaire and interviews reveal a great deal of enthusiasm and innovation within the beekeeping community; however, this potential cannot be realized unless beekeepers and crop growers on the island can be guaranteed a safe and certain supply of honeybees from season to season.

Finally, in Chapter 13, Debnath and Keske offer a scientific account of the technological advances in the production of four native berry crops (blueberries, partridgeberries/lingonberries, cranberries, and bakeapples/cloudberries) with high nutritional properties and distinctive flavours. They note that berry crops are rich in vitamin C, cellulose, and pectin, and that these berries produce anthocyanins, which are believed to have important therapeutic values (including fighting inflammation). Advancements in production techniques and marketing opportunities make these crops ripe for the picking, so to speak, although whether berries can contribute to food security and food sovereignty goals will depend on local control over berry production and distribution. Blueberries are already Canada's top exported berry crop, and the robust, distinctive flavours of the NL berries offer room for market expansion. Debnath and Keske describe various aspects of berry crop improvement and their propagation using plant tissue culture methods at the St. John's Research and Development Centre of Agriculture and Agrifood Canada in Newfoundland and Labrador, and also provide several photographs illustrating progress being made in the cultivation of these crops and the beauty they provide to native landscapes.

In summary, though several common themes presented in this book resonate nationally and internationally, the material is focused on Newfoundland and Labrador and the province's unique and distinctive culture, history, land, and people. *Food Futures* presents a modern-day context for recurrent themes that have faced Newfoundlanders and Labradorians for generations with an optimism that wisdom has been gained from experience. These insights, along with recent innovations, serve as a reminder of the promise that sustainable food systems and agrifood production can be achieved within the province and in similar climates across the globe. We believe that multidisciplinary collaboration is imperative to bringing these issues to the forefront.

Although food security and the right to food are recognized as among the top humanitarian priorities of the twenty-first century, the inhabitants of Newfoundland and Labrador, as elsewhere in the world, have confronted these issues for generations. Self-reliance in the presence of obstacles gave rise to the Newfoundland and Labrador food culture. In other words, food systems research has recently gained momentum in academia, although navigating the elements to ensure that food is available to all is a timeless issue for the province.

Optimism for the advancement of a sustainable Newfoundland and Labrador food system is perhaps best depicted in the words of a young entrepreneur encountered through casual conversation, who recently started a catering company in western Newfoundland (Ellsworth, 2016): "I believe that, with humble ingredients and a little bit of hard work, you can cook with delicious, nutritious food. It's amazing what you can do with just a turnip."

ACKNOWLEDGEMENTS

I'd like to extend a warm thank you to Lynne Phillips and Sharon Roseman for their hours of input, reviews, and conversations that supported the Editor's Introduction. Paul Foley also provided high-quality editing and insight about the fisheries and Fogo Island process. Jennifer Dare's contributions as a research assistant are also noted and appreciated.

NOTES

1. In their seminal article, "Human health implications of environmental contaminants in Arctic Canada: A review," Van Oostdam et al. (1999: 6) use the terms "country foods" and "traditional foods" interchangeably to describe food sources among Indigenous persons in the Canadian Arctic that are harvested from local stocks. These terms are contrasted with "imported" or "market" foods that are usually purchased at a store. The authors also note that Indigenous cultural groups use specific terms for traditional foods related to their cultures; in the context of Labrador that might be "Inuit" or "Innu food"; elsewhere in Canada, it could be described, for example, as

"Dene food" or "Métis food." Other scholars, including those contributing to this book, reinforce the use of the term "country foods" for Indigenous foods secured by hunting/fishing and gathering. Of course, the distinctions are complicated by distinct and sustained periods of European contact. As Hanrahan (2008) indicates, some agricultural practices among Indigenous people in Atlantic Canada (such as the production of root vegetables) were influenced, even adopted, from contact with Europeans over centuries. Furthermore, increased awareness of the Mi'kmaw of eastern Canada — and the formation of the Qalipu Mi'kmaw First Nation band office in western Newfoundland — invites intriguing conversation about country food in the context of the heritage of the widely scattered population of Mi'kmaw persons, many of whom are only recently learning of their heritage and embracing it wholeheartedly (Robinson, 2014). Thus, for purposes of clarity, this book incorporates the perspective of Van Oostdam et al. that "country foods" reflect Indigenous food sources, while acknowledging that there have been periods of contact, disruption, and assimilation over the centuries.

It is worth mentioning that some "traditional" settler foods also reflect Indigenous foods, though of course there are nuances between this usage of the term and the traditional/country foods as articulated by Van Oostdam et al. Furthermore, "traditional" Newfoundland foods include imports like molasses, though these would be distinguished as "market foods" according to the definition presented by Van Oostdam et al., as well as moose, which is an introduced species (Broders, Mahoney, Montevecchi, and Davison, 1999). Further compounding the conversation is the growing awareness of the food sovereignty movement (often simultaneously described as "La Vía Campesina" [2002]), which provides increasing awareness of the importance of a balanced agroecology and environmental goals that are ostensibly enhanced through local food production, as opposed to large-scale corporate production. This has brought awareness to "peasant foods" that may be produced locally, illuminating the knowledge of locals (or "peasants") who may seemingly present sustained knowledge that can lead to a balanced, long-term outlook for food production when agroecological factors like soil health are considered (Rosset, Sosa, Roque Jaime, and Ávila Lozano, 2011). Using this threaded example, a peasant food could reflect a "traditional" Newfoundland food, such as root vegetables used in Jiggs dinners, which may or may not be viewed as a country or traditional food to Indigenous persons.

In summary, we acknowledge that complex situations and interactions have given rise to the food consumed over the centuries within the province of Newfoundland and Labrador, elsewhere in Atlantic Canada, and across the country. We respect these different perspectives, and we do our best to rely on the prevailing literature to incorporate consistent language for discussing the multidisciplinary research studies presented in this book.

2. Newfoundland Jiggs dinner consists of salt beef or roast turkey, potatoes, cabbage or turnip greens, carrots, and a bag of pudding, all typically boiled in a pot. While this is considered a Newfoundland dish, not surprisingly, there are several regional or household variations. Traditional Jiggs dinner contains salt beef. Peas porridge pudding made of yellow peas is also traditional. Blueberry or figgy (raisin) duff, essentially dough wrapped in cheesecloth and cooked in the pot, may be substituted. Cranberry sauce, mustard pickles, and pickled beets may be served on the side.
3. Many organizations have conflicting definitions of food sovereignty. Examples from Desmarais and Wittman (2014) include L'Union des producteurs agricoles (UPA), the Coalition Souveraintè Alimentaire, and the Union Paysanne. Some literature considers the "international peasant movement" and the practice of maintaining local, small-scale, traditional food systems that are accessible to marginalized persons to be synonymous with food sovereignty.
4. Several examples include support and accommodations for breastfeeding, improving work space to support home meal preparation and care of elderly parents, and promotional activities that encourage gardening and home food production. A more thorough discussion is presented in Keske and Phillips (2017).

REFERENCES

Bartlett, G. 2017. "DFO abandoning tags for Newfoundland and Labrador food fishery: DFO looking at licensing system for 2018 that won't include tags." CBC News, 19 May. At: http://www.cbc.ca/news/canada/newfoundland-labrador/food-fishery-tags-abandoned-newfoundland-labrador-1.4123257.

Baumann, S., A. Engman, and J. Johnston. 2015. "Political consumption, conventional politics, and high cultural capital." *International Journal of Consumer Studies* 39, 5: 413–21. doi:10.1111/ijcs.12223.

Binimelis, R., M. Rivera-Ferre, G. Tendero, M. Badal, M. Heras, G. Gamboa, and M. Ortega. 2014. "Adapting established instruments to build useful food sovereignty indicators." *Development Studies Research* 1, 1: 324–39. doi:10.1080/21665095.2014.973527.

Bird, L. 2015. "An island orchard: How one couple is working towards Newfoundland food security." CBC News, 7 Nov. At: http://www.cbc.ca/news/canada/newfoundland-labrador/crow-brook-orchard-increasing-food-security-1.3303953.

Brinkley, C. 2013. "Avenues into food planning: A review of scholarly food system research." *International Planning Studies* 18, 2: 243–66. doi:10.1080/13563475.2013.774150.

Broders, H. G., S. P. Mahoney, W. A. Montevecchi, and W. S. Davidson. 1999. "Population genetic structure and the effect of founder events on the genetic variability of moose, *Alces alces,* in Canada." *Molecular Ecology* 8, 8: 1309–15. doi:10.1046/j.1365-294X.1999.00695.x.

Butler, L., E. Dabrowska, B. Neis, and C. Vincent. 2017. "Farm safety: A prerequisite for sustainable food production in Newfoundland and Labrador." Unpublished manuscript in review.

Cadigan, S. T. 1992. "The staple model reconsidered: The case of agricultural policy in northeast Newfoundland, 1785–1855." *Acadiensis* 21, 2: 48–71.

———. 2009. *Newfoundland & Labrador: A History*. Toronto: University of Toronto Press.

Canadian Food Inspection Agency. 2016. "Organic products." At: http://www.inspection.gc.ca/food/organic-products/eng/1300139461200/1300140373901.

CBC News. 2014. "Lester's Farm opens new food truck utilizing fresh crops." 3 Aug. At: http://www/cbc.ca/news/canada/newfoundland-labrqador/lester-s-farm-opens-new-food-truck-utilizing-fresh-crops-1.2726207.

Charlebois, S., F. Tapon, M. von Massow, E. van Duren, P. Uys, E. Fraser, . . . M. McCormick. 2016. *Food Price Report 2016*. University of Guelph Economic Brief. At: http://canadianbudgetbinder.com/wp-content/uploads/2016/01/Food-Price-Report-2016-English.pdf.

Daniels, J. "California almond harvest may break records despite drought." CNBC, 29 Aug. At: http://www.cnbc.com/2016/08/29/california-almond-harvest-may-break-records-despite-drought.html.

Desjardins, E. 2016. "Inspiring and informing through food studies." *Canadian Food Studies* 2, 1: 1–3. doi:10.15353/cfs-rcea.v3i1.156.

Desmarais, A. A. 2012. "La Vía Campesina." *The Wiley-Blackwell Encyclopedia of Globalization.* Toronto: John Wiley & Sons.

———, N. Wiebe, and H. Wittman. 2011. *Food Sovereignty in Canada: Creating Just and Sustainable Food Systems.* Halifax: Fernwood.

——— and H. Wittman. 2014. "Farmers, foodies and First Nations: Getting to food sovereignty in Canada." *Journal of Peasant Studies* 41, 6: 1153–73. doi:10.1080/03066150.2013.876623.

Ellsworth, B. 2016. Personal communication, 25 Sept.

Everett, H. 2007. "A welcoming wilderness: The role of wild berries in the construction of Newfoundland and Labrador as a tourist destination." *Ethnologies* 291, 2: 49–80.

Food First NL. 2015. "Everybody eats." At: http://www.foodfirstnl.ca.

Hanrahan, M. 2008. "Tracing social change among the Labrador Inuit and Inuit-Metis: What does the nutrition literature tell us?" *Food, Culture & Society* 11, 3: 315–33. doi:10.2752/17517-4408X347883.

Hochedez, C., and J. LeGall. 2016. "Food justice and agriculture." *Spatial Justice* No.9 (Jan.), special edition on food justice and agriculture. At: http://www.jssj.org/issue/janvier-2016-dossier-thematique/.

Iacovetta, F., V. J. Korineka, and M. Epp. 2012. *Edible Histories, Cultural Politics: Towards a Canadian Food History.* Toronto: University of Toronto Press.

Inglis, D. 2009. "Comment on Josée Johnston and Shyon Baumann/3. Towards the Post-Foodie." *Sociologica* 1: 1–6.

Johnston, J., and S. Baumann. 2009. "'Tension in the kitchen': A response to the comments, the politics of foodie discourse: idealized, ironic, materialist?" *Sociologica* 1: 1–10. doi:10.2383/29569.

——— and ———. 2015. *Foodies: Democracy and Distinction in the Gourmet Foodscape*, 2nd ed. New York: Routledge.

Keske, C. M. H. 2014. "Home grown: Trends in agricultural economics and small scale agricultural production." Symposium: Our Food, Our Future, Research That Feeds Newfoundland and Labrador. Sponsored by Newfoundland and Labrador Department of Natural Resources Agrifoods Development Branch. Corner Brook, NL, 6 Nov. At: www.ourfoodourfuture.ca/pastsymposia.

———. 2015. "Food security, food sovereignty, and the agricultural supply chain in Newfoundland and Labrador." Department of Economics Visiting Speaker Series, Support for Scholarship in the Arts, by Vice President Academic, St. John's, 30 Jan. doi:10.13140/RG.2.1.3909.9764. At: https://www.researchgate.net/publication/303518068_Food_security_food_sovereignty_and_the_agricultural_supply_chain_in_Newfoundland_and_Labrador.

———. 2016. "How does growing NL's agricultural sector make the regional economy bloom?" Symposium: Our Food, Our Future: Research That Feeds Newfoundland and Labrador." Sponsored by Newfoundland and Labrador Department of Natural Resources Agrifoods Development Branch, St. John's, 2 Nov. At: www.ourfoodourfuture.ca/pastsymposia.

———, J. B. Dare, T. Hancock, and M. King. 2016. "The connectivity of food security, food sovereignty, and food justice in boreal ecosystems: The case of Saint-Pierre and Miquelon." *Spatial Justice* No. 9 (Jan.). Special Issue on Food Justice and Agriculture. At: http://www.jssj.org/article/la-connexion-entre-la-securite-alimentaire-la-souverainete-alimentaire-et-la-justice-alimentaire-dans-les-ecosystemes-boreals-le-cas-de-saint-pierre-et-miquelon/.

——— and L. Phillips. 2017. "An equitable, sustainable food system vision for Newfoundland and Labrador." In R.E. Ommer, B. Neis, and D. Brake, eds., *Asking the Big Questions: Reflections on a Sustainable Post Oil-dependent Newfoundland and Labrador*, Section 1.5, Royal Society Canada, Atlantic Reflections, 55–63.

Kneen, B. 1995. *From Land to Mouth: Understanding the Food System*, 2nd ed. Toronto: NC Press.

Koç, M., J. Sumner, and A. Winson. 2012. *Critical Perspectives in Food Studies*. Toronto: Oxford University Press.

Lavallée-Picard, V. 2016. "Planning for food sovereignty in Canada? A comparative case study of two rural communities." *Canadian Food Studies / La Revue canadienne des études sur l'alimentation* 3, 1: 71. doi:10.15353/cfs-rcea.v3i1.73.

La Vía Campesina. 2002. "Declaration NGO Forum FAO Summit Rome+5," 13 June. At: https://viacampesina.org/en/declaration-ngo-forum-fao-summit-rome5/.

Lionais, D. 2015. "Social enterprises in Atlantic Canada." *Canadian Journal of Nonprofit and Social Economy Research* 6, 1: 25.

Loopstra, R., N. Dachner, and V. Tarasuk. 2015. "An exploration of the unprecedented decline in the prevalence of household food insecurity in Newfoundland and Labrador, 2007–2012." *Canadian Public Policy* 41, 3: 191–206. doi:10.3138/cpp.2014-080.

Low, C. 1967. *The Founding of the Co-operatives*. National Film Board of Canada. At: http://www.nfb.ca/film/founding_of_the_cooperatives/.

Lowitt, K. 2009. *A Community Food Security Assessment of the Bonne Bay Region*. CURRA, Memorial University of Newfoundland. At: http://www.curra.ca/documents/CFS%20Assessment%20Report_%20Final_Oct%2009.pdf.

———. 2014. "A coastal foodscape: Examining the relationships between changing fisheries and community food security on the west coast of Newfoundland." *Ecology and Society* 19, 3: 48. doi:10.5751/ES-06498-190348.

McMichael, P. 2014. "Historicizing food sovereignty." *Journal of Peasant Studies* 41, 6: 933–57. http://dx.doi.org/10.1080/03066150.2013.876999.

Murray, H. C. 2002. *Cows Don't Know It's Sunday: Agricultural Life in St. John's*. St. John's: ISER Books.

Murton, J., D. Bavington, and C. Dokis, eds. 2016. *Subsistence under Capitalism*. Montreal and Kingston: McGill-Queen's University Press.

Omohundro, J. 1994. *Rough Food: The Seasons of Subsistence in Northern Newfoundland*. St. John's: ISER Books.

Paddon, W. A. 2002 [1989]. *Labrador Doctor: My Life with the Grenfell Mission*. Toronto: James Lorimer & Company.

Pottle, H. 1979. *Newfoundland Dawn without Light: Politics, Power, and the People in the Smallwood Era*. St. John's: Breakwater.

Robinson, A. 2014. "Enduring pasts and denied presence: Mi'kmaw challenges to continued marginalization in western Newfoundland." *Anthropologica* 56, 2: 383–97.

Rosset, P. M., B. M. Sosa, A. M. Roque Jaime, and D. R. Ávila Lozano. 2011. "The *Campesino-to-Campesino* agroecology movement of ANAP in Cuba: Social process methodology in the construction of sustainable peasant agriculture and food sovereignty." *Journal of Peasant Studies* 38, 1: 161–91. doi:10.1080/03066150.2010.538584.

Shaw, D. J. 2007. "World Food Summit, 1996." *World Food Security*, 347–60.

Sobal, J., L. Kettel Kahn, and C. Bisogni. 1998. "A conceptual model of the food and nutritional system." *Social Science & Medicine* 47, 7: 853–63. doi:10.1016/S0277-9536(98)00104-X.

Song, A. M., and R. Chuenpagdee. 2015. "A principle-based analysis of multilevel policy areas on inshore fisheries in Newfoundland and Labrador, Canada." In S. Jentoft and R. Chuenpagdee, eds., *Interactive Governance for Small-scale Fisheries: Global Reflections*, vol. 13, 435–56. Basel, Switzerland: Springer International Publishing. doi:10.1007/978-3-319-17034-3_23.

Statistics Canada. 2009. *2006 Census of Agriculture*. At: http://www.statcan.gc.ca/ca-ra2006/indexeng.htm.

———. 2010. *Farm Cash Receipts*, Catalogue no. 21-011. At: http://www.statcan.gc.ca/pub/21-011-x/2011002/t031-eng.pdf.

———. 2011. "Farm and farm operator data." Catalogue no. 95-640-X. At: http://www.statcan.gc.ca/pub/95-640-x/2011001/p1/prov/prov-10-eng.htm#Farm_area.

———. 2015. "Fruit and vegetable consumption, 2014." At: http://www.statcan.gc.ca/pub/82-625-x/2015001/article/14182-eng.htm.

———. 2016. "Body mass index, overweight or obese, self-reported, youth, by sex, provinces and territories (number)." At: http://www.statcan.gc.ca/tables-tableaux/sum-som/l01/cst01/health84a-eng.htm.

———. 2017. *Statistics Canada Census Profile, 2016 Census*. At: http://www12.statcan.ca.ca/census-recensement/2016/dp-pd/prof/details.

Tanner, Adrian. 2014. *Bringing Home Animals*, 2nd ed. St. John's: ISER Books.

Tarasuk, V., A. Mitchell, and N. Dachner. 2015. *Household Food Insecurity in Canada, 2013*. Report No. 3. Toronto: PROOF (Research to identify policy options to reduce food insecurity). At: http://proof.utoronto.ca/wp-content/uploads/2015/10/foodinsecurity2013.pdf.

Tye, D. 2010. *Baking as a Biography: A Life Story in Recipes*. Montreal and Kingston: McGill-Queen's University Press.

Van Oostdam, J., A. Gilman, E. Dewailly, P. Usher, B. Wheatley, H. Kuhnlein, . . . V. Jerome. 1999. "Human health implications of environmental contaminants in Arctic Canada: A review." *Science of the Total Environment* 230, 1 (July): 1–82. doi:10.1016/S0048-9697(99)00036-4.

Wald, N., and D. Hill. 2015. "'Rescaling' alternative food systems: From food security to food sovereignty." *Agriculture and Human Values* 33, 1: 203–13. doi:10.1007/s10460-015-9623-x.

Wall, L. 2016. "Recreational fishery extends season with added weekends." CBC News, 20 May. At: http://www.cbc.ca/news/canada/newfoundland-labrador/cod-fishery-extended.1.3590805?.

Wein, E. E., J. H. Sabry, and F. T. Evers. 1991. "Food consumption patterns and use of country foods by native Canadians near Wood Buffalo National Park, Canada." *Arctic* 44, 3: 196–205.

Windfuhr, M., and J. Jonsén. 2011. *Food Sovereignty: Towards Democracy in Localized Food Systems*. Rugby, UK: ITDG Publishing.

Part I

Gardening and Local Food Production

Tomatoes © Dave Howells

1

Food Literacy and Home Economics in Twentieth-Century Newfoundland and Labrador

Lynne Phillips

INTRODUCTION

This chapter examines home economics as a food literacy movement in Newfoundland and Labrador from the 1920s to the 1960s. "Food literacy" is defined here as knowledge of parts or all of the food system that potentially enables food security.[1] While the concept of food literacy was not employed by home economists — or anyone — during this time period, they aimed to address a wide range of food-related issues, including malnutrition, deficiency diseases, cooking and nutritional standards, gardening, and the overall well-being of people living in the outports and in northern communities.

Home economists were not alone in expressing concern about such issues. Alarm about the poor dietary situation of the rural population in Newfoundland and Labrador came from a variety of quarters, including the nursing profession, religious missions, the adult education movement, the Jubilee Guilds,[2] and Memorial University College in St. John's (now known as Memorial University of Newfoundland, or MUN). Indeed, in the first half of the twentieth century, there was some degree of collaboration among these groups as they worked to reach out to the outport and northern rural populations.

From the 1920s to the 1940s, perception of widespread rural poverty in Newfoundland provided new opportunities for young women to play a larger role in their country's future. These opportunities often involved pursuing post-secondary education and moving beyond the comfort zone of one's own community. There was a sense of adventure linked to these opportunities (English, 2011) but also a chance to make a difference, as the country had emerged from World War I with the idea that future peace could be forged through education.

This was a good time to become a home economist. Home economics was a relatively new profession, and was considered a respectable alternative to teaching and nursing for women. Home economists in Newfoundland emphasized the home as a place around which education about food and household management could take hold. They spread this knowledge widely by teaching in schools (in St. John's and beyond), but also by training nurses, teachers, domestic servants, housewives, mothers, and the general public. These diverse audiences learned of food's nutritional content and its importance for health, but also of food preparation, food preservation, food storage, and meal planning.

In Newfoundland, outport fishing communities depended, when they could, on supplementary agriculture and other land-related activities such as tending to goats and sheep, berry picking, and hunting (Cadigan, 2002; Murray, 2010; Omohundro, 1994). The focus in this chapter is on gardens, given their potential to yield "sufficient, safe, nutritious food" with the application of appropriate food knowledge. "Common kitchen gardens" have been cultivated in Newfoundland since the 1700s, when "overwinterers" created gardens to access fresh food (MacKinnon, 1991: 34). Gardens endured into the twentieth century, accompanied by frequent debates about their utility vis-à-vis commercial agriculture (cf. Province of Newfoundland, 1956; Whitaker, 1963; Inglis, 1976). Kitchen gardens (and their associated root cellars) persisted as a feature of many rural households such that early home economists took them for granted in the first half of the century; gardens were an assumed component of their food literacy work.[3]

Placing a food literacy lens on home economics not only reveals important contributions of an often-maligned field, but it also helps to show how food knowledge was deeply undercut by the complex processes of "modernization" in mid-century Newfoundland. The dominant narrative about gardens in Newfoundland is that they were deserted in the post-World War II period, when modernization was embraced by all and women abandoned their productive activities to take on a consumer role (Ommer, 2007). For the rural population, this meant turning "their spades into can openers," to borrow a phrase from Herbert Pottle (1979: 74). The assumption was that people were keen to leave their "precarious" rural existence for the convenience and apparent security of modern food and modern life.[4] In this narrative the "can" is imagined as a better source of food security than the "spade," and food literacy is essentially unnecessary. Closer examination permits another story to be told, however, one that identifies the strengths of the early work of home economists in a way that champions the spade and retrieves the importance of food literacy.

To examine this process, I follow the activities of one Newfoundland home economist, Edna Baird, who established the Household Science program at Memorial University College in 1933.[5] I argue here that food literacy work undertaken by people like Edna Baird was sidelined in the second half of the twentieth century, and the marginalization of food literacy served to reduce avenues for addressing food insecurities in Newfoundland and Labrador.[6]

EARLY FOOD LITERACY EFFORTS

World War I revealed the poor health of much of the country's population: almost half of the Newfoundlanders and Labradorians signing up to enlist were considered medically unfit for duty (Sharpe, 2014). Yet, this was not exactly news in 1914. Wilfred Grenfell, a British medical missionary, had already begun his work in Labrador and northern Newfoundland to address rural poverty and what he saw as a shocking absence of medical care. Outbreaks of "deficiency diseases" like tuberculosis, scurvy, rickets, and beriberi, along with reports that people lived on little more than "bread and tea," led to a growing recognition that there was a significant relationship between poor diet and disease. It was during the first quarter of the century that Grenfell began to encourage families to cultivate their own gardens. Grenfell knew of the highly successful gardening experiments of the Moravian Missions in Labrador, more than 100 years before (H. Rollman, personal communication, 2015; Wilson, 2015). Grenfell's efforts were greatly challenged by soil and climatic conditions, and the vegetable gardens they established in Brig Bay, Flowers Cove, and St. Anthony were abandoned. Yet, as we shall see, such gardening initiatives persisted here and elsewhere on the island.

Newfoundland's major food literacy effort in the mid-1920s emerged from the unlikely place of Memorial University College (MUC). MUC was established in 1925 as a memorial to those who fell in World War I. J. L. Paton, born in England and a renowned educator, was appointed its first President. Paton was a "feisty reformer" (MacLeod, 1999) who embraced the vision of future peace through the hard work of education. Paton was also very supportive of female students and encouraged them to continue their studies elsewhere after graduating from MUC. From the start, Paton was inclined towards public outreach, helping to publish weekly columns of "College Notes" in the St. John's *Daily News* to inform the community about MUC activities (Paton, 1926), including home economics (called Household Science and sometimes Domestic Science at MUC).

Paton's goal was to bring *education to the people.* He had close ties with the World Association for Adult Education and the Newfoundland Adult Education Association (NAEA). Both MUC and NAEA were financially supported by the Carnegie Foundation and were championed by the same people, including Paton (English, 2011; Overton, 1995). Adult education was viewed as distinct from regular teaching in grade schools; the central idea was to bring education to the people rather than the other way around. Early on, the "Opportunity School" method was adopted to promote adult literacy: instructors (all female) went to the outport communities to teach night school and to hold home visits with women during the day. Home economists – young women from away or Newfoundlanders who trained elsewhere – were brought into the communities as guest instructors. These instructors stayed in the community for one or two months, teaching "a curriculum strong on household science and literature" (English, 2011: 28). In this way, improving literacy was not only about the three Rs of reading, writing, and arithmetic. According to Paton: "Students learn how to care for the body, how to figure out the daily budget, how to dress properly and to sew, how to prepare food, to learn something of the wide world and to work for the common good of all" (as cited in Overton, 1995: 260).

In the early years, household science was taught as part of Normal School; that is, teachers who were to enter into the school system were instructed in household science as a teachable subject. However, there was no Household Science *program* at MUC, which was understood to offer a two-year preparation for attaining a full degree elsewhere; it allowed for two years of university background without having to go "overseas." Edna Baird was one of MUC's first graduates in 1927 (MUC Calendar, 1928–30). Baird went on to obtain a BA from Dalhousie University in Halifax (1929), a Bachelor of Home Economics from Montreal's McGill University (Macdonald College), and a graduate diploma in Dietetics from Johns Hopkins University in Baltimore (Balsom, 1989).

In 1933, just before leaving his position as President, Paton invited Baird to come on board as a Memorial University College faculty member and to create a Household Science program. Baird accepted and began her position by immediately engaging the larger public on food literacy issues, an orientation that was the *modus operandi* throughout her life. That year (1933) she not only developed the curriculum and taught her day classes, but she held non-credit evening courses open to the public, trained teachers, connected with the Grace Hospital Nursing School, and gave a talk to the Rotary Club on dietetics.

Baird designed MUC's Dietetics and Home Economics program to support the work of nurses, teachers, housewives, mothers, and domestic help, in addition to meeting the needs of students who wanted to pursue a career in Home Economics. She offered classes in a wide range of themes, including cooking, kitchen management, the nutritional value of food, disease and diet, nutrition of the child, hospital diets, and the aesthetics of the meal. For example, in 1934, the Thursday evening class in "Cookery Arts and Kitchen Management" taught the following topics: Care and Cleanliness of Equipment; Methods of Measuring; Temperatures; Basic Recipes for Batters and Doughs; Uses of Sour Milk; Soups; Fresh Meats (how to clean and prepare with various methods of cooking); Deep Fat Frying; Desserts (milk, eggs, cornstarch, gelatin); Salads and Salad Dressings; Meal Planning; and Table Service (Baird, 1934). Baird was kept busy as the only faculty member appointed to the program.

It is worth noting that MUC also had a "Farm and Garden group," supported by then President A. G. Hatcher and the Department of Natural Resources. This group, started in 1934, focused on improving knowledge about agriculture and gardening. Speakers were brought in to talk about the challenges of growing beets, carrots, and parsnips in Newfoundland soil. The public was invited to these discussions through notifications in the local newspaper. For example, one clip in the *Evening Telegram* (1940) states that

"Anyone interested in improving his or her knowledge of using profitably a piece of land, big or little, is cordially invited." The uncharacteristic language used here ("his or her") and the reference to big or small property may indicate that there was productive conversation between the Household Science program and the Farm and Garden group at MUC.[7]

Edna Baird's food literacy work was complemented by other efforts to work with the outport communities at this time. In 1935, the very successful Jubilee Guilds of Newfoundland (later, the Women's Institute) was established. Spurning charity approaches, the Guilds' mission was to support rural communities, particularly women, "to help themselves" through handicrafts and home economics (Cullum, 2014). Though the Guilds are sometimes thought of as a movement that replaced food production with craft production, the home economics aspect of the Guilds' work was always significant. The organizing secretary position in the Guilds was consistently held by a home economist. For example, Anna Templeton, who graduated with a bachelor's degree in Home Economics from Macdonald College at McGill, opted to apply her skills to the Jubilee Guilds in 1937. Under her care, the Guilds became "a training school in the field of Home Economics" (Richard, 1989: 54). Significantly, the methodology of the Guilds was very similar to that of the NAEA Opportunity Schools: fieldworkers went to the outport communities, when requested, to help organize branches and to hold classes in cooking home produce, preserving, sewing, and knitting. There were many requests for new Jubilee branches in the 1930s and 1940s, such that the central office sometimes had a difficult time keeping up (Cullum, 2014).

The global economic crisis in the 1930s also prompted calls, from various sources, for people to grow their own vegetables. The *Evening Telegram* supported vegetable gardening through several editorials and articles, calling on people to grow potatoes, for example, complete with instructions (MacLeod, 2001). The newspaper also launched a weekly column on agriculture, "Let the Land Help us Out," dealing with the special conditions of farming in Newfoundland (MacLeod, 2001). The Department of Public Health broadcast messages urging residents to "grow your own vegetables."

When Newfoundland came under the Commission of Government in 1934, improving diet through advancements in agricultural production became a focus of government policy. The government sought to "rehabilitate" Newfoundlanders through the discourse of self-help, encouraging them into land settlement schemes so they could grow their own food (Overton, 1995). Heavy-handed messages about the importance of thrift and the negligent behaviour of the slothful shifted responsibility for food security to the colonial population, though the government's messages were not always successful.[8]

The important point here is that the Newfoundland public was bombarded with opportunities to become "food literate" in the 1930s and 1940s. Outreach efforts by home economists, the Guilds, and other groups highlighted the importance of homegrown produce to diet, and of diet to well-being. This extensive dissemination of food knowledge, however, was soon to be challenged.

THE SCIENCE AND ART OF FOOD LITERACY

As an academic discipline, Home Economics was a hybrid: it was inclined towards science and scientific methods (especially in the area of food testing), and yet it had a strong commitment to manual skills, practical knowledge, and aesthetic values. This hybridity was both its strength as a food literacy movement and its Achilles heel as a profession in the modern academy.

In responding to an inquiry from a Mrs. Northcroft in 1935, President Hatcher at Memorial University College concludes by saying: "As to any prospects for Domestic Science teachers in this country, I do not think I can write very hopefully; our facilities here, are, of course, limited by our financial resources" (Hatcher, 1935).[9] While this comment may be understandable given that it was written during the Depression, it is interesting that no comments are to be found about financial constraints for the other sciences during this time. A review of the requested costs of the different MUC courses taught in 1937–38 shows that Household Science had limited equipment needs compared to Biology, Chemistry, and Engineering. The Household Science budget included requests for what we might consider the required platform for home economists: items such as bowls, cups, preserving jars, jelly moulds, cloths, candle holders, and, in one year, a sewing machine. These requests contrasted, both economically and symbolically, with departmental requests for dozens of microscopes, chemicals, slides, lanterns, barometers, and museum cases (Memorial College Evening Classes, n.d.). Claims of resource limitations directed at Household Science hint at future difficulties for the field.

Baird consistently asked for improved lab conditions to facilitate teaching and research. For example, in her annual report in 1945 Baird requests – in addition to her "long-hoped-for plan" to extend Home Economics to male teachers – an updated laboratory so that she can undertake food testing. Referring to the RCAF food lab in Guelph, she writes:

> I was impressed with the valuable work which was carried out in this laboratory not only in planning well balanced dietaries, but also with the frequent analysis of their daily food. Such a laboratory in the Memorial University College would be invaluable to Newfoundland and with the cooperation of the other Science departments of the College much work could be done in obtaining better health for our people. (Baird, 1945)

Neither MUC nor, later, MUN was able to meet Baird's laboratory requests,[10] making it difficult for her to fulfill the science requirement of this field's expectations regarding knowledge production. This lack of support was also one way to ensure that Home Economics, and its food literacy efforts, remained marginalized as a science.

In 1946, by then an associate professor at MUC, Baird took a sabbatical. During this time she surveyed Home Economics programs in eastern Canada and the United States; she also went to Europe to assess the post-World War II food situation.[11] While she was away, it was proposed (perhaps from within the Department of Education) to make Home Economics optional in teacher training. When Baird heard this news she wrote to President Hatcher from Montreal: "I can find no precedence for making Household Science optional in the training of teachers. Here and in Nova Scotia the tendency is to make it a definite and full time subject. When I come back I hope to be prepared to oppose it very strongly" (Baird, 1946). Baird clearly saw her sabbatical as a mission to gather enough information from other parts of the world to bolster the Household Science program in Newfoundland and Labrador.

There were parallel concerns about food literacy – and similar biases – in Canada at this time. The Canadian Council on Nutrition (CCN) was created in 1938, along with the country's first Canadian nutrition program, producing Canada's Official Food Rules in 1942, 1944, and again in 1949 (Mosby, 2014; Canada's Food Guide was not created until 1961).

Ian Mosby's work on the impact of World War II on Canadian kitchens reveals how the politics and "science" of nutrition effectively sidelined women's professions concerned with such issues, including the Canadian Home Economics Association and Canadian Dietetics Association. Mosby demonstrates that a deep misogyny was woven into the federal nutritional programs. For example, in response to a notice regarding the establishment of a nutritional division in the federal government in 1941, E. W. McHendry (a University of Toronto biochemist who dominated the CCN) appealed to the deputy minister of the time, recommending that "For the sake of getting things done sensibly, I hope that the senior appointment goes to a man and not to a dietician" (cited in Mosby, 2014: 33). It did.

A Nutrition Council was established in Newfoundland in 1943 by the Newfoundland Medical Association. If there were tensions between the Medical Association and home economists, they were not obvious, as the Council always had a nutritionist or home economist on board and maintained a public orientation. The Nutrition Council suggested to the government the implementation of an educational program on "helpful dietary habits and methods of using and preparing local food stuffs" (Overton, 1998: 22), and launched a nutrition campaign to promote the production, consumption, and preservation of local food. Home economist Baird made a major contribution to this initiative, serving as the dietician for the Newfoundland Council of Nutrition between 1943 and 1946 (Lush, 1999). In that capacity, and in addition to her day job, she broadcast over 220 radio shows, wrote a weekly column, "Food and Your Health," for local newspapers, and produced two booklets for distribution on food and nutrition.

For home economists like Baird, who ventured beyond St. John's and into the outport communities, a holistic approach to food literacy meant embracing the larger sense of home that was integral to the fishing household economy, i.e., including supplementary land-based activities. From a food literacy perspective, it was considered important to educate "homemakers" about *all* of the skills and knowledge required for a healthy meal, from the practical to the aesthetic. Baird advocated taking pride in one's home management skills and appreciating art and beauty. The aesthetic wisdom that might be brought to cultivating gardens is striking in this sense. Writing of his ethnographic studies in northern Newfoundland in the 1980s and 1990s, John Omohundro (1995: 164) explains that:

> In fact, for some country gardens (those along roadsides as much as 40 km from the house) as well as kitchen gardens, a kind of housekeeping aesthetic exists. One's reputation as a gardener depends partly on how neat one's garden is. Many Newfoundlanders perceive a beauty in a highly regular arrangement of square and rectangular raised beds, carefully "lined out" during planting with string and stakes.

Taking pride in a well-laid-out and productive garden aligned with the well-planned and well-prepared meal — an orientation that the early home economists could applaud as part of food literacy. This view also suggests that gardening could be undertaken for reasons other than survival or poverty. The garden could have a "housekeeping aesthetic" — and qualify as a kind of "folk art" (Omohundro, 1987) — and was capable of fostering emotions such as guilt, rather than relief, when one is unable to tend to it. This point is lost when only the science or productivity component of home economics and agricultural expertise is emphasized.

That there may be something empowering about gardening — whether due to pride in producing one's own food or delight in working with a particular aesthetic — is not acknowledged in the second half of the twentieth century by those keen to modernize Newfoundland's food production. Early home economists in Newfoundland, on the other hand, were always able to recognize the potential of the garden. This is in part because they employed a "ground-up" methodology that respected local foodways. So, while they took on the role of reformers, with all the power that the term implies, they employed an ethnographic methodology that potentially challenged the dominant paradigm. For example, they did not advocate a middle-class division of labour or a foreign model of dietary change. Their approach, rather, "was to work within the culture and financial experiences of their new clients" (Lush, 2008: 230). Lush finds that the focus was not about "forcibly chang[ing] the local diet" from above. Rather, it was about recognizing the potential of the existing diet and trying "to shape the dietary reform campaign around the [fishing] family work schedule" (Lush, 2008: 231). Home economists took the time to understand local food preferences and to determine how people's diets could benefit from the available resources around them. However, while this respect for people's local circumstances may have helped to produce an audience for this early version of food literacy, the role of the home economist was about to radically shift.

THE CAN OPENER: CONFEDERATION, MARKETS, AND MUN

Joey Smallwood, Newfoundland's first premier after Confederation, set the stage for change. For him, modernizing Newfoundland was necessary to support "our toilers... whose greatest toil and endurance could not provide their families with enough to eat or wear" (1948; cited in Overton, 1998: 1). In this comment, Smallwood aims to relegate rural "toil" and the spade to the past, disconnecting them from a future geared towards prosperity. Celebrating all that was modern, Smallwood's vision facilitated the circulation of new ideas about convenience, hygiene, consumption, and work.

When Newfoundland entered Confederation with Canada in 1949, much change occurred to the food systems of the new province. Tariffs no longer protected vulnerable local markets. "Baby bonus" cheques gave women money to purchase goods, while corporate advertising was directed at women as consumers (Phillips, 2014). Investment in roads and supermarkets facilitated food imports and made available a whole new range of commercial foods. Ideas about "modern" life spread quickly, along with new views about food literacy.

The extent to which food literacy became compromised by this process is seen in a 1959 pamphlet on how to cook with fish, written by the Home Economics Section in the federal Department of Fisheries for a demonstration at the Newfoundland Agricultural and Homecrafts Exhibition. Significantly, almost all of the recipes call for canned or processed goods: canned fish (salmon, tuna, lobster), processed food (potato chips for "tuna crunch" and corn flakes or graham wafer crumbs for "crispy baked fillets"), and other canned ingredients (e.g., tinned mushroom and tomato soup for sauces). In this version of food literacy, all that is required is opening a can and heating its contents. No need to tend the garden or prepare the meal; it is just a matter of having cash. And Newfoundland quickly and abruptly became a cash economy geared towards this kind of consumption.

Ideas about modern life also impacted the academy. The transition from Memorial University College to the Memorial University of Newfoundland was expedited by Smallwood, who envisioned the university as a centrepiece for modernization of the new province. His legislature quickly approved the university's creation in 1949. Becoming a university presented an opportunity for academic departments and the professions to develop and expand. The curriculum of the Household Science program in the early 1950s was rigorous, still helmed by Baird. The program retained its hybrid approach, with required courses in foods and cookery, applied art, household administration, physiology, nutrition, and diet therapy, in addition to English, chemistry, a foreign language, and math or physics, and, in the second year, economics. Course content was demanding.[12]

On the other hand, the shift to university status also made the Household Science program vulnerable: Associate Professor Baird was still its only faculty member, and student numbers were small. As competition for resources and professional status increased, some departments were supported over others. Questions as to Household Science's fit and relevance in a modern university were couched in terms of the small numbers of students signing up. Baird argued that student numbers would return if her lab conditions improved and if a full degree rather than a two-year program could be offered. However, some shared the sentiment that, with the move to university status, there was a need to "get rid of the dead wood" (MacLeod, 2001).

As long as Hatcher remained President of Memorial University, it appears that Household Science was protected. When he stepped down in 1952, the program lost an advocate and was left vulnerable to those who felt it did not belong. The Dean of Arts during this period, A. C. Hunter, was apparently hostile towards Household Science: "The Hunter clique thought certain studies — such as language and literature — were like orchids and should be valued in the garden, while others, like domestic science, should be weeded out" (MacLeod, 1990: 81).[13]

Though never expressed explicitly in written documents, Memorial University clearly followed the pattern of many other universities that declined the inclusion of what were viewed as college-level, practical fields in order to focus on "pure research" and matters of public importance (Rossiter, 1982; Stage and Vincenti, 1997). Not only was home economics a gendered science, but food literacy — a cornerstone of the profession — was simply not regarded as a matter of intellectual concern in the 1950s.

Perhaps it is not surprising, then, that Baird moved from her academic position to help launch MUN's new Extension Division in 1959. There is no evidence that she was pushed out of her academic position, but, significantly, the Household Science program was "confidentially" suspended the same year she left her academic post (MUN, 1959). Moreover, the university rubbed salt in the wound by taking away her professorial title of associate professor and paying her a smaller salary as an Extension employee. In March 1960 she was sent a letter from the Board of Regents stating that as an assistant to the Director of Extension, Baird would receive a salary of $6,700 with no incremental scales, as would normally be the case in an academic position. As an academic, she would make between $8,000 and $8,500, so this was a substantial decrease. The lengthy disagreement between Baird and the Board of Regents drew in the new Director of Extension, S. J. Coleman. A strong supporter of Baird and her work, Coleman appealed to then President Gushue, but to little effect.[14]

Despite these setbacks, Baird took on the new Extension Services position with her usual enthusiasm, playing a key role in disseminating food and nutrition information via Extension's network throughout the province. Baird's job category as administrative assistant to the director belied the range of work she undertook, since home economics figured strongly in the early years of MUN Extension.[15]

Extension's 1961–63 Annual Report highlights 20 teaching modules titled "At Home with Edna Baird," a weekly half-hour television program that Coleman had asked her to write and produce. Baird's hybrid approach is still evident in this project. Subjects covered were: Better Breakfasts, New Ways of Cooking Fish, Meat Cutting for Economy and Nutritive Value, Frozen Foods, How Safe is Our Food, Arts in Home and Dress, Chats about Children, Consumer Buying Standards, A Way of Looking at and Choosing Pictures for the Home, and Homemakers of Tomorrow. The series ended with an international buffet showcasing food from around the world. Throughout the series, Baird hosted important guest speakers, including the Chief Health Instructor, the Chief Medical Health Inspector, artist Mary Pratt, the home economist with the federal Department of Fisheries, and home economist Anna Templeton, the organizing secretary for the Jubilee Guilds at that time. That year, Edna Baird also sat on the board of the Jubilee Guilds. Together, Baird and Templeton made a formidable food literacy team.

In its 1963–65 Annual Report, MUN Extension Director Coleman notes again Baird's significant contributions to outreach:

> Miss Baird spent three weeks in Corner Brook (with visits to Stephenville and Deer Lake) to confer with the 15 tutors in those places and visit classes. She gave three television programs on nutrition in Corner Brook and three in Stephenville. . . . Her two booklets on food and nutrition were sent out in response to more than one hundred requests arising from these programs. She also addressed various Women's groups. Response to her questionnaire on nutrition reveals that although Newfoundland diets are still far from ideal, the situation is improving, especially in the consumption of milk. 200 more booklets were sent out in response to requests returned with the questionnaire.

Through Extension services, the university was an agent of modernization for the province, spreading its tentacles to the outports, to the west coast, and to Labrador. Using communications technology to enable outreach, MUN Extension could "shape the public sphere" (Webb, 2014: 86). This is precisely the approach that Edna Baird had always taken — though without the technology — to increase her reach in enhancing food literacy. She did this effectively despite working in environments not always friendly to women. By some accounts, MUN Extension was no exception, particularly after the departure of Coleman in 1965. Extension Services had a strong reputation as "Pioneers in Community Engagement" (Webb and Bishop-Sterling, 2012) that brought international attention to MUN, but there is evidence that MUN Extension was not enlightened about gender issues. Webb (2014) writes of the case of Laura Jackson, who fought hard to do fieldwork with women farmers in the Labrador Straits. She was apparently told by another fieldworker, Margaret Davis: "be prepared for a rough ride 'cause they [the men who run Extension] don't like to hire women, they don't see the need. They see the traditional community, the traditional power structure, and they are not thinking outside that box" (cited in Webb, 2014: 96). Another Extension staff member, Linda Cullum, echoes Davis's view: "I felt like we were exporting to community groups, perhaps a very fine class analysis, and a political analysis, but very conventional gender norms" (cited in Webb, 2014: 96).

Baird could be accused of proliferating "conventional gender norms." Many home economists have been likewise accused. The details of her career have been provided here precisely because they challenge any attempt to dismiss her work on this basis. Baird confronted convention on many levels. People like Baird bravely engaged the public about issues that more powerful others would rather ignore. Placing a *public* emphasis on self-knowledge about food preservation, storage, and preparation – on food literacy for everyone – not only "helped people to help themselves," but also challenged "the way it's always been." Moreover, Baird's work with MUN Extension garnered a considerable amount of attention from the public precisely during the period when the food system in the province was being radically altered.

While Baird was actively engaging the public about the importance of food literacy, the world was changing around her. By the mid-1960s, there were not only new modes of accessing food, but changing views on the role of home economics and the home economist.

In the mid-1960s, home economics began to morph into "Family Studies" in Newfoundland and elsewhere. Family Studies emphasized personal grooming and family relationships over meal preparation. Without a home in the university, home economics agendas could be driven by government interest. According to *Home Economics: A Teaching Guide, Grades VII–XI*, a booklet produced by the Department of Education, Grade 7 students learned: "cleanliness, voice, posture, courtesy, manicure, care of shoes, care of socks, hair care, comb and brush care – pressing a skirt, attaching a button," as well as knowledge about body type and eating habits (Department of Education, 1965: 5, 9). The Grade 9 course aimed to "teach girls to work cheerfully, to work at home willingly, and to be considerate of other family members." The guidebook, assuming an all female audience, proposed that girls must "learn to be neat and clean and to set high standards for themselves" (1965: 19), to know the "value of becoming a gracious hostess," and to understand "social customs pertaining to afternoon tea" (1965: 26). These goals not only indicate a decidedly middle-class orientation that was much more muted in the earlier period, but they also suggest a certain dismissal of the more holistic food literacy approach previously practised.

An apparent disdain of the "old" Home Economics continued into the 1970s. In a 1974 brief from the Department of Education requesting the university to make Home Economics a university entrance subject, a distinction is clearly made between the "inadequate" old idea of Home Economics "as an option for non-achieving girls; to teach them the homely [*sic*] skills of cooking, serving and home management" and the contemporary understanding of Home Economics, which should do more than cooking and sewing. According to one presentation on the new Home Economics to the Newfoundland Home Economics Association:

> In this day and age when families have to deal with the complexities of consuming rather than producing, when people are bombarded with problems caused by social and technological change, when the very structure and function of families is undergoing rapid and often perplexing and frightening change the school must have a commitment to fit the student with fundamental competencies which will be effective in personal and family living. (Ghory, 1974: 40)

The commitment to "fit" students with such competencies marks a radical change from the commitments of earlier years.

CONCLUSION

This chapter has reframed home economics as a food literacy movement, as it was practised in the first half of the twentieth century in Newfoundland and Labrador. During this period, home economics had four key characteristics: it adopted a holistic, land-to-plate approach to food literacy in the home; straddled arts and science orientations to food issues; learned about and respected local foodways; and engaged effectively with the public to increase food literacy beyond the household. I draw on these characteristics to identify how this historical analysis might yield potential lessons for us today.

The holistic approach of home economics during this early period, while perhaps articulated more through practice than as theory, indicated how a robust food literacy requires a range of knowledge and skills along the food chain, from harvesting produce in the garden and preparing, preserving, and storing food for the future, to cooking and presenting food in a meal. It has been suggested here that the garden, as an extension of the household, was viewed as an important source of both proper diet and aesthetic pleasure. This point challenges the argument that people only garden out of necessity, as a matter of survival, and that gardening is something from which they crave escape. It is worth noting that the garden as a source of food has persisted in Newfoundland and Labrador despite the availability of food by other means, including in store-bought cans, and despite the dominance of consumer culture. We perhaps owe this legacy of garden practices at least in part to the home economist.

Edna Baird married and retired in 1970, initially without a pension (Balsom, 1989). By this time the "old" Home Economics was all but erased from the academy, including from MUN Extension activities.[16] Transformation of the economy, the university, and the profession itself meant that opportunities to bring skills and critical knowledge about food to the public were all but eclipsed. While some might debate whether Household Science belonged in the sciences or the arts, the important point is that its exclusion marked the loss of the only voice for food literacy within the university — at a time when it was arguably most needed.

While Edna Baird always promoted beauty and the aesthetic side of food settings, she struggled to maintain her footing in food science. Food science requires labs, and labs were (and still are) expensive. However, behind this problem was always the question of whether food itself was a valid topic within the university context. With its connections to domestic concerns and women, Household Science would perhaps always be viewed as a "lesser" science. Baird's efforts to have Household Science relevant to the arts and science were admirable, but ultimately she lost. In this light, it seems ironic that we have recently created a food studies program that crosses departments and faculties.[17]

That home economists in Newfoundland and Labrador had a deep respect for local foodways is what drew my interest to this topic in the first place. In my review of historical documents, I found no evidence of this profession denigrating the local diet; instead, there was an insistence on listing the unique foods that people in Newfoundland and Labrador consumed, a listing that reflected close attention to how people lived. This "ethnographic" approach to food reform surprised me, especially since it took place long before MUN Extension's outreach to outport communities. Starting from a local understanding of how and what people eat is surely an important lesson for food literacy initiatives today.

Finally, I have highlighted the extensive public outreach of home economics in Newfoundland in these early years. Home economists engaged the public with a dizzying level of energy. Edna Baird's story makes clear that teaching at Memorial University College was just one of a broad suite of food literacy activities she undertook. During one year alone, while teaching 43 Household Science students, she also taught student nurses at the Grace General Hospital in dietetics, food preparation, and cooking; conducted experiments on cooking with skim milk for the Commission of Government; liaised with the Adult Teachers Association; taught a course in the Theory of Foods and Practical Cookery for the Women's Division of the Royal Canadian Air Force and Navy; and, with her Household Science students, created a public exhibition on nutrition. The determination to occupy multiple spaces for spreading the word speaks to the need to be flexible and strategic in how we educate people about food skills and knowledge.

In 1949, when Memorial became a university and Newfoundland became a province, the food literacy orientation promoted by people like Edna Baird was obscured and sidelined. The market fundamentally altered how food was accessed and the profession of home economics realigned its interest away from the place of food knowledge in society.

By focusing on the historical changes in home economics and its often-troubled relationship to the academy, insight can be gained into how a form of food literacy, underpinned by knowledge creation and skill, was eclipsed in favour of a weakened food literacy based on deskilling and knowledge reduction. This process of "modernization" undoubtedly diminished the food literate public in Newfoundland and Labrador in the second half of the twentieth century, which in turn closed pathways for achieving food security.

This historical research sheds light on the importance for food security of supporting food literacy efforts today that are holistic in scope, hybrid in conceptual orientation, ethnographically informed, and publicly engaged. All four of these features work best when supported by university research open for many professions and for many views of the food world.

ACKNOWLEDGEMENTS

I'd like to thank Mel Baker, Pauline Cox (MUNFLA), Jeff Webb, Diane Royal, and the CNS at the QEII library, St. John's, for their support of this research. Thank you to Catherine Keske for her fine editorial skills, and to Sally Cole for reviewing the manuscript.

NOTES

1. As a relatively recent concept, there is no common definition of "food literacy" from which to draw. "Food security" is usually defined as: "when all people, at all times, have physical and economic access to sufficient, safe and nutritious food that meets their dietary needs and food preferences for an active and healthy life" (FAO, 1996). It is assumed in this chapter that food literacy facilitates food security insofar as food-related skills and knowledge impact success in getting food from land to plate. Low skills and knowledge about food preparation, for example, can translate into a situation of food insecurity, despite the availability of food.
2. The Jubilee Guilds were created in 1935, initiated and supported by prominent St. John's women. Cullum (2014: 180) writes: "These women aimed to rehabilitate what they saw as economically and socially depressed rural communities, and to build a stronger nation of Newfoundland through the 'uplift' of its rural women and families." The organization joined the Women's Institutes in 1951, and in 1968 changed its name to the Women's Institutes of Newfoundland and Labrador ("Women's Institutes and Jubilee Guilds," 1994: 606–07).
3. A section on home gardening in a *Newfoundland Health Bulletin* (1939: 7) states that "one of the greatest mistakes of farm life" is not growing "larger quantities of vegetables . . . as one of the attractions of any home is the garden and its table products." While this bulletin has no author, its contents — focusing on recipes, food preparation, and attractive table settings — indicate that it was likely written by a home economist.
4. This is the Newfoundland version of the classic modernization tale that disparages the rural past to make way for the future. It was, for example, the story told by Henry Ford — "the man who freed the farmer" — when he envisioned the automobile as an escape from the drudgery of rural life (Phillips, 2012). As indicated later in this chapter, Joey Smallwood and Henry Ford had this in common as champions of modernization.
5. Edna Baird was born in Botwood, Newfoundland, in 1908 to William H. and Bertha Baird. She had three sisters. Her father is listed as a millman at the time of his marriage in 1902. The family moved to St. John's in 1916, where Edna later attended Methodist College and, in 1925, MUC.
6. Reference to Newfoundland alone is appropriate to the historical time period covered in this chapter (the province was not renamed Newfoundland and

Labrador until 2001). However, in some cases it seems more fitting to use the full name of the province, and I have done so. The specific role of food literacy in twentieth-century Labrador is work that remains to be done, although see Schiff and Bernard, Chapter 7.

7. Given that many home economists were trained through McGill's Macdonald College, an agricultural college, one wonders whether there was general collaboration between Household Science and the Farm and Garden group at MUC, and to what extent the latter may have influenced the former in taking a holistic, land-to-plate approach to food literacy. However, I have found no evidence of collaboration.
8. Overton (1998) offers a good example. The consumption of whole grain flour ("brown flour") was made compulsory by the Commission for anyone on government assistance, though it had been shown by then that its consumption could improve health for all. The significant stigma that became attached to brown flour — referred to by some as "cattle feed" — ensured that the Commission's plan convinced few, "even the hungriest dole recipients," to eat it (cited in Overton, 1998: 18).
9. Mrs. Northcroft's inquiry into the status of Home Economics programs in the country indicates a growing international interest in the field in the 1930s.
10. Neither is there evidence to suggest that she was successful in introducing household science to male teachers.
11. Baird was warned against going so soon after the war, but she went anyway (*Evening Telegram*, 1947) — just one of many clues that she had a deep curiosity about the world and was open to learning new approaches to food literacy.
12. For example, the description of the Foods and Cookery course was: "A study of foods, their composition, economic selection and nutritive value; the general principles of cooking and their application to food preparation treated from an experimental point of view; the planning, preparation and serving of meals" (MUN, 1950: 33).
13. I feel compelled to point out the considerable irony in this use of a garden metaphor here, given what has been argued above.
14. Coleman (1960) wrote: "She was invaluable to me, in my first months, in giving local information. Then, after Christmas 1959, the Nutrition and Housekeeping course was held at Gander — she made a notable success of

it." On 17 November 1961 the Board agreed to pay Baird $7,000 per annum, plus $1,000 per annum as "a special payment" in recognition of her lengthy service as an associate professor (Board of Regents). However, Baird was not able to retain her professorial status, though this was apparently promised to her when she transferred to Extension.

15. Coleman specifically mentions Home Economics in a 1960 *Evening Telegram* article on MUN's Extension Services. See also Coleman (1960) for a discussion of Baird's work in Gander.
16. This does not mean that no home economists were advocating food literacy during this time. Olga Anderson's contributions in the 1970s and Mary Mackey's work in the 1980s appear to be exceptions, but this point awaits future research.
17. Establishing a food studies program is one of a number of projects recently launched by Food Advocacy Research at Memorial (FARM).

REFERENCES

Baird, E. 1934. "Outline of work for evening classes, Domestic Science, 1934–1935." Memorial University of Newfoundland, President's Office Files, Box PO-9.

———. 1945. *Report of the Department of Household Science, 1944–1945.* St. John's: Memorial University College.

———. 1946. Correspondence to A. G. Hatcher. President's Office Files, Box PO-5.

Balsom, D. 1989. Interview with Edna Baird and Joan Bown, 8 Feb. MUN Folklore and Language Archive, Accession No. 89-061.

Board of Regents. 1961. Correspondence to E. Baird, 17 Nov. President's Office Files, Board of Regents Files, Box PO-23.

Cadigan, S. 2002. "The role of agriculture in outport self-sufficiency." In Rosemary Ommer, ed., *The Resilient Outport: Ecology, Economy, and Society in Rural Newfoundland*, 241–62. St. John's: ISER Books.

Coleman, J. 1960. *Report on the Experimental Extension Course on Home Economics Held in Gander.* President's Office Files, Box PO-37, Extension — General, 1953–1960.

Cullum, L. 2014. "'It's up to the women': Gender, class, and nation building in Newfoundland, 1935–1945." In L. Cullum and M. Porter, eds., *Creating This Place: Women, Family and Class in St. John's, 1900–1950*, 179–201. Montreal and Kingston: McGill-Queen's University Press.

Department of Education. 1965. *Home Economics: A Teaching Guide, Grades VII–XI*. St. John's.

English, L. 2011. "Adult education on the Newfoundland coast: Adventure and opportunity for women in the 1930s and 1940s." *Newfoundland and Labrador Studies* 26, 1: 25–54. At: https://journals.lib.unb.ca/index.php/NFLDS.

Evening Telegram. 1947. "Woman of the Week," 18 Oct., 10–11.

———. 1960. "MUN Extension Director has long-term program," 26 Feb., 3.

Food and Agriculture Organization of the United Nations (FAO). 1996. *Rome Declaration on World Food Security and World Food Summit Plan of Action*. Proceedings from the World Food Summit, Rome, Nov.

Ghory, M. 1974. "The new Home Economics curriculum for Newfoundland schools." *Newfoundland Home Economics Association* 7: 39–41.

Hatcher, A. G. 1935. Correspondence to D. M. Northcroft, 7 Oct., University Archives, Box PO-9.

Inglis, G., ed.. 1976. *Home Gardening in Newfoundland: Proceedings of the Centre for the Development of Community Initiatives*. St. John's: Memorial University of Newfoundland.

Lush, G. 1999. "Newfoundland women pioneering careers in science, 1880–1949." BA Honours thesis, Memorial University of Newfoundland.

———. 2008. "Nutrition, health education and dietary reform: Gendering the 'New Science' in Newfoundland and Labrador, 1893–1928." Master's thesis, Memorial University of Newfoundland.

MacKinnon, R. 1991. "Farming the rock: The evolution of commercial agriculture around St. John's, Newfoundland, to 1945." *Acadiensis* 20, 2: 32–61.

MacLeod, M. 1990. *A Bridge Built Halfway: A History of MUC, 1925–1950*. Montreal and Kingston: McGill-Queen's University Press.

———. 1999. "John Lewis Paton." At: http://www.mun.ca/memorial_history/publications/MacLeod/Paton.html.

———. 2001. "Making friends and enemies: Public relations at MUC, 1925–50." *History of Intellectual Culture* 1, 1: 1–15. At: http://www.ucalgary.ca/hic/.

Memorial College Evening Classes. n.d. President's Office Files, Box PO-5.

Memorial University of Newfoundland (MUN). 1950. *MUN Calendar, 1950*. St. John's.

———. 1959. "Teaching of household science." Proceedings from the Senate Meeting, Feb. St. John's.

Mosby, Ian. 2014. *Food Will Win the War: The Politics, Culture and Science of Food on Canada's Homefront*. Vancouver: University of British Columbia Press.

MUC Calendar, 1928–30. President's Office Files.

Murray, H. C. 2010 [1979]. *More Than 50%: Woman's Life in a Newfoundland Outport, 1900–1950*. St. John's: Flanker Press.

Newfoundland Health Bulletin. 1939. President's Office Files.

Ommer, R., ed. 2007. *Coasts under Stress*. Montreal and Kingston: McGill-Queen's University Press.

Omohundro, J. 1987. "The folk art of the raised bed." *Garden Magazine* (May/June): 10–15.

———. 1994. *Rough Food: The Seasons of Subsistence in Northern Newfoundland*. St. John's: ISER Books.

———. 1995. "'All hands be together': Newfoundland gardening." *Anthropologica* 37, 2: 155–71. doi:10.2307/25605808.

Overton, J. 1995. "Moral education of the poor: Adult education and land settlement schemes in Newfoundland in the 1930s." *Newfoundland Studies* 11, 2: 250–82. At: http://www.mun.ca/nls/.

———. 1998. "Brown flour and beriberi: The politics of dietary and health reform in Newfoundland in the first half of the twentieth century." *Newfoundland Studies* 14, 1: 1–27. At: http://www.mun.ca/nls/.

Paton, J. L. 1926. *Report of Newfoundland Memorial University College, 1925–1926*. President's Annual Report. St. John's: Memorial University of Newfoundland. At: http://www.mun.ca/memorial_history/Presidents_Report.html.

Phillips, L. 2012. "Alternative mobilities." Paper presented at the annual meeting of the Centre for Diaspora and Transnational Studies, "Foodways: Diasporic Diners, Transnational Tables, and Culinary Connections," University of Toronto, Oct.

———. 2014. "'Women not like they used to be': Food and modernity in rural Newfoundland." In J. Paige-Reeves, ed., *Women Redefining the Experience of Food Insecurity: Life off the Edge of the Table*, 243–60. Lanham, Md.: Lexington Books.

Pottle, H. 1979. *Newfoundland Dawn without Light: Politics, Power, and the People in the Smallwood Era*. St. John's: Breakwater.

Province of Newfoundland. 1956. *Report of the Newfoundland Royal Commission on Agriculture, 1955*. St. John's: Queen's Printer.

Richard, A. 1989. *Threads of Gold*. St. John's: Creative Publishers.

Rossiter, M. 1982. *Women Scientists in America*. Baltimore: Johns Hopkins University Press.

Sharpe, C. 2014. "The 'Race of Honour': An analysis of enlistments in the armed forces of Newfoundland, 1914–1918." In *Essays on the Great War: Papers Published in Newfoundland and Labrador Studies*, 9–41. Special publication of *Newfoundland and Labrador Studies* in Commemoration of the 100th Anniversary of WWI. St. John's: Faculty of Arts Publications, Memorial University of Newfoundland.

Stage, S., and V. Vincenti, eds. 1997. *Rethinking Home Economics: Women and the History of a Profession*. Ithaca, NY: Cornell University Press.

Webb, J. 2014. "The rise and fall of Memorial University's Extension Service, 1959–91." *Newfoundland and Labrador Studies* 29, 1: 84–116. At: http://www.mun.ca/nls/.

——— and T. Bishop-Sterling. 2012. "Pioneers in community engagement: MUN Extension, 1959–1991." *Newfoundland Quarterly* 104, 4: 45–47. At: http://www.mun.ca/nq/.

Whitaker, I., ed. 1963. *Small-scale Agriculture in Selected Newfoundland Communities: A Survey Prepared for the ARDA Administration*. Report No. 14024, Rev. No. 1013. St. John's: ISER Books.

Wilson, L. 2015. *The Labrador Garden Project: An Annotated Bibliography and Research Notes*. Intangible Cultural Heritage Inventory. St. John's: Agricultural History Society of Newfoundland and Labrador.

"Women's Institutes and Jubilee Guilds." 1994. In C. F. Poole and R. H. Cuff, eds., *Encyclopedia of Newfoundland and Labrador*, vol. 5, 606–07. St. John's: Harry Cuff.

2

Commuting to Garden: Subsisting on Bell Island

Sharon R. Roseman & Diane Royal

INTRODUCTION

On a windy afternoon in August of 2015, Fred Parsons and his daughter Cheyenne showed us their greenhouse made of recycled lumber – the muffled sound of wind chimes fading as we moved away from their back porch. The greenhouse was filled with colours and smells of tomato, cucumber, carrot, beet, and mint. Outside, there were more plants, including green beans, rhubarb, strawberries, and raspberries. The year before, they had also cultivated potatoes and onions. In late summer and early fall, the two go berry picking. Their devotion to self-provisioning was evident in their thriving plants as well as winter preserves. Beyond their own tight-knit family unit, Fred, his wife Miranda, and Cheyenne are generous neighbours. When a friend fell ill and requested fresh blueberries, Fred picked five buckets for her in one day. As Fred put it, during the summer months "Every day I'm at this. Every day."

This chapter draws on ethnographic fieldwork with Bell Island residents, such as Fred Parsons, to explore the links between subsistence gardening, place attachment, and commuting mobilities in this part of Newfoundland and Labrador. The "mobility turn" perspective foregrounds "the movement of people, ideas, objects, and information" (Urry, 2010: 17) to counter earlier frameworks that overemphasized the boundedness of social units. In the Newfoundland context, the term "gardening" distinguishes between labour-intensive horticultural activities in smaller planted beds or greenhouses for the purpose of self-provisioning essential for food security and intensive agricultural production of mainly commercial crops on larger expanses of land (e.g., Antler, 1977; Murray, 1979; Felt, Murphy, and Sinclair, 1995; Omohundro, 1985, 1994, 1995; Porter, 1983: 87). We are employing an approach central to humanistic ethnography of highlighting and honouring the experiences and thoughts of several individuals who represent a range of patterns, in this case a group of six adults[1] who were the children, in-laws, and relatives of men who worked in the Bell Island iron ore mines. These individuals embody different mobilities trajectories related to the post-closure period.

Located in Conception Bay across from the Avalon Peninsula, Bell Island is comprised of the municipality of Wabana and the unincorporated settlements of Lance Cove and Freshwater. An influx of workers seeking employment at six surface and submarine mines that operated at different times from the 1890s to 1966 led Bell Island to have one of the largest populations in Newfoundland by the mid-twentieth century (Martin, 2003; Neary, 1973: 111; Neary, 1975; Weir, 2006). However, when operations ceased at the last functioning iron ore mine in 1966, there was extensive permanent out-migration to places with manufacturing jobs such as Galt, Ontario (now part of Cambridge), mining areas such as Wabush, Labrador, and other parts of North America. As with most other areas of the province, Bell Island's current population is aging and not being replaced by incoming residents. The 2016 Census of Population registered under 2,500 inhabitants (Statistics Canada, 2017), many of whom commute to the Newfoundland mainland daily to work. To reach services in the St. John's area, Bell Islanders are reliant on public ferry transportation across the "Tickle" to the wharf in Portugal Cove.[2] They then travel by road in privately owned vehicles or taxis since there are currently no buses or other public transportation services in Bell Island or the municipality of Portugal Cove-St. Philips.

This chapter is organized in five remaining sections. We next outline our conceptual framework and then discuss our methods and introduce the six Bell Island gardeners who participated in the research. We subsequently summarize the historical context for gardening practices on Bell Island. This is followed by a three-part section where we discuss gardening as a place-making

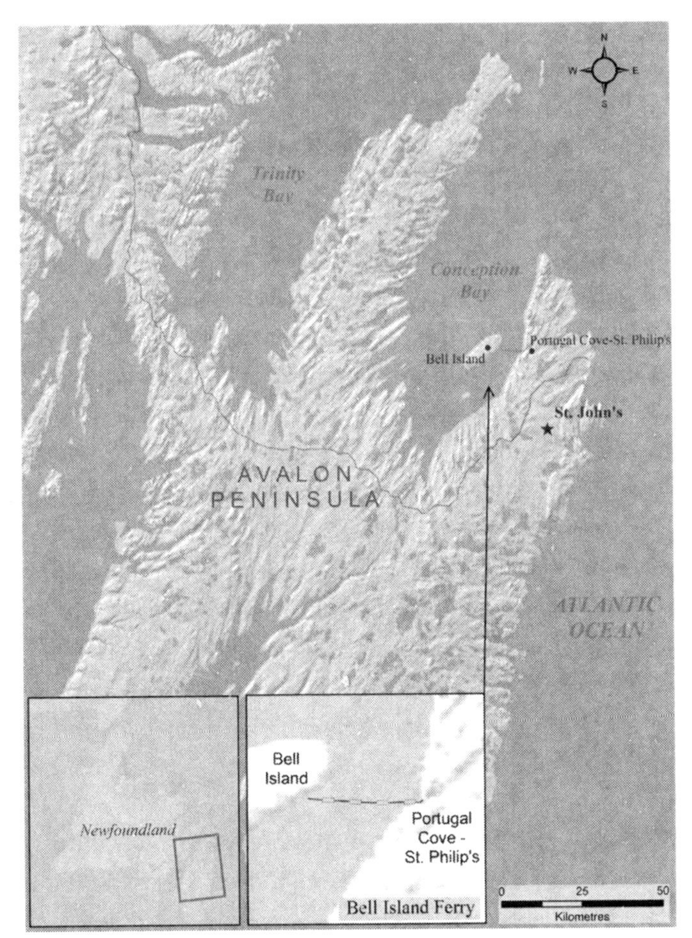

Figure 2.1. Map of study area. Bell Island, Newfoundland and Labrador, showing ferry crossing to Portugal Cove. (Cartography by Myron King)

activity, in terms of people sharing strong memories of their own childhoods and their making of garden places in the present period. Next, we discuss the flows of people, things, capital, and information that are part of Bell Island gardening mobilities. We end with the chapter's conclusion.

CONCEPTUAL FRAMEWORK

Our argument is organized around an exploration of the interplay between place-making and mobilities in relation to subsistence gardening on Bell Island. We view gardening as one important way some Bell Islanders are remaking place in the wake of reverberations decades after an industrial closure. We thus consider the interconnections between food production rooted in specific spaces and the geographical movements fed by and feeding into this activity (Tuan, 1977: 152–60), as places are "acted out" (Certeau, 1984: 97–98) or made (Casey, 1996; Feld and Basso, 1996; Pink, 2008a).

Place attachment is often particularly strong in areas characterized by extensive mobilities, including in contexts of migration and industrial closures (Clifford, 1997; Mah, 2012; Olwig, 2007; Stack, 1996; Winson and Leach, 2002). Place attachment involves people developing, through practices, emotional bonds to locations associated with specific sets of meanings, memories, social relationships, and activities across various spatial and temporal scales (Low and Altman, 1992). Place attachment, therefore, occurs as part of active, ongoing processes. Bonds between people and places are reinforced through a variety of activities; these often include unpaid social reproductive labour such as housecleaning, home renovations, and food conservation and preparation (Massey, 1994; Roseman, 2002). Keeping a clean, orderly, and pleasant home, for example, can be "a strategy for providing a highly mobile population with a sense of stability" (Jones, 1985, in Boland Ahrentzen, 1992: 124). So, too, can subsistence gardening. Like unpaid labour within the home, places and people's attachments to them are often made and remade through gardening, whether on individually owned or leased plots or in community garden spaces (e.g., Halperin,1990: 68–69, 81–82; Milbourne, 2012; Stocker and Barnett, 1998).

As Mah (2012: 153) found in her research on the strength of place attachment in different locations impacted by industries shutting down or being scaled back, "neither mobility nor fixity creates a sense of loss," it is rather "limited [economic] choice." David Ralph and Lynn A. Staeheli (2011: 524) have emphasized "the importance of understanding home as simultaneously mobile and sedentary, as localised and extensible." They see home as being "like an accordion, in that it both stretches to expand outwards to distant and remote places, while also squeezing to embed people in their proximate and immediate locales and social relations" (see also Gustafson, 2001; Milbourne and Kitchen, 2014). Despite deeply questioning sedentarist assumptions, leading thinkers in the "mobility turn" literature also emphasize the links between people's mobilities and the locational, infrastructural places that enable and interact with them (e.g., Adey, 2010; Verstraete and Cresswell, 2003; Urry, 2003). The gardens of Bell Island can be understood as examples of the sort of material, institutional, and symbolic places-in-the-making that have a dialectical relationship with the movements of those who tend to them. This perspective on gardens parallels Tornaghi's consideration of how urban agriculture can be viewed as both an important livelihood activity and a form of "place-making" (Tornaghi, 2014). We examine how the continuity and revitalization of subsistence gardening not only contribute to food security, but more broadly constitute one among a series of ways in which residents enact their commitment to Bell Island's history and current context. Like other routine activities that occur in specific locations, gardening provides a conduit for reinforcing people's emotional attachments to their own properties, kin, and the island as a whole. As part of this process, it strengthens existing local conceptualizations and generates new sets of meanings that define their individual and shared cultural identities as Bell Islanders.

The interdisciplinary field of mobility studies emerged in the early 1990s and focuses on a wide range of mobilities, including movements of people and objects through geographical space and information through virtual space (Cresswell, 2006; Urry, 2010). Both are central to gardening on Bell Island, as we illustrate below. We are following a feminist approach to mobilities, labour, and livelihood that highlights how wages and other cash remittances such as pensions and income support cannot be extricated from unpaid tasks such as subsistence gardening — tasks that allow individuals to reproduce themselves and other members of their households and communities daily, weekly, and annually as well as through the generations (Luxton, 2006: 36–37; Brodkin Sacks, 1989; Roseman, Barber, and Neis, 2015). We are interested in the social reproduction of individual households, the community of Bell Island itself, and the broader kin and friendship networks that extend beyond. The unpaid labour of producing food for one's household and as gifts for extended family members and neighbours encompasses social reproduction at these various scales. As part of a commitment to remaining on the island in the wake of a devastating industrial closure, gardening is one example of residents' unpaid labour contributions to making and remaking Bell Island's place within the regional and global political economies (Tornaghi, 2014; Lefebvre, 1991).

In the mid-nineteenth century, the term "commuting" was employed in reference to American suburban dwellers using public transportation to reach their jobs in urban centres, which led to "the 'commutation' of their daily fares to lower prices, when purchasing tickets in monthly quantities"

(Muller, 2004: 64; Gregory et al., 2011). The term came to be used for various modes of transportation, from cars to bicycles, and for purposes other than reaching employment. Moreover, researchers highlighted how commuting journeys are often not one-purpose or one-stop trips (e.g., Hanson and Hanson, 1981; Hanson, 1985).

To highlight the various distances, itineraries, and travel modes related to gardening, the commuting mobilities in this case study involve commuting off-island to paid employment, commuting off-island to purchase tools and inputs used in gardening, and commuting locally within Bell Island as part of gardening. The idea of "commuting to garden" is empirically descriptive but also has broader metaphorical significance. Along with many of their neighbours, some of the six individuals portrayed here were continuing to commute and garden. If they and many of their neighbours had not begun commuting off-island to earn cash wages in the 1960s, they might not be living on Bell Island in the 2010s. However, as with the case of some of the gardeners discussed below, commuting by ferry and working full-time off-island creates extra challenges for keeping up the intensive work required for gardening. Some Bell Island gardeners delayed or constrained how much gardening they did earlier in life. Their years of commuting, in effect, allowed them to garden once retired or when their work schedules shifted from permanent and full-time to either temporary contracts or full retirement.

METHODS AND PARTICIPANTS

Our primary research method was to ask participants to take us on audio tours of their gardens and to participate in semi-structured interviews about their gardening memories and work during and after these guided visits. Our adoption of this method of interviewing follows Sarah Pink's use of "video tours" (Pink, 2006: 101). Each garden tour was documented using a digital audio recorder as our research participants showed us what they were growing in their gardens in 2015 and explained how they had built up their garden spaces, reflecting on links to earlier periods in their lives. This chapter also draws on participant observation that includes visiting, gardening work, and participation in community events. Two of the gardeners (George Hickey and Harriett Taylor) also took us on lengthy audio tours by car and explained pertinent information about Bell Island's gardening and agricultural history, among other topics. Glenda Tedford also accompanied us on a number of lengthy walking tours, during which she explained the island's history, including aspects related to food production. To study the history of gardening, we also consulted key sets of archival and secondary sources.

By asking people to take us on "garden tours," we reinforced the role of both gardening labour and narratives in the making and remaking of place (Tuan, 1991).[3] Through our questions, we elicited accounts of residents' memories of gardening from earlier periods in their lives and of their current gardening practices. The information relayed to us on these occasions parallels the kind of garden talk that Bell Island gardeners regularly share with kin and neighbours.

Two of the gardeners who participated in our study, Des McCarthy and Harriett Taylor, were long-term commuters who had travelled back and forth across the Tickle to work for many decades and were semi-retired at the time of the fieldwork. Des's garden in the West Mines area of the island sits next door to his childhood family home. In retirement, he expanded the size and range of his beds significantly, trying out new crops and techniques he learned about from talking to other gardeners, from reading, and from experimenting. In 2015, he grew a long list of vegetables including kale, multi-coloured Swiss chard, cauliflower, broccoli, and brussel sprouts. He also tended to fruit-bearing plants (rhubarb and strawberries), as well as cherry, pear, and apple trees. An enthusiastic cook, Des prepared many meals for his family using ingredients from the garden.

Also a career-long commuter, Harriett Taylor started to focus on vegetable gardening in retirement. When co-workers asked about her plans for retirement she answered, "gardening" — adding that she hoped to be a "green thumb on Bell Island." At her retirement party she received an envelope from her colleagues. "I looked inside and there was a picture of a greenhouse from Costco. It was a real surprise." But when post-tropical storm Leslie struck in September of 2012, it took down Harriett's new greenhouse. Together with her husband, Frank, they managed to salvage most of the structure and they rebuilt it the following year. In 2015, Harriett's greenhouse provided a warm, peaceful space where bright red tomatoes poked out from greenery — a mate for the lettuce, beets, and carrots that also grew there. Harriett regularly exchanged seeds, plants, and food within a close-knit network of extended family members and neighbours. They also gathered together to help with bottling and preserving.

Two additional gardeners, Fred Parsons and George Hickey, had also commuted off-island by ferry but over shorter periods. Along with many other Bell Islanders, when he was an adolescent, Fred worked in the Portugal Cove fish plant during capelin season. Fred was also mobile on the ocean and land in other jobs, including fish harvesting on his father's boats when he was young and later on vessels he and his brother owned, as well as employment in agriculture and carpentry. Fred's current gardening occurs in the greenhouse and beds adjacent to the home built by his grandfather, which Fred has lovingly renovated. Like many men, his grandfather came to Bell Island to work as a carpenter when the mines were operating. Fred attributes the centrality of subsistence gardening and preserving to his Bell Island upbringing. He grew up helping his parents manage the garden and livestock and he is now teaching his young daughter.

Like Fred, George did not commute by ferry for very long. Although he commuted across the Tickle for six years prior to retirement, most of his career was spent working in the hospital on Bell Island. George learned to garden roughly 50 years ago when, as a young boy, he worked for a Bell Island farmer. There, he learned lessons about food subsistence that were passed down to his four children, grandchildren, as well as other Bell Islanders when he was a volunteer with the Bell Island Development Association Farm Project (discussed below). When his children were young, he kept a large garden with potatoes, turnips, and cabbage – or, as he phrased it, "Your vegetables for a Sunday meal." He also raised livestock including cattle, pigs, and chickens. In 2015, George's garden was much smaller than in the past, but he was still growing potatoes and keeping chickens. He shared plans for expansion in 2016 because, as he put it, "I kind of misses it a bit."

The final two gardeners who participated in our study, Glenda Tedford and Dorothy Clemens, both held various wage jobs for many years in other parts of North America as well as on Bell Island. As a young girl growing up on The Front,[4] Glenda Tedford, formerly Bennett, remembers hearing about a downshift in her immediate family's gardening when her father started working in the mines. When the final mine closed in 1966, the Bennett family was just one of many out-migrant families to leave Bell Island for mainland Canada. In 2008, Glenda and her husband Bob retired back to Bell Island. Glenda's kitchen garden in 2015 consisted of two raised beds as well as a large planter box for herbs, built by her son. Glenda regularly grew vegetables, including potatoes, asparagus, turnip tops, and cabbage. Having a kitchen garden reconnected Glenda with early childhood memories of her grandparents' farm and parents' gardens. Dorothy Clemens is a long-term resident of Lance Cove, having moved there with her Bell Island-born husband at the beginning of the 1980s. She grew up a "farmer's girl" on the outskirts of Galt, Ontario, where her father owned a small greenhouse and her parents also kept a kitchen garden and livestock, including chickens and pigs. Dorothy's husband, Gerald, worked for a few years as a fish harvester. Although Dorothy and Gerald used to maintain a larger garden on their property, as well as on Gerald's parents' land, they have more recently "scaled back." Even so, her 2015 greenhouse played host to impressive-looking tomatoes, green beans, a grapevine, lettuce, herbs such as basil, as well as an array of flowers. Other plants were found growing in outside beds. These comprised a large assortment of vegetables, including lettuces, onions, radish, tomatoes, potatoes, carrots, garlic; fruits such as strawberries, rhubarb, raspberries, as well as black and red currants; and flowers. "I really love my garden," Dorothy maintained. "I look forward to spring."

A HISTORY OF GARDENING ON BELL ISLAND

Gardening, agriculture, and animal husbandry were central activities on Bell Island for centuries. Although the island's important mining history is often highlighted in writings about Bell Island and is very present in public representations of its past, islanders also emphasize the historical importance of other industries such as shipbuilding as well as the dominance of agrarian, fishing, and hunting activities.

Although we are not aware of any recorded archaeological surveys or excavations from the pre-colonial period, Bell Island would have almost certainly been a site for myriad subsistence activities by Indigenous populations. The island was used at least as early as the seventeenth century as a fishing station by mariners from England, Ireland, France, and the Channel Islands, as well as temporarily by settlers who later moved to places such as Bay de Verde (Hammond, 2004: 1–3). The first European considered to have been a permanent settler was the fisherman and farmer Gregory Normore from the Jersey Islands, who began living on Bell Island in the late eighteenth century together with his wife, Catherine Cook from Harbour Grace (Hammond, 2004: 3). Over the following centuries, alongside fish harvesting and processing, seasonal travel to the seal hunt, shipbuilding, brick work, carpentry, commerce, and mining, vegetable cultivation and animal husbandry for commercial as well as subsistence purposes remained central features of Bell Island's economy and society (Coxworthy, 1985, 1996; Hammond, 2004; Neary, 1975: 211; Rennie, 1998; Weir, 2006). The endurance of a pattern of multiple forms of employment and self-provisioning through much of the twentieth century on Bell Island was not uncommon in the Newfoundland context (Neary, 1975: 206). As Cadigan notes, on the northeast coast, "As early as 1785, . . . subsistence agriculture became essentially a subsidization of the mercantile fishery" (Cadigan, 1992: 52).

In different historical periods, governments have recognized and even encouraged commercial agriculture in Newfoundland to boost the availability of local food supplies and to promote economic diversification. For example, when 30-year leases for agricultural production in the vicinity of St. John's were first allowed in 1813, a registry indicated that approximately 1,000 acres were under cultivation illegally (Shaw, Drummond, and Murray, 1956: 25). Much of the arable area around St. John's came to be farmed (Murray, 2002).

Records from 1814 show that some early nineteenth-century agricultural activity on Bell Island was on rented land (Hammond, 2004: 9, 10, 41). By 1836, the census indicated that 359 people lived on Bell Island, 260 acres were owned, and 148 acres were being cultivated. These efforts to produce food for humans and livestock included 6,570 "bushels of potatoes" a year, 152 "tons of hay," and 120 "Meat Cattle" (Hammond, 2004: 11). By 1891, just a few years before mining began, Bell Island's population was 709. A late nineteenth-century account by Reverend Lloyd Rees provides a vivid portrait of farming homesteads in Lance Cove where "Everyone had a strawberry patch, a row or two of gooseberry and black current [*sic*] trees, and a few drills of 'small seed'" (Rees in Hammond, 2004: 14). Bell Island produce was well known in St. John's, as can be seen in newspaper accounts such as this one from the *Evening Telegram* in 1897: "The finest potatoes which came to the city these months past came from Bell Isle this morning." It was "[t]he farming and fishing family of John and Jabez Butler [that] sold their rights to the land to the Nova Scotia Steel Company, which started a mine in 1894" (Cadigan, 2009: 162; Martin, 2003: 53–54).

Figure 2.2. Bell Island. Houses and gardens, the Beach, c. 1904. (Geography Collection, Historical Photographs of Newfoundland and Labrador, Centre for Newfoundland Studies)

Many Bell Island men from fishing and farming families began working in the mines to supplement the livelihood that they and other members of their households continued to pursue with these other activities. Similarly, it was not unusual for residents whose households were established on the island only because of mining jobs to harvest wild seafood, hunt birds, and to pick berries as well as plant basic subsistence crops such as potatoes, turnips, and cabbage and to raise livestock. This economic mix was also practised by the families of miners who commuted weekly to work in the mines from homes in other communities, mainly from other parts of Conception Bay. Although the iron ore mines of Bell Island provided industrial employment, the work was seasonal for much of its history and layoffs occurred at various junctures. Even when miners had more steady shifts throughout the year, aside from engineers and executive staff, the wages of the men and boys who worked underground and of the few women who worked in the offices and in other functions were insufficient to fully support themselves and their families (Weir, 2006, also Martin, 2003).

The extent and purpose of food production on Bell Island have shifted over time, depending on both general and individual households' circumstances. For example, as authors such as Bown (n.d.) and Sheppard (2011: 50) point out, during the Great Depression, despite Dominion Steel and Coal Corporation's announced intention in 1930 to expand operations, two of the four mines were closed down completely. Meanwhile, work in the remaining two changed from full- to part-time, with operations being restricted to only two days each week. This pushed many mining families to turn to fishing and agriculture: "In order to survive, Bell Islanders returned to their agricultural heritage, growing crops to sell on the mainland or catching the resources of the sea around them" (Sheppard, 2011: 50). Sheppard cites Bown's report about 1934 indicating that "12,000 barrels of potatoes were grown and 66,000 gallons of milk were produced on the island" (Sheppard, 2011: 50; Bown, n.d.). Martin (2003: 58) notes:

> Those miners lucky enough to retain their jobs worked for a fraction of the normal salary and spent their spare time tending vegetable plots on land leased without fee from DOSCO. Unemployed miners compensated for lost pay by catching seals, seabirds and rabbits.

Like our research participants' accounts, Kay Coxworthy's collections of rich narratives about Bell Island contain descriptions that underscore the centrality of food production. In 1985, for example, Howard Dyer recounted: "Everyone I knew had a small farm — we had one ourselves. We had pigs, hens, a few cows and a horse. I remember in the Fall of the year, digging sixty or more barrels of potatoes and lots of turnip and cabbage to be used by our family during the winter" (Coxworthy 1985: 92). Luke Roberts Jr. described how gardening often preceded work in mining for male children:

> most families were large, so a man had to do his shift at the mines and then tend his garden, because they needed a lot of vegetables to feed them during the winter and spring . . . they had no automatic equipment then, it was all done by hand. So a man might keep his sons home, after grade eight, to work in the

> garden, because he knew that once a boy reached a certain age, he would go to work in the mines anyway. (Coxworthy, 1996: 52)

Bonds between kin and neighbours were strengthened through the necessity of shared food production, as Rosemarie Farrell explained about growing up in West Mines: "living next door were her father's brother Uncle Matt, Aunt Catherine and their children . . . the families were close, the brothers shared a vegetable garden and all of the kids would get together with their parents to plant and harvest the crops" (Coxworthy, 1996: 94).

By the mid-twentieth century, while Bell Island farmers and gardeners still supplied many of the local food needs, many families purchased staples as well as vegetables from producers or stores, with some coming from farms in the areas around St. John's (Murray, 2002: 156, 158). Various efforts encouraged agriculture and animal husbandry after the mines closed. These included the Bell Island Development Association Farm Project that provided employment and organized a community pasture as well as vegetable planting and storage (Bell Island Development Association, 1986, 1993; Bell Island Economic Development Committee, 1990). In recent years, Tourism Bell Island instituted a small community garden, planted garden boxes to supply the Keeper's Café located in the former lighthouse keeper's house, and began to grow vegetables and flowers in a greenhouse established and formerly used by the Operation Sunshine Garden Centre (Tourism Bell Island, 2015).

However, as occurred in other parts of Newfoundland and Labrador, despite the continuing operation of some commercial farms and gardens on the island, there has been a major decline in food cultivation and animal husbandry, and the population of Bell Island as a whole has become increasingly reliant on imported food purchased from stores (e.g., Kindl, 1999: 137; Food Security Network of Newfoundland and Labrador, 2014; Whitaker, 1963).

We now return to the six gardeners' accounts. The section is divided into three parts. The first highlights our research participants' memories of earlier gardening experiences and the second turns to the place-making that continues to occur through their current gardening activities. The third part details the diverse mobilities patterns associated with gardening.

GARDENING MEMORIES, PLACE-MAKING, AND MOBILITIES

Gardening Memories

As is the case for other Bell Island residents, gardening for our six participants involves both producing food and being connected with the past. During the garden tours, we asked about childhood memories of gardening and discovered that the majority were raised in families that maintained a garden as well as livestock. There were both commonalities and diversity in their memories of childhood gardening. Most referenced the significance of the garden and livestock for their family's food security. As Fred Parsons put it, "I wasn't reared up with a silver spoon or nothin', but I always had a bite to eat." Harriett Taylor remembered her parents sharing a garden with her grandparents: "Kept two families going, hey? For sure." Glenda Tedford recalled how important gardens and livestock were when the ice came into the bay:

> It was harder to get supplies because in winter the ice used to move in regularly; like in the bay. I have a picture, my brother does, of my dad pulling a sled with supplies on it across the ice in the bay. So it was harder to get supplies — so people had to be more self-sufficient. A lot of families were large.

Although George Hickey's family did not grow vegetables, he learned about vegetable cultivation from a local farmer. A few years later, when his father was laid off from the mines and went to work on the other side of Newfoundland in Port aux Basques, George recounted the centrality of both his wages and access to food from the farm where he worked: "I know for seven weeks we were waiting on a cheque. We lived off of my five dollars, plus all the vegetables and milk." Similarly, Des McCarthy started his first vegetable garden as a newlywed living with his in-laws. Although his father had maintained a garden in his youth, he "rarely spent a summer home" due to various activities (such as cadets), as well as paid labour. As an adult, it was his father-in-law who taught him how to plant and maintain a garden. He explained:

> Teresita's father had a little garden. When we first got married we stayed with them for a year, year and a half or so, and I put a little garden in. That was the first time I ever had a vegetable

garden. I must say that grew well. He had me started on beet, potato, carrot — that was it. That was it.

For those whose families had gardens on Bell Island, all referenced childhood gardening and other subsistence chores. As Fred recalled, "We used to have to take turns, go out and clean up the pigs, wash 'em, and everything." Much like Fred, Dorothy explained that everyone in her family in Ontario had garden chores: "It was a family thing. We all pitched in — Dad was fair. I mean everyone had a little job to do, right? Every family is like that." Des echoed her thoughts: "Everybody had their chores. Doesn't matter, you know, if you got two or three girls or two or three boys. They're all out there [in the garden]. They all got to do their duties. They all have their duties." In contrast, Glenda felt that birth order determined her relatively lesser chore-load in the garden: "I remember going around throwing scratch at the hens," she said, "But being the second youngest, the older ones were doing it [the chores]." Glenda's large family of 12 was not unusual on Bell Island in the mid-twentieth century.

In speaking about their chores in relation to their siblings, the interviewees also highlighted the many hours of hard work their parents and grandparents put into food production. This was always alongside many other paid and unpaid work obligations. As Des put it, "The women tended to be doing the wash, the dishes, stuff like that. While the guys were out there weeding or trenching. But the later years, it was kind of unisex." This statement mirrors patterns found by other researchers looking at the division of household labour, particularly in relation to food provisioning, within the broader Newfoundland context (Murray, 1979; Felt, Murphy, and Sinclair, 1995; Pocius, 1991; Porter, 1988; Sinclair, 2002). The oral histories in our Bell Island study suggest that, while there may have been a primary parent or grandparent gardener, the related tasks were managed regardless of age or gender. So even in families where there was a "point person" for gardening, many others helped out. This would have been particularly prevalent within island families — such as Glenda's grandparents' — who had large amounts of land and numerous responsibilities related to agriculture as well as taking care of livestock. As she explained: "My grandparents had their hay — they used to get it by the old graveyard up over the Beach Hill. That one. They had gardens and animals and really were quite self-sufficient. So it's always been there — although I was so young I didn't get to appreciate it or anything."

Among our research participants, the extent to which specific individuals in their families were involved in gardening varied over their lifetimes. As was the case in other mining households on Bell Island, Glenda recalled: "I remember Dad used to work double shifts, if he'd ever get them. So when you came home from that you didn't really want to be out gardening. You just didn't have the energy or the time and lots of times you didn't have the daylight to do it." When her father began working in the mines, Glenda's parents scaled down their garden and shifted responsibilities, despite having once maintained large amounts of land and livestock with her grandparents. Glenda's mother then kept a smaller kitchen garden. Similarly, Harriett's parents also shared a larger garden and livestock with her grandparents. But Harriett's mother moved into the role of primary gardener when her father was laid off from the mines and became increasingly mobile for work. As she described, "Well my mother kept a garden then because Dad was working. Where he was not home because he was on the boats, sailing and things like that." In retirement, however, her father planted "a nice patch behind the house" with potatoes, beets, onions, and dahlias.

We now turn from these vivid memories of gardening in the past to an analysis of how the six Bell Islanders who participated in this research performed place-making through gardening in outside beds and greenhouses near their homes, as well as through various mobilities that took them to and from these food production spaces.

Place-making through Gardening

To return to the opening story, like people's homes, businesses, and places of study and worship, Fred's greenhouse and neighbouring garden beds are examples of places being made and remade in a mobile world (Casey, 1996; Feld and Basso, 1996; Pink, 2008b; Tuan, 1977). Such gardening spaces have been built and are cultivated through various levels of mobilities both on and off Bell Island. Indeed, these gardens are often the starting or ending points in the journeys of mobile gardeners. Harriett, for example, often started a summer morning in her greenhouse, watering and tending to plants before using the ferry to commute to work across the Tickle on the Newfoundland mainland. Likely because we were not, by any means, the first to be shown people's gardens, the garden tours we were taken on constituted eloquent combinations of gesture and narration (Pink, 2006). The six profiled gardeners described the spaces where they carry out gardening activities, placing an emphasis on strength, recycled materials, security, and their own feelings of satisfaction from working in the garden.

As part of making place in and through food growing, the Bell Island gardeners spoke about the importance of building strong greenhouses, which are especially vulnerable in the coastal weather of a small island located in the waters of the North Atlantic. Structures must sustain year-round high winds, frequent tropical storms, and occasional hurricanes, as well as heavy winter snowfalls. Dorothy named her 2015 greenhouse "Igor" after the 2010 hurricane that "totally flattened" its predecessor. As noted earlier, Harriett's brand new greenhouse – a retirement gift from her colleagues across the Tickle – was blown down into her neighbour's garden after tropical storm Leslie. "It was just twisted metal," Harriett described. "I was heartbroken." Dorothy and Harriett both spoke about placing high importance on salvaging what they could from the wreckage – ultimately rebuilding stronger structures. But extreme weather is not the only risk to gardens on Bell Island. In comparing his 2015 garden to the previous year's, Des recounted, "There is no comparison between last year and this year. Last year I had nothing. I grew everything, but the cows got in. That's why you have the fence." As Des told it, several escaped cows from a nearby pasture enjoyed a garden feast at the end of summer: "I heard a bang. They were finished, they were on their way. They were going for dessert somewhere, I don't know," he said laughing. "They had their main course here, let me tell ya." Just as Des rebuilt a stronger fence around his garden, Harriett and her husband replaced the greenhouse's aluminum uprights with a wooden frame: "So we recycled everything off of the original, except for the aluminum uprights which the panels were on. Now he's just got them stapled to the wood." Harriett's greenhouse is now multi-purpose as it converts into a shed for firewood storage at the end of autumn.

The use of recycled materials in the gardens ranges from large structures to small knick-knacks – items obtained from Bell Island and across the Tickle. Fred described using recycled wood to rebuild his greenhouse in 2014 because of normal rotting over time: "This [greenhouse] come out of a place that was in a store – buddy tore it out and give me the lumber. I cleaned it up and made a greenhouse out of it. All recyclable lumber. Never went to the store and bought new stuff. All recyclable." In Dorothy's garden, her beds were lined with brick, rather than the more standard wooden frames found

Figure 2.3. Harriett Taylor's late summer harvest. (Photo by Diane Royal)

in most other island gardens. She explained, "I was going to get Gerald to make me some wooden frames, but then we had all the brick from the two chimneys off the old house roof and I said, 'I can use the brick.'" She also pointed out an old bathtub – relocated during a renovation and mostly used to store recycled water.

Dorothy noted the recycled parts that serve both functional and decorative purposes. In her greenhouse, she used pieces of what were once dark-blue blinds on the greenhouse's exterior, as well as hangers and curtain rods to keep plants off the ground. Inside, she pointed out a cleverly trellised grapevine: "He's just growing up through and around – and old curtain rods to hold it up." She added, "I like my ornaments too. I do all these things with bells. They are flea market finds. That's basically all they are." Yet Dorothy's "flea market finds" personalize her greenhouse and garden in a way that highlights the aesthetic contribution that food growing, like interior home decoration, brings (see Pocius, 1991: 99). All six of the interviewees spoke about the multi-sensory appeal of their gardens, also a draw for visiting relatives and neighbours (Milbourne, 2012).

Figure 2.4. Dorothy Clemens's mid-summer gardening. (Photo by Diane Royal)

Another form of place-making using recycled elements is soil creation. Both Des and George described the process of making their own soil as both active and ongoing – a garden activity that takes many years to perfect. Although Des also spoke about commuting across the Tickle to occasionally purchase soil, he described how he also makes his own – primarily through composting. The soil in his 2015 garden had been created "over the years, and with lots of manure in it." George, as well, described making garden soil using manure from his son's Bell Island farm: "My young feller got sheep up there. I go up – I got a bike and trailer and that – I go up and get sheep manure."

He uses a particular system of planting and manure management that he has honed over many years:

> I make my drills and then I set the potatoes. Then I bury it in the manure. Then I put the clay over the manure. Some people put the manure in the drills, then the potato. But I'm different – I does stuff a little. Because my mind is, the manure on the potatoes. When it rains and everything, it washes all the juice over the potato. But if the potato is on top and it rains, it's still there. The manure is still beneath the potatoes. I do it that way. I don't know if it's right. I don't know if it's wrong. But I'm having success with it.

Figure 2.5. George Hickey's after-harvest potato plot in fall. (Photo by Diane Royal)

For several gardeners in our sample, their gardens served as a form of food security for their own and other households, both on and off Bell Island. In this chapter, we are following the 1996 World Food Summit's definition of food security as "when all people at all times have access to sufficient, safe, nutritious food to maintain a healthy and active life" (WHO, 2015). This is particularly important on a small island where gardening can offer both an alternative and a higher-quality food source. As Fred described, gardening is a form of food security for his family of three: "We've been here weeks without a boat. Still survived. Because, with me, I stock a lot of stuff. Like meats. I makes tomatoes, has me beet, has me potatoes. Has enough stuff to do me for a month or two, right? If the boat don't run you got stuff on hand so you won't go hungry." When Fred and his wife Miranda moved into his family home, from an apartment where they did not have access to gardening space, it dramatically changed their ability to develop some self-sufficiency through food production, preservation, and storage.

Figure 2.6. Fred Parsons: raising tomato seedlings to prepare for spring planting. (Photo by Sharon R. Roseman)

We also believe that the Bell Island gardeners who took us on tours are acting partly from a food sovereignty impulse. Like the concluding statement that came out of the FAO World Food Summit in 2002, the six gardeners we spoke with believe in their "right to food and to produce food" (La Vía Campesina, 2002). For example, although Glenda regularly purchases groceries on the island as well as across the Tickle, having a garden means she can also feed her family organic vegetables. As she described:

> I'm very into organic or non-sprayed. Doesn't have to be totally certified organic, but I don't use any sprays on my garden. I haven't had it soil tested or anything; I'm not that extreme, but I just try and make it a bit better than you would in the grocery store. So that's a good part of the reason I started the garden.

Figure 2.7. Glenda Tedford's potatoes in progress. (Photo by Sharon R. Roseman)

The gardeners in our study described feelings of satisfaction from having a garden. As Glenda noted, "I just like going out into the garden and seeing something grow. It's just a nice feeling." She added, "It's so nice to pick potatoes because you just go under the soil and there's so many. It's like finding a treasure. It's amazing, I love it. I love that part of the gardening." But finding satisfaction in the garden is not only about "seeing something grow." It's also the full process of planning, planting, nurturing, and harvesting — all aspects of place-making. As Dorothy described, "It's a passion to get out here and clean up all the winter debris and start planning where I'm going to put this or that or the other thing. It's true." Des as well spoke about enjoying putting a lot of thought — including hours of research — into his garden. For all six of the interviewees in our study, the garden is a place of contentment. The derived satisfaction is, at least partially, due to the gardeners' hard work and immense dedication to the making of the garden space. The major effort of maintaining a vegetable garden was an aspect underscored by all six interviewees. In Dorothy's words, this involves "a lot of work; it's really a lot of work."

Some of the work that Dorothy and others identified involves not just the many hours of gardening labour but also the time, energy, and cost involved in moving to and from gardens in order to procure materials needed to grow food as well as to share food and information with others. We now turn to a discussion of the kind of mobilities associated with gardening on Bell Island.

Gardening Mobilities

In contrast to the garden as a physical and symbolic aspect of home and a place-making "anchor," we also focus on the mobilities to and from people's gardens (Urry, 2010). As in the past, in addition to ferry-dependent commuting across the Tickle, there are many "local" mobilities on Bell Island itself. The six Bell Island gardeners highlighted here move between their gardens and other spaces on the island as well to locations across the Tickle. Although many Bell Islanders, like Des and Harriett, still sometimes commute to work, we also emphasize that the ferry is used for purposes of social reproduction — including tasks related to gardening (Roseman, Barber, and Neis, 2015). Some Bell Islanders frequently use the ferry to gather gardening supplies and to exchange information and island-grown food. We consider the variability among their multi-modal trips, as well as the ways in which things and capital, people and information, travel back and forth across the Tickle (Urry, 2010).

As is common among highly mobile populations, trips across the Tickle varied widely among the profiled gardeners. Of the six interviewees, three described making regular ferry trips — roughly once a week or more. They spoke about picking up gardening inputs while simultaneously managing other tasks of social reproduction, such as banking. Des and Glenda referred to including gardening supplies on their already existing "to do" lists: "I'm in town a couple times [a week] anyway," Des commented. Glenda explained her strategic approach to commuting: "Whenever I go over, I always link stuff up. On my way back I'll get the gardening stuff." As well as occasionally crossing for gardening supplies, chiefly seeds, Harriett also described bringing garden-grown food and preserves across the Tickle for relatives.

Several of the gardeners spoke about bringing food, as well as plants, across the Tickle as a part of visiting practices occurring at increasingly broader mobility scales – related, in part, to out-migration from Bell Island to the Newfoundland mainland. Des described making plans to cross the Tickle to help his daughter, who grew up on Bell Island but now lives in Paradise, put in her first garden. He also spoke about gathering up garlic to get her started: "I got to get there next week because my daughter is starting a garden and I got to start that now – around the 20th of September. So she'll grow some garlic over the winter." Harriett also spoke about the importance of being able to share homegrown food and plants with relatives and friends, many of whom now live on the other side. Last year, for example, she got together with her island sister to bottle sauce using her garden's tomatoes. They then shared it with their children who live across the Tickle.

Self-provisioning requires the purchase of many items, both on Bell Island and across the Tickle. In addition to smaller, individual items such as seeds, several interviewees described having to organize larger loads. Such shopping patterns are an additional form of mobility related to gardening. For the most part, the gardeners in our sample used their own vehicles or travelled with relatives and neighbours. Although Glenda usually makes supply trips across the Tickle in her car, she spoke about her husband Bob having to go get soil with their truck and trailer: "We get a truckload, or Bob will go bring his trailer over and get a load when we were putting in our gardens. So he'd bring his trailer over to town and get a load of soil at that place on Portugal Cove Road." Although he mostly creates his own, Des also spoke about using his truck to haul soil back to Bell Island from across the Tickle – often a much bigger and more involved trip than when picking up seeds or plants.

Supplies, including gardening materials, also come over on the ferry to Bell Island, including those sold at local Bell Island stores. Fred especially described procuring most of his gardening supplies on Bell Island, rarely needing to cross the Tickle. Aside from the recycled lumber on his greenhouse, he spoke in detail about the process of buying the greenhouse's plastic covering, as well as other gardening items, locally. In the case of Harriett's original greenhouse, she explained how it was purchased in St. John's, "came in boxes," and was brought over on the ferry – delivered to her home by a local service. Although much of the gardening supplies are purchased on the Avalon Peninsula – mostly due to the prevalence of large nurseries – all of the gardeners expressed the crucial role that local, island options provide. One gardener's comment – "I get little things if I'm stuck for something" – was a prevalent description of island shopping patterns within our study.

As briefly noted above, the casual sharing of food across the Tickle and around the island is commonplace. On Bell Island, Dorothy spoke about exchanging vegetables and homemade food. Her husband still "shares some potatoes with his buddy up the road. So we usually trade back and forth." She added, "If I'm baking muffins there's half a dozen, I'll go down to the neighbour." Although Des described preparing warm lunches using food from the garden for his daughter who commuted from the mainland to Bell Island for work, he also spoke about the centrality of growing food to share with neighbours and friends. His next door neighbour always "gets a feed of broccoli," and a friend who bikes around the island for exercise makes frequent stops:

> she'll come up and she'll grab – she's on the bike every day, right? – and she'll go and have a few strawberries. A little energy, whatever. And if she sees anything she likes, she'll call me and say, "you know your kale is going well" and I'll end up picking some kale for her.

As is common throughout Newfoundland and Labrador, in cases of illness or death, Bell Islanders have long provided food support. The story of Fred picking five buckets of blueberries for a sick friend is just one of many examples. As Harriett explained:

> Even now, if somebody dies, if we know the family, we'll cook and send a pot of soup, or a pot of stew. Make a cake. Something they can have when they come home or if they have a large crowd. Something that can help them through so they don't have to cook. That's very common here.

Information about gardening also freely flows on the island as well as back and forth across the Tickle. Several of the gardeners in our sample described gathering information from the owners and others working in nurseries on the Avalon Peninsula while purchasing inputs. A few, Glenda especially, spoke about getting advice from relatives who garden. These included her sister Doris, who also lived on the island, her sister Joan in St. John's, and her sister Leona, who lived in Aurora, Ontario, but spent part of her summers on Bell Island. Des and Harriett spoke about gardening as a topic of conversation at their places of employment across the Tickle in St. John's. For Harriett, gardening discussions in the workplace helped motivate her to make gardening a priority as she looked toward retirement.

Advice can also come from people who cross the Tickle from mainland Newfoundland and elsewhere to visit Bell Island. Glenda, through her small business and volunteer work, interacted with many tourists and island visitors. She described often conversing with them about gardening. The plastic cover on her garden, she explained, resulted from a visitor suggesting she put plastic over the newly planted seeds to provide protection and heat up the soil. She maintained, "Anyone that can give me any advice, I take advice from them." George recalled consulting with visiting mainland gardeners in the 1980s when he volunteered with the Bell Island Development Association. Information about gardening is also widely circulated around the island. George has been gardening for almost six decades and is frequently asked to share tips with his Bell Island relatives and neighbours. Harriett referenced visits to the gardens of her neighbours and relatives. "It peaks my interest; seeing what they've got out growing," she said. Des, in turn, described people coming over to ask him about his plants while he worked in the garden. This form of visiting and sharing of information was common among all six of our interviewees. Receiving and sharing information virtually was also routine. Several gardeners, especially Des and Glenda, spoke about searching the Internet when they had questions. As Des put it, "I Google everything." Harriett exchanged gardening tips with former colleagues over e-mail: "Usually I e-mail now since they're retired. I e-mail them or take a picture of something. And say to them, 'How do I get rid of this?' 'Have you seen this before? What's my problem here?' And they are good like that." Dorothy discussed listening to talk radio, sometimes even calling in: "I've got a lot of treasures from the CrossTalk [CBC] radio station," she chuckled.

Figure 2.8. Grumpy keeping an eye on things. (Photo by Desmond McCarthy)

The last two sections of this chapter have examined how gardeners such as the six individuals profiled in this chapter have kept alive the practice of subsistence gardening on Bell Island, a practice that contributes significantly to place-making and place attachment in and through food production and forms of gardening commuting.

CONCLUSION

This chapter has focused on six Bell Island gardeners whose lives have been filled with numerous obligations, including both waged and unwaged labour tasks. Among the activities that have competed for their attention, they have all dedicated significant effort and time to cultivating food during the short growing season. Like elsewhere, continuing or returning to subsistence production in rural spaces in particular can be seen as part of food security efforts, food sovereignty politics, and even broader processes of reclaiming rural histories and identities (Roseman, 2002, 2008).

As in other areas of the province, Bell Islanders have participated for centuries in food harvesting and production activities for both subsistence and commercial purposes. For much of the 70-year period of the operation of the iron ore mines, many families living on and off Bell Island combined gardening with wage work in the mines. Although the boost to employment options represented by what became a large mining operation was welcomed on Bell Island and beyond, the wages that could be earned from most of these new jobs were not sufficient to support the often sizable families that were common in the early and mid-twentieth century. Therefore, our research participants and other Bell Islanders highlight how crucial it was to continue to have access to spaces where food could be grown and animals could be raised. It was also true that, in some households, mining employment and subsequent layoffs had a significant impact on the extent and nature of subsistence gardening in specific periods. In the latter decades of the twentieth century, reflecting a wider pattern of increased reliance on imported and commercial products, in combination with extensive out-migration and increased daily commuting, there was a decline in gardening on Bell Island. Bell Islanders did not, however, leave behind their memories of extensive gardening and farming — a core aspect of their history. Many, including the main research participants in our study, plan to continue to garden through their lives. Some have returned to more intensive gardening in recent years. Others described plans to expand their food production activities in the coming years.

On Bell Island, place-making through gardening is reinforced by the various associated mobilities, including the frequent movement of capital and things, people, and information. These mobilities involve travelling on foot, bicycle, and motorized vehicle around Bell Island itself, as well as crossing the Tickle by ferry to purchase gardening tools, seeds, and other inputs and to bring relatives homegrown products. Information about gardening similarly circulates both locally and throughout the Avalon Peninsula through visiting, the radio, the Internet, and analogue reading materials.

This research reinforces the point that the sense of security and feelings of attachment that come with continuous forms of place-making should not be viewed separately from the many mobilities that feed into these processes of commitment to place. Recent engagements with the "mobility turn" literature have explored this non-contradiction (e.g., Ralph and Staeheli, 2011: 524), an insight that has long been reinforced in foundational works by scholars representing a range of theoretical perspectives (e.g., Adey, 2010; Certeau, 1984; Clifford, 1997; Feld and Basso, 1996; Lippard, 1997; Urry, 2010: 253–70).

Bell Island has always constituted an interesting example of this conjunction, given its small island status and a shifting but always present reliance on marine mobilities to be connected to mainland Newfoundland and elsewhere. However, the challenges associated with striving to be firmly rooted in one's home space while living with myriad mobilities are particularly salient in cases of industrial closure. The use of the ferry for employment and social reproduction became increasingly fundamental after the final operating iron ore mine ceased production. This includes travel for the purpose of purchasing seeds, plants, tools, and other gardening inputs during or alongside commuting for employment as well as visits to kin for the purpose of food and meal sharing – mobilities that have contributed to strong extended family networks across the Tickle. As it always has been, producing food is a central aspect of Bell Islanders' commitment to place, history, and family – a commitment played out through the commuting mobilities that allow individuals such as Dorothy Clemens, George Hickey, Des McCarthy, Fred Parsons, Harriett Taylor, and Glenda Tedford to literally put down roots year after year.

ACKNOWLEDGEMENTS

First and foremost, we wish to thank the six Bell Island gardeners who generously agreed to be profiled in this chapter: Dorothy Clemens, George Hickey, Des McCarthy, Fred Parsons, Harriett Taylor, and Glenda Tedford. Our study would not have been possible without them. Their dedication to growing food for their families is inspiring, and we are grateful that they were willing to share their experiences and memories. This study is part of a large multidisciplinary team project. On the Move Partnership: Employment-Related Geographical Mobility in the Canadian Context is a project of the SafetyNet Centre for Occupational Health & Safety Research at Memorial University. On the Move is supported by the Social Sciences and Humanities Research Council through its Partnership Grants funding opportunity (Appl ID 895-2011-1019), Innovate NL, Government of Newfoundland and Labrador, the Canada Foundation for Innovation, and numerous university and community partners in Canada and elsewhere.

NOTES

1. This is a sub-sample of our larger, ongoing study on Bell Island commuting. Participants were recruited from participant observation and snowball sampling.

2. The term "tickle" is used for various places in the province where there is a "narrow difficult strait" (Story, Kirwin, and Widdowson, 1999: 565).
3. Quantitative methods have also been used to study populations' sense of place in different contexts (for example, see Cross et al., 2011).
4. "The Front" is a local place name referring to the area facing the "Tickle."

REFERENCES

Adey, Peter. 2010. *Aerial Life: Spaces, Mobilities, Affects.* West Sussex, UK: Wiley-Blackwell.

Antler, Ellen P. 1977. "Women's work in Newfoundland fishery families." *Atlantis* 2, 2: 106–13.

Bell Island Development Association. 1986. *The Bell Island Development Association Newsletter,* Jan.

———. 1993. "Across the Tickle." *The Bell Island Development Association Newsletter* 3, 4.

Bell Island Economic Development Committee. 1990. *Developing for the '90s.* Town of Wabana.

Boland Ahrentzen, Sherry. 1992. "Home as a workplace in the lives of women." In Irwin Altman and Setha M. Low, eds., *Place Attachment,* 113–38. New York and London: Plenum Press.

Bown, Addison. N.d. "Early history of Bell Island." At: http:/www.bellisland.net/history/earlyhistory.

Brodkin Sacks, Karen. 1989. "Toward a unified theory of class, race, and gender." *American Ethnologist* 16, 3: 534–50. doi:10.1525/ae.1989.16.3.02a00080.

Cadigan, Sean T. 1992. "The staple model reconsidered: The case of agricultural policy in northeast Newfoundland, 1785–1855." *Acadiensis* 21, 2: 48–71.

———. 2009. *Newfoundland & Labrador: A History.* Toronto: University of Toronto Press.

Casey, Edward. 1996. "How to get to space in a fairly short stretch of time: Phenomenological prolegomena." In Steven Feld and Keith Basso, eds., *Senses of Place,* 13–52. Santa Fe, NM: School of American Research Press.

Certeau, Michel de. 1984. *The Practice of Everyday Life,* Steven Rendall, trans. Berkeley: University of California Press.

Clifford, James. 1997. *Routes: Travel and Translation in the Late Twentieth Century.* Cambridge, Mass.: Harvard University Press.

Coxworthy, Kay. 1985. *Bell Island: Memories of an Island*. St. John's: Jesperson.
———. 1996. *"The Cross on the Rib": One Hundred Years of History*. Bell Island, NL: Creative Concepts Plus.
Cross, Jennifer E., Catherine M. Keske, Michael G. Lacy, Dana L. K. Hoag, and Christopher T. Bastian. 2011. "Adoption of conservation easements among agricultural landowners in Colorado and Wyoming: The role of economic dependence and sense of place." *Landscape and Urban Planning* 101, 1: 75–83. doi:10.1016/j.landurbplan.2011.01.005.
Doyle, Emily. 2014. *The Benefits and Sustainability of School Gardens: A Case Study of St. Francis School Greenhouse (Harbor Grace, NL)*. St. John's: Leslie Harris Centre of Regional Policy and Development. At: http://www.mun.ca/harriscentre/reports/arf/2012/12-13-TSP-Final-Doyle.pdf.
Evening Telegram. 1897. "Bell Isle produce," 20 Mar.
Feld, Steven, and Keith Basso, eds. 1996. *Senses of Place*. Santa Fe, NM: School of American Research Press.
Felt, Lawrence, Kathleen Murphy, and Peter R. Sinclair. 1995. "'Everyone does it': Unpaid work and household reproduction." In Lawrence Felt and Peter R. Sinclair, eds., *Living on the Edge: The Great Northern Peninsula of Newfoundland*, 39–56. St. John's: Institute of Social and Economic Research, Memorial University of Newfoundland.
Food Security Network of Newfoundland and Labrador. 2014. "10 ways to take action." At: http://www.foodsecuritynews.com/fact-sheets.html.
Gregory, Derek, Ron Johnston, Geraldine Pratt, Michael Watts, and Sarah Whatmore, eds. 2011. *The Dictionary of Human Geography*. Oxford: John Wiley & Sons.
Gustafson, Per. 2001. "Roots and routes: Exploring the relationship between place attachment and mobility." *Environment and Behavior* 33, 5: 667–86. doi:10.1177/00139160121973188.
Halperin, Rhoda H. 1990. *The Livelihood of Kin: Making Ends Meet "The Kentucky Way."* Austin: University of Texas Press.
Hammond, Rev. John W. 1982. *Wabana: A History of Bell Island from 1893–1940*. Grand Manan, NB: Print'N Press.
———. 2004 [1979]. *The Beautiful Isles: A History of Bell Island from 1611–1896*. St. John's: Pentecostal Assemblies of Newfoundland Printing Department.

Hanson, Susan, and I. Johnston. 1985. "Gender differences in work-trip length: Explanations and implications." *Urban Geography* 6, 3: 193–219. doi:10.2747/0272-3638.6.3.193.

Harvey, David. 1989. *The Condition of Postmodernity*. Oxford: Blackwell.

Kindl, Rita. 1999. "Change and continuity: Three generations of women's work in North West River, Labrador." MA thesis, Department of Anthropology, Memorial University of Newfoundland.

Jones, Jacqueline. 1985. *Labor of Love, Labor of Sorrow: Black Women, Work and the Family from Slavery to the Present*. New York: Basic Books.

La Vía Campesina. 2002. "Declaration NGO Forum FAO Summit Rome+5," 13 June. At: http://viacampesina.org/en/index.php/main-issues-mainmenu-27/food-sovereignty-and-trade-mainmenu-38/398-declaration-ngo-forum-fao-summit-rome5.

Lefebvre, Henri. 1991 [1974]. *The Production of Space*, Donald Nicholson-Smith, trans. Oxford: Blackwell.

Lippard, Lucy R. 1997. *The Lure of the Local: Senses of Place in a Multicentered Society*. New York: New Press.

Low, Setha M., and Irwin Altman. 1992. "Place attachment: A conceptual inquiry." In Irwin Altman and Setha M. Low, eds., *Place Attachment*, 1–12. New York and London: Plenum Press.

Luxton, Meg. 2006. "Feminist political economy in Canada and the politics of social reproduction." In K. Bezanson and M. Luxton, eds., *Social Reproduction: Feminist Political Economy Challenges Neo-liberalism*, 11–44. Montreal and Kingston: McGill-Queen's University Press.

Mah, Alice. 2012. *Industrial Ruination, Community, and Place: Landscapes and Legacies of Urban Decline*. Toronto: University of Toronto Press.

Martin, Wendy. 2003 [1983]. *Once Upon a Mine: Story of Pre-Confederation Mines on the Island of Newfoundland*. Special Volume 26. Canadian Institute of Mining and Metallurgy. Ste-Anne-de-Bellevue, Que.: Harpell's Press Cooperative.

Massey, Doreen B. 1994. *Space, Place, and Gender*. Minneapolis: University of Minnesota Press.

Milbourne, Paul. 2012. "Everyday (in)justices and ordinary environmentalisms: Community gardening in disadvantaged urban neighborhoods." *Local Environment* 17, 9: 943–57. doi:10.1080/13549839.2011.607158.

——— and Lawrence Kitchen. 2014. "Rural mobilities: Connecting movement and fixity in rural places." *Journal of Rural Studies* 34: 326–36. doi:10.1016/j.jrurstud.2014.01.004.

Muller, Peter O. 2004. "Transportation and urban form: Stages in the spatial evolution of the American metropolis." In S. Hanson and G. Giuliano, eds., *The Geography of Urban Transportation*, 3rd ed., 59–85. New York: Guilford Press.

Murray, Hilda Chaulk. 2002. *Cows Don't Know It's Sunday: Agricultural Life in St. John's.* St. John's: ISER Books.

———. 2010 [1979]. *More Than 50%: Woman's Life in a Newfoundland Outport.* St. John's: Flanker Press.

Neary, Peter. 1973. "'Traditional' and 'modern' elements in the social and economic history of Bell Island and Conception Bay." *Historical Papers/ Communications historiques* 8, 1: 105–36. doi:10.7202/030763ar.

———. 1975. "The epic tragedy of Bell Island." In Joseph R. Smallwood, ed., *The Book of Newfoundland*, 201–30. St. John's: Newfoundland Book Publishers.

Nelson-Hamilton, Laura. 2011. "Growing together: The intersections of food, identity, community, and gardening in St. John's, Newfoundland." MA thesis, Department of Gender Studies, Memorial University of Newfoundland.

Olwig, Karen Fog. 2007. *Caribbean Journeys: An Ethnography of Migration and Home in Three Family Networks.* Durham, NC: Duke University Press.

Omohundro, John T. 1985. "Efficiency, sufficiency, and recent change in Newfoundland subsistence horticulture." *Human Ecology* 13 (Sept.): 291–308. doi:10.1007/BF01558253

———. 1994. *Rough Food: The Seasons of Subsistence in Northern Newfoundland.* St. John's: ISER Books.

———. 1995. "Living off the land." In Lawrence Felt and Peter R. Sinclair, eds., *Living on the Edge: The Great Northern Peninsula of Newfoundland*, 103–27. St. John's: Institute of Social and Economic Research, Memorial University of Newfoundland.

Pink, Sarah. 2006. *Doing Visual Ethnography*, 2nd ed. London: Sage.

———. 2008a. "An urban tour: The sensory sociality of ethnographic place-making." *Ethnography* 9, 2: 175–96. doi:10.1177/1466138108089467.

———. 2008b. "Mobilising visual ethnography: Making routes, making place and making images." *Forum Qualitative Sozialforschung/Forum: Qualitative Social Research* 9, 3.

Pocius, Gerald L. 1991. *A Place to Belong: Community Order and Everyday Space in Calvert, Newfoundland.* Montreal and Kingston: McGill-Queens's University Press.

Porter, Marilyn. 1983. *Place and Persistence in the Lives of Newfoundland Women.* Aldershot, UK: Avebury Press.

———. 1988. "Mothers and daughters: Linking women's life histories in Grand Bank, Newfoundland, Canada." *Women's Studies International Forum* 11, 6: 545–58. doi:10.1016/0277-5395(88)90108-2.

Ralph, David, and Lynn A. Staeheli. 2011. "Home and migration: Mobilities, belongings and identities." *Geography Compass* 5: 517–30. doi:10.1111/j.1749-8198.2011.00434.x.

Rennie, Rick. 1998. "Iron ore mines of Bell Island." Newfoundland and Labrador Heritage Website. At: http://www.heritage.nf.ca/articles/economy/bell-island-mines.php.

Roseman, Sharon R. 2002. "'Strong women' and 'pretty girls': Self-provisioning, gender, and class identity in rural Galicia (Spain)." *American Anthropologist* 104, 1: 22–37. doi:10.1525/aa.2002.104.1.22.

———. 2008. *O Santiaguiño de Carreira: o rexurdimento dunha base rural no concello de Zas.* A Coruña: Baía Edicións.

———, Pauline Gardiner Barber, and Barbara Neis. 2015. "Towards a feminist political economy framework for analyzing employment-related geographical mobility." *Studies in Political Economy* 95: 175–203. doi:10.1080/19187033.2015.11674951.

Sheppard, Nath W. 2011. *Historic Bell Island: Dawn of First Light.* St. John's: Flanker Press.

Sinclair, Peter R. 2002. "Narrowing the gaps? Gender, employment and incomes on the Bonavista Peninsula, Newfoundland, 1951–1996." *Atlantis* 26, 2: 131–45.

Stack, Carol. 1996. *Call to Home: African Americans Reclaim the Rural South.* New York: Basic Books.

Statistics Canada. 2017. "Census profile, 2016 census." At: http://www12.statcan.gc.ca/census-recensement/2016/dp-pd/prof/index.cfm?Lang=E.

Stocker, Laura, and Kate Barnett. 1998. "The significance and praxis of community-based sustainability projects: Community gardens in Western Australia." *Local Environment* 3, 2: 179–91. doi:10.1080/13549839808725556.

Story, George M., William J. Kirwin, and John D. A. Widdowson, eds. 1999. *Dictionary of Newfoundland English*, 2nd ed. Toronto: University of Toronto Press.

Tornaghi, Chiara. 2014. "Critical geography of urban agriculture." *Progress in Human Geography* 38, 4: 551–67. doi:10.1177/0309132512542.

Tourism Bell Island Inc. 2015. "Projects." At: http://www.tourismbellisland.com/project_details.html?id=9&sub_id=13#a.

Tuan, Yi-Fu. 1977. *Space and Place: The Perspective of Experience*. Minneapolis: University of Minnesota Press.

———. 1991. "Language and the making of place: A narrative-descriptive approach." *Annals, Association of American Geographers* 81: 684–96. doi:10.1111/j.1467-8306.1991.tb01715.x.

Urry, John. 2003. *Global Complexity*. Cambridge: Polity Press.

———. 2010. *Mobilities*. Cambridge: Polity Press.

Verstraete, Ginette, and Tim Cresswell, eds. 2003. *Mobilizing Place, Placing Mobility: The Politics of Representation in a Globalized World*. Amsterdam: Rodopi.

Weir, Gail. 2006 [1989]. *The Miners of Wabana: The Story of the Iron Ore Miners of Bell Island*, 2nd ed. St. John's: Breakwater Books.

Whitaker, Ian. 1963. *Small-scale Agriculture in Selected Newfoundland Communities*. St. John's: Institute of Social and Economic Research, Memorial University of Newfoundland.

Winson, Anthony, and Belinda Leach. 2002. *Contingent Work, Disrupted Lives: Labour and Community in the New Rural Economy*. Toronto: University of Toronto Press.

World Health Organization (WHO). 2015. "Food security." At: http://www.who.int/trade/glossary/story028/en/.

3

Living Lessons of the School Food Environment: A Case Study of a School Greenhouse in Newfoundland and Labrador

Emily Doyle & Martha Traverso-Yepez

INTRODUCTION

In Canada and around the world, there is growing awareness of interconnections among school food, health, learning, and environmental sustainability (Ashe and Sonnino, 2013; Bagdonis, Hinrichs, and Schafft, 2009; Black et al., 2015; Heart and Stroke Foundation, 2013; Henderson, 2011; Morgan and Sonnino, 2008; Robert, 2011; Rojas et al., 2011). A farm-to-cafeteria program is one example of a school food intervention that emphasizes the positive interconnections of food, health, and learning. It brings local produce onto the school plate by connecting students and schools to local food systems. Another is a school gardening program that exposes children to healthy food and teaches both curriculum and ecological awareness via food cultivation. These two examples of interventions, here defined as actions taken to improve the state of the school food environment, attempt to integrate school food with health, the curriculum, and the environment. They can be contrasted with more restricted (and somewhat unsuccessful) interventions that promote the consumption of healthy food through health education messages or changes in food offerings (Olstad, Raine, and Nykiforuk, 2014). A complex view of school food takes into account the interaction of individual, social, and physical features that influence what type of food is eaten at school, what is learned about food in the curriculum, and how schools interact with the surrounding environment and food-related practices (Olstad et al., 2014; Robert, 2011; Rojas et al., 2011). This view of the school food environment is founded on an ecological health framework.

Within the fields of both food studies and school health, ecological (or systems) frameworks are increasingly being used to define the healthfulness of the school environment. The Comprehensive School Health (CSH) framework is one such example. It is advocated by the World Health Organization (WHO), the International Union of Health Promotion, and in Canada, the Joint Consortium for School Health, of which the province of Newfoundland and Labrador is a member (International Union for Health Promotion and Education, 2009; Joint Consortium for School Health, 2013; WHO, 2016). The CSH model recognizes the interdependence of education, health, and the community in creating a healthy school environment. Its underlying values can be traced back to the philosophy of ecological health introduced by the Ottawa Charter for Health Promotion (1986). Ecological health promotion focuses on individuals, the community, and institutional and public policy levels to make the healthy choice the easy choice (McLeroy, Bibeau, Steckler, and Glanz, 1988; WHO, 1986).

Between 2005 and 2009, the Newfoundland and Labrador government invested over $6 million to create the "Healthy Schools Healthy Students" (HSHS) initiative. This partnership between the Department of Health and Community Services and the Department of Education was a progressive endorsement of Comprehensive School Health. While almost a decade has passed since the introduction of Newfoundland and Labrador's CSH program, there has been little publicly available monitoring or evaluation of this initiative (Card, 2008). Research about the food environment of schools in Newfoundland and Labrador has also been scarce (Coalition for School Nutrition, 2001; Goss Gilroy Inc., 2013; Hanrahan and Ewtushik, 2001; Health Research Unit, 2012).

The CSH framework is used to structure this case study of a school greenhouse program in the rural community of Harbour Grace, Newfoundland and Labrador. We consider how student health is a reflection not only of their individual decisions but also of the influence of a complex interaction of community, organizational, and environmental factors (McLeroy et al., 1988; Rayner, 2012). A number of recent investigations of the food environment use an ecological framework (Ashe and Sonnino, 2013; Black et al., 2015; Blay-Palmer, 2016; Lang, 2009; Olstad et al., 2014; Rojas et al., 2011).

Different manifestations of this systematic approach to understanding the food environment exist in Canada (Black et al., 2015; POWER UP!, 2015; Rojas et al., 2011).

One example is the Think and Eat Green @ School (TEG@S) project in British Columbia, which aims to understand how the public school system can reconnect food, health, and the environment (Rojas et al., 2011). As part of this project, Black et al. developed a tool inspired by the CSH framework. This conceptual framework is used to measure the level of a school's engagement in food systems. It takes into account the individual, social, economic, and ecological factors and processes that may influence the health of children at school (Black et al., 2015).

Research on the implementation of CSH reveals that one of the greatest challenges is sustaining comprehensive co-ordination among multiple stakeholders and across differing contexts (Bassett-Gunter, Yessis, Manske, and Gleddie, 2015; Deschesnes, Martin, and Hill, 2003; McIsaac, Read, Veuglelers, and Kirk, 2013). This case study adds to the literature by showing how the CSH framework may be implemented successfully to build a sustainable school food environment. The St. Francis School greenhouse in Harbour Grace also offers an opportunity to see the development of a food environment intervention in the context of the province of Newfoundland and Labrador. The long history of the greenhouse facilitates an assessment of what factors enabled co-ordination and what factors may have inhibited the development of this intervention.

CASE STUDY METHODOLOGY AND DATA PRESENTATION

The data included in this case study were collected throughout 2013. Initially, this study was framed as a pilot investigation into the health benefits of school gardens in Newfoundland and Labrador. An ecological model was adopted as way to gauge the multiple potential health benefits of this school garden (Ozer, 2007). The case study was considered to be the ideal method for engaging with this particular school about the topic. It permitted investigation into how multiple stakeholders experienced and perceived unique circumstances that led to the development of the St. Francis School greenhouse (Denzin and Lincoln, 2011; Flyvbjerg, 2006). In total, 14 open-ended interviews were held with teachers, administrators, community members, and government officials who were connected to the St. Francis greenhouse throughout the 20 or so years of its development. The interviews were followed by a focus group during which initial findings were presented and discussed among participants. The interviews, focus group, and continued contact maintained with the school community helped to inform the process, and are the foci of this study.

Initial data analysis about the case and continued exposure to growing research on the school food environment led to the adoption of the CSH model to define the full range of interactions among potential ecological health factors and outcomes. In the following section, the greenhouse is considered from multiple angles. This includes a brief description of the development of the greenhouse, the policy environment, the social and physical environment, the teaching and learning environment, and community partnerships. The goal is to understand how the introduction of and use of a school greenhouse transformed the school food environment and how these changes influenced student's learning and health.

Figure 3.1. The St. Francis School and greenhouse. (Photo by Emily Doyle)

ST. FRANCIS GREENHOUSE: TWO DECADES OF EXPERIENCE WITH A SCHOOL FOOD INTERVENTION

St. Francis School is in Harbour Grace, a town in Conception Bay, NL, with a population of about 3,000 (Statistics Canada, 2012). In the early twentieth century Harbour Grace had the second largest population in the province, and therefore it was the site of some of the province's first schools (Pitt, 2015). St. Francis opened in 1961 as a Catholic school, becoming non-secular in 1998. The school has since transitioned from a high school to a junior high and now accommodates students from kindergarten to the eighth grade.

The inspiration to build a greenhouse is said to have originated with a previous school custodian, Gustav Reinhart, to whom the greenhouse was officially dedicated. Beginning work at the school in 1974, Reinhart was an inspiration to the staff and students, meticulously caring for the school grounds by growing flowers and shrubs. He also kept a wooden greenhouse in the back of the schoolyard (Bowman, 1994). The St. Francis greenhouse, built in 1992, is a 2,400-square-foot polycarbonate structure (Figure 3.1). At the time, it was a model for technological innovation, featuring automated watering, ventilation, and lights. The vice-principal at this time, identified as RB, was a strong driving force behind the greenhouse. He felt automation was necessary in order to manage the greenhouse in the busy school environment. In addition, students would be exposed to the latest technology in the horticulture industry and would learn as much about computers as they would about plants and entrepreneurship. The original financial plan for the greenhouse was that students would be able to generate enough money from the sale of tree seedlings grown in the greenhouse to sustain the cost of its operation.

Since the beginning, the greenhouse has been used for different purposes and has relied on different sources of funding and support. These sources include the Eastern School District (now the Newfoundland and Labrador English School District), the Lower Trinity Regional Economic Development Board (now defunct), the Department of Health and Community Services, the Department of Natural Resources (which used the greenhouse to feed the 2012 NL Summer Games athletes), Environment Canada, and the provincial Department of Environment and Conservation. Today, the greenhouse shows some signs of depreciation and it is in need of a new influx of funding to reinvigorate both the infrastructure and programming. However, there are promising signs of a partnership involving the school, a nearby farm, and a local catering company, the latter of which would supply the school cafeteria with produce that students would grow in the greenhouse (Robinson, 2015). While the greenhouse was not initially planned as a "school food intervention," it is described in this way both because for much of its history the greenhouse has been a site of food production and also because this appears to be the direction of this facility in the years to come (Figure 3.2).

Figure 3.2. The greenhouse in the 1990s. (Photo by St. Francis schoolteacher)

POLICY, LEADERSHIP, AND MANAGEMENT OF THE ST. FRANCIS SCHOOL GREENHOUSE

This section explores how leadership and management practices have interfaced with shifting policy directions throughout the history of the St. Francis School greenhouse. Broadly speaking, some significant policy-related issues influenced the creation and development of the greenhouse. In the 1990s, the cod moratorium had a provincial-wide impact on the Newfoundland and Labrador education system. At this time funding became available for projects that would help to expand the economy outside the fishery. In the early 1990s, there was a strong push from both the federal and provincial governments to find alternative employment opportunities and education for young people (Sheppard, 2003). RB, the vice-principal, was looking for something that would make "teaching as real as possible." RB promoted the idea of a school greenhouse as a way to give students the experience of operating a business. The business plan was to grow tree seedlings to sell throughout the province. He contacted greenhouse manufacturers across North America to learn about potential designs and approached Human Resources Development Canada (HRDC) and the Atlantic Canada Opportunities Agency (ACOA) with his proposal, which was subsequently approved. The concept was innovative for its time, offering students considered likely to drop out of school ("at-risk students") cross-curricular, hands-on instruction to increase their interest in school.

The transition from secular to non-secular schools in 1998 reflected another important political change. One of the participants remarked that this transition resulted in an increasing amount of bureaucracy and a decreasing amount of control at the school level. This is noteworthy given that one of the primary policy objectives of this educational reform was to increase local involvement in decision-making by transferring full control of the provincial education system from the churches to the provincial government (Kelly, 1997). As part of the shift from a denominational to a non-denominational system, St. Francis was converted from a high school to a junior high. One of the consequences of this decision was that courses designed for implementation in the greenhouse – horticulture and entrepreneurship – no longer neatly fit into the junior high curriculum. Those interviewed also noted that the school experienced a high turnover rate of staff after this transition. This meant that whole-staff projects, previously maintained from year to year, were harder to sustain as there were fluctuations in the staff each year.

Another significant policy that influenced the greenhouse intervention was the Healthy Schools Healthy Students initiative. The greenhouse found its second revival as a complement to the Department of Education's "Healthy Living" curriculum for schoolchildren in Newfoundland and Labrador (Public Health Agency of Canada, 2009). The enactment of this new vision of the greenhouse came from the Lower Trinity South Regional Development Association (LTSRDA), which continued to use the greenhouse to engage children in entrepreneurial activities. This time, however, children were growing food themselves, thereby empowering them to learn about healthy food and how to grow and process it (Public Health Agency of Canada, 2009; Sullivan, 2011). Throughout the early 2000s, much of the programming was executed by the regional economic development organization. In 2013, Newfoundland and Labrador's regional economic development organizations lost both national and provincial funding, which undercut the services that had been available to children.

Leadership for the greenhouse has come from administration, and the school council has also played a critical role in sustaining and supporting the greenhouse throughout its existence. When the school district considered dismantling the greenhouse, the school council rallied to its support. Research participants described the early years of greenhouse development as "more flush with resources" and they stated that, in recent times, there have been more cuts at the school and community levels. This has made it challenging to maintain the high level of support and human resources needed to fulfill the initial vision.

Shifts in policy, leadership, and management have influenced both the development and the attrition of the greenhouse. The bottom line is that, despite these shifts, the greenhouse still stands more than 20 years after being built and it is still an inspiration for the school and the surrounding community. Almost everyone interviewed mentioned the substantial amount of time, patience, and energy people had to invest to ensure greenhouse projects were running smoothly. This constant upkeep often required visiting the greenhouse after hours and on weekends. Greenhouse programming also required knowledge about plant production, curriculum development, and extra supervision of students — even project development and management skills — "and resources weren't there for that." In the past, the school had a committee responsible for keeping the projects moving, and there is currently interest in setting up a committee to oversee the greenhouse program. The reality is that even establishing a committee takes extra time and resources. Participants commented on the fact that opportunities depend on resources and leadership: "The opportunities are endless. If somebody had the resource to take it and run with it. A lack of resources has led to underutilization."

TEACHING, LEARNING, AND THE ST. FRANCIS SCHOOL GREENHOUSE

In the early years, learning about the growth of tree seedlings, the operation of the greenhouse, and the development of the business offered students a hands-on and cross-curricular learning opportunity. Active engagement was often a natural component of teaching and learning in the greenhouse and had a positive influence on the greater teaching and learning environment in St. Francis School.

In those early days of the greenhouse enterprise, staff members were encouraged to design a curriculum that would allow them to use the greenhouse. This initial push from administration led to the development of a horticulture course and also helped the school become a leader in the area of technology education. The greenhouse was equipped with video cameras that allowed students from neighbouring communities to learn about greenhouse operations from a distance. The video cameras led to teaching video production and marketing and the early adoption of smart boards (i.e., interactive touch-sense white boards) in the school. The impact of these changes on the learning environment was positive. Several years after the greenhouse's debut, a newspaper article reported the success of students involved with the greenhouse: "Even though one third of the students . . . are considered to be at risk of dropping out, none of the enterprise students have dropped out since the course began two years ago and it has recorded a 100 per cent pass rate" (Bowman, n.d.).

When the school transitioned into a junior high and the Lower Trinity Economic Development Board began to use the greenhouse to grow food rather than tree seedlings, the benefits of hands-on and cross-curricular learning persisted (Figure 3.3). The teachers who were interviewed believed that hands-on learning led to enhanced memory formation and also to an enhanced sense of ownership and a positive sense of involvement for students. "It was amazing to see how much pride those students showed in terms of what they were able to accomplish in the greenhouse." Giving students the chance to engage in a project with a tangible purpose proved to be highly meaningful and beneficial — especially, teachers observed, for those who may not have excelled at traditional pencil-and-paper activities. By providing an alternative learning environment, the school was able to accommodate different learning styles, thus promoting inclusive education. Learning in the greenhouse classroom also enabled a caring environment as students learned to work better together:

> The ones who came in fast and furious had to slow down because there are hoses and things to trip over. You have to get the plant in the centre of the pot and you can't rip out the root. So that type had to slow down. And the one that was shy, meek, and mild did end up saying "My turn in there, my turn!"

Figure 3.3. Food growing in the St. Francis greenhouse. (Photo by St. Francis schoolteacher)

Teachers were also motivated to learn, as expressed by one respondent: "Part of it was for me to learn a little bit more, like, can we grow corn? How do we grow corn? I think the whole thing has been a really good learning process, not only for the kids, but for me and the student assistants." Teachers often took pride in the greenhouse and shared positive memories from their experience: "I do remember the looks on their faces and how exciting it was to be outdoors and to be gardening in October."

Some examples of project-based learning or hands-on learning organized around the investigation and resolution of real-world problems (Smith and Gruenewald, 2008) took place in the greenhouse. In one instance children transplanted lettuce started in the greenhouse to a local farm. When the lettuce was mature, they harvested, bagged, labelled, and sold it at the local grocery store. In another case, students grew tomatoes, then processed the tomatoes into salsa and sold it. These projects offered links to subjects taught in the curriculum, such as the chemistry and biology of growing and preserving plants, the mathematics of selling produce, and the language of marketing.

As the literature shows, there is a link between active engagement and students' health and well-being. That is, the more freedom students are given to participate in decision-making and the learning process, the more likely they are to develop intrinsic motivation that can positively enhance academic achievement and overall well-being (Rowling and Jeffreys, 2006). Yet the planning, execution, and evaluation of active engagement in schools can be challenging. The development of project-based learning requires extra teaching resources, which have been achievable during times of funding. However, a number of participants in this study pointed out that classroom teachers do not generally have enough time on top of their current workload to easily design these potentially cross-curricular, project-based experiences within the established curriculum and current evaluation schemes.

THE INTERRELATED SOCIAL AND PHYSICAL ENVIRONMENTS OF THE ST. FRANCIS SCHOOL GREENHOUSE

At times there was a noticeable effect of the greenhouse on the social environment of the school and the surrounding community. The greenhouse was often cited as a source of pride for those interviewed, and it seemed to have the effect of drawing people together. Examples of this effect occur at a number of levels: first, as noted above, students learned to co-operate in new ways; second, the formation of a greenhouse committee and a greenhouse after-school club increased involvement and interaction between teachers and the community; third, the greenhouse led to projects that connected St. Francis with other schools throughout the province; and finally, the greenhouse depended on the involvement of a long list of partners, which will be discussed in further detail below.

Although the physical environment is considered an important aspect in developing a healthy school, it is often given less attention than factors such as leadership and governance (McIsaac, Storey, Veugelers, and Kirk, 2015). One of the study participants believed that the physical attributes of the greenhouse contributed to its success and longevity:

> Twenty-five years ago, whoever had the foresight to say I'm spending a lot of money and I'm getting the best facility I can [got it right]. I'm not going with wood and plastic, I'm going with glass and steel. And that greenhouse has stood pretty solid for up to 20 to 25 years — when I walked in my first impression was: Wow! This is an expensive facility, and that they had the foresight to say, we're going with the top-notch, we're not going half in, we're not going slap-happy. We're investing in a solid structure. And to me, that was one of the best decisions that they made.

This participant's comment supports the notion that the built environment is important in creating healthy school communities. With the physical existence of the greenhouse came new learning possibilities for teachers, students, and the surrounding community. Students were also motivated by the aesthetics of the greenhouse, as commented by another teacher: "They really liked playing in the dirt. I think they thought, sensory-wise, it was probably very sensory fulfilling. Plus, there's something about the light out here and the heat and the building structure itself — they really seem to enjoy that part of it."

Teachers noted that the greenhouse had a therapeutic effect on them. It was also used as leverage for grant proposals: "I was applying for different grants, and people saying all right, what he's doing out there, he's pretty innovative, so we'll give him that grant." In addition, there is an important interaction between the physical environment and the social environment in determining the health of students. One participant remarked:

> Anything that helps with the coming together of people is going to help improve the collective health of the group because we know that sense of connection and diminishing of the isolation, sharing of skills and ideas, the sense of camaraderie, [it] helps if

> people have that connection, the chances are that they're going to feel better about themselves.

However, the difficulty of measuring these health benefits was also noted by a respondent:

> Can you say, "Okay, we build a garden and chronic disease goes down?" Probably not, but if you want to be able to make that environment, then do you want to make that investment too, because eventually things become mainstream? They become the common.

This comment highlights the difficulty of researching or evaluating health within an ecological framework. It also provides insight into how the "built environment" helps to define the baseline that becomes a common or acceptable way of living. This, in turn, leads to questions about current approaches to school health investment. Currently, most such investment comes in the form of small project grants. However, this type of short-term investment would not yield the significant change to the built environment that occurred at St. Francis School.

COMMUNITY PARTNERSHIPS AND THE ST. FRANCIS SCHOOL GREENHOUSE

At different points throughout its history, the school greenhouse has partnered with the Department of Health, the Department of Natural Resources, the Department of the Environment and Conservation, the Regional Economic Development Board, the municipality, local businesses, parents, and the school council. This diversity of partners can be viewed as a sign of healthy engagement with the community and provides evidence of the type of community/school interaction that is a critical component of an ecological approach to health. These partnerships can also be interpreted as having developed out of necessity. As voiced by one participant, "I know with our current budgets in terms of the school board, we wouldn't be able to sustain or keep the greenhouse going, so without the community support, it probably wouldn't be able to continue or exist any longer." The potential of the greenhouse to have a favourable influence on the community was also mentioned:

> You get a return on it, for goodness sake. You get a big return. And we're not just talking a return financially. Once it [growing food] gets in the school system, then the kids will become involved [and] then they'll go home and tell their mom and dads, "This is what I'd like to do," right? I mean, and that's the way we get the best support, from the kids telling their grandparents.

This was apparent following a project in which five students were working together in the greenhouse: "And I have to say, two kids out of the five had actually convinced their parents to either break ground or put up a greenhouse."

Engagement and support from the community were central to the operation of the greenhouse. However, a fine balance exists between building community capacity and unloading burdens and responsibility onto the community. Today, community engagement is further complicated by liability issues. For example, in response to the question of whether organizing the greenhouse as a community garden would help to sustain it, one participant addressed the liability issue: "You can't do an open community access garden because it is attached to a school and that's just for liability." The tension at the root of school–community partnerships is a theme revisited in the following section.

DISCUSSION: THE GREENHOUSE AS A RESPONSIVE YET UNINTENDED SCHOOL FOOD ENVIRONMENT TRANSFORMATION

This case study is unique for two reasons: first, the St. Francis School greenhouse, built in the early 1990s, was a novelty for the province (and perhaps even for Canada); and second, its more than 20-year history has granted a broader perspective to examine how the multiple components of an ecological health framework combine and interact. The Comprehensive School Health framework helped to structure this case study, bringing to light how a number of factors combined to either facilitate or inhibit an ecological approach to school health.

As mentioned in the introduction, one of the main challenges to the successful implementation of CSH is being able to sustain a co-ordinated relationship among multiple stakeholders, each of whom brings their own agenda and own way of communicating. Some of these stakeholders include health and education policy-makers, teachers and public health professionals, community members (from farmers to town councils to parents), and students. Due to the interconnectivity, untangling student health, teaching and learning, the social and physical environment, and community partnerships is a challenging task.

When respondents were asked how the greenhouse impacted student health, the most common answer was that it was through the increased exposure to healthy food (the physical environment). Emphasis on the direct link between student health and healthy food is important yet limited. Understood more holistically, "health" extends beyond physical wellness to include emotional, spiritual, and mental well-being. In this perspective, the positive influences on child health from the teaching and learning environment and from the physical environment became clearer through the ecological framework. Community involvement also had a positive yet indirect influence on student health. One participant describes this nicely:

> Benefits are expanding way beyond what kids have to learn. If you're going to bring in parents' involvement or other community-partner involvement, then you've got a municipal thing happening. And you've got a skill set and a knowledge base being built in the community. Then you've got your health promotion piece that goes with it.

The aesthetics of the greenhouse also had a positive impact on the health of students. According to the teachers it fulfilled sensory needs. Other positive effects were inclusiveness and teacher engagement.

Despite this array of potential benefits, there were significant barriers to the full utilization of the greenhouse. At no point was there a policy that explicitly advocated the use of the greenhouse to support broad-scale healthier school practices. The lack of institutional support meant that the school community has had to continually reinvent the greenhouse in order to continue to sustain benefits. An amazing degree of multi-stakeholder co-ordination evolved despite the lack of any overarching institutional support. Sustaining this co-ordination has been a constant struggle, which has made it challenging to maintain positive impacts on student health and learning. Shifting to an ecological view accommodates the perspective that health and learning are indivisibly connected. However, this connected view of health and learning must contend with the current purpose and design of the educational system ("how do you test it?") and from commonly accepted understandings of health. As one interviewee put it: "it's the hospitals and the dialysis machines and all that stuff."

Perceptions of the types of learning or actions expected to occur in schools and how "health" is defined impact the potential use of the greenhouse. For example, while growing food at school was viewed as positive for children's health due to the increase in active learning, there could be unintended negative consequences from fostering unfair competition with small, local growers. Another example is that while growing food at school was seen as positive for children's health due to the increase in community partnerships, the increased resources required from both school and community could be viewed as negative as well.

This case study can be used to shed light on the interrelationships that affect both school health, as framed by the CSH model, and the school food environment (Robert, 2011). Looking at the St. Francis greenhouse through a CSH lens also exposes some of the tensions that arise from competing or differing interests. Conflicting positions must be negotiated in order to foster potential benefits from food environment interventions (Black et al., 2015; Olstad et al., 2014).

USING "FOOD" TO FOSTER CONNECTIONS BETWEEN SCHOOL AND COMMUNITY

If the greenhouse had been planned as a school food environment intervention from the outset, this may have led to more harmony among policy, the teaching and learning environment, the physical environment, and the social environment. Investigating the role of urban school gardening in the twenty-first century, Gaylie (2011) has said that perhaps the most important benefit of school gardens is how they lead us to question the assumptions made about the "place" that schools have in society. This is increasingly important as students are spending more time today in school and are perhaps less healthy than ever before (Ogilvie, 2016). For these reasons, the lessons from this case study add to the recognition of how school food environment interventions can help us to challenge how children are educated today, as the "school garden is potentially the start of a groundswell of movement for teachers ready to engage students in experiential, transformative environmental learning" (Gaylie, 2011: 7). In Newfoundland and Labrador, as elsewhere, there has been drastic change over the past 50 years in how food is produced, who produces it, and where it is produced (Nesheim, Oria, and Yih, 2015). An important contribution of this case study is the acknowledgement from educators that important lessons, which were developed from an intimate connection with the land and sea and which were incorporated into the cultural fabric of society, are now not often being taught.

The initial investment in the school's physical infrastructure resulted from a need to diversify the economy when the fishery collapsed. This political and economic investment might not have been thought of as relating to food policy at the time, but in the case of the St. Francis greenhouse it allowed those connections to be made.

Concerns about the present-day food system and an interest in preserving the tradition of sustainable food practices in Newfoundland and Labrador were guiding principles that motivated many key actors in the community. Teachers who were active in the St. Francis greenhouse program were motivated by the opportunity to reconnect children with the food system. This began with RB, who described his early childhood as a source of inspiration for his involvement with the greenhouse project:

> Yeah, it came from my father. I mean, we always had our own vegetables. So every summer we had to do the weeding and all that stuff. And in the fall, we'd have to do the harvesting and we used to have to put it all away in the cellar. And every now and then father would say to me, RB, go up to the cellar now and bring down some vegetables and a side of pork. He'd have a pig up there hung up, and I'd go up and cut off a piece of that and go down the cellar and get the cabbage and the turnip and the potatoes. Then out behind our house was a small little plot of land about 10 feet by 20 feet, so I'd start picking away at that, putting potato seed in and all of a sudden I saw it growing and turning it out. So I got interested that way.

RB's vision for the development and use of the school's land resonated with community members, inspiring them to revisit and identify with traditional food practices. Some participants mentioned that their concern about the degree of pesticides and preservatives present in produce was a reason to grow food: "We've gotten away from growing our own vegetables as a family over the last 50 or 60 years. So I think that's something we should all get back to."

Participants asserted that students became conscious of the fact that agriculture is possible in Newfoundland and Labrador, "A lot of them didn't have the faith that it could be done in Newfoundland." This exposure is critical for developing agricultural capacity at the provincial level (Quinlan, 2012). Also, the greenhouse exposed students to farming and increased their understanding of the province's food system, a subject not typically learned in school. In doing so, they also learned that farming is a viable occupation:

> Once they're in level one, they've got their first set of courses picked, their teachers and their counsellors are talking about, well this is the way you're going — you're going to go engineering, you're going to be a teacher, nurse, doctor, lawyer. Nobody mentions farmer, you know, unless your dad and grandfather and your Uncle Tom, whoever, was a farmer. You really didn't have that exposure.

The initial intent to diversify the economy was one factor that led to the building of the greenhouse. From that grew a number of unforeseen connections to the Newfoundland and Labrador food system: (1) students were exposed to new foods; (2) teachers felt enhanced motivation to engage students with the greenhouse because of their concern about the lack of food knowledge and also concern about the nature of "modern" food; and (3) students learned that local food production was possible. This case study helps demonstrate how the St. Francis greenhouse, while not planned as a school food environment intervention, was well positioned to help participants respond to some of the complex challenges that characterize the Newfoundland and Labrador food system — an increasing dependency on industrialized, processed, and imported food, loss of food production and food system knowledge, an aging farmer population, and low fruit and vegetable consumption (Food First NL, 2015). Food was a means to tie together the fragmented worlds of school, community, agriculture, health, culture, and politics (Wallinga, Story, and Hamm, 2009).

The fact that there have been few successful interventions that have led to broad-scale improvements in students' health and well-being and the fact that the province's food system is in need of improvement are the main incentives for further investigating what factors may enhance or diminish the ability to nurture comprehensive health outcomes at school (Food First NL, 2015; Olstad et al., 2014; Stuckler and Nestle, 2012). Future research could explore the systems thinking that is behind an ecological view of health and the current understandings of the school food environment. Another future study could examine how this fits into the purpose and design of the Newfoundland and Labrador education system. However, in future investigations it will be interesting to analyze how educational reforms that have occurred in the province over the last few decades have impacted long-term health and learning outcomes. Previous research echoes observations made by participants in this case study that a school's ability to promote health is mitigated by such factors as changes to curricula and pressures to raise standardized assessment results (McIsaac, Read, Veuglelers, and Kirk, 2013). Implementation and promotion of holistic school health, such as through the CSH framework, necessarily require more detail about the decision-making process, which calls for reflection from a teacher's or a student's perspective (Samdal, 2013). More time, energy, and resources are needed to understand and document how student learning and health are influenced by adaptations to the school food environment, such as by providing children with an opportunity to connect to communities and to the surrounding food system (Rojas et al., 2011).

FINAL CONSIDERATIONS

The essence of an ecological approach to public health lies in gaining an acceptance and understanding of the complex connections among people, places, communities, and the environment. These connections determine the quality of our lives and how people live together in sharing our resources and infrastructure, such as air, water, soil, and food (Rayner, 2012). However, making and understanding these connections is challenging. Our goal in presenting this case study was not to minimize or eliminate the complexity of the school food environment but to link aspects of it that may have previously seemed disconnected. This was accomplished by outlining how the various components of the comprehensive model of school health (CSH) interacted throughout the development of the St. Francis greenhouse. Weaving the stories and experiences of the people who have been connected to the greenhouse over the past 20 years revealed that the passing of time is a significant factor when researching ecological public health. It allows a deepening of perspective necessary for an observation of people and the systems they live in (Sturmberg, 2013).

This chapter ends with a story told by one of the study participants. The individual described the importance of filling the entire greenhouse with plants in order for the greenhouse to flourish:

> If you leave the greenhouse half empty, you don't fill it up right, you never get the humidity built up because you don't have enough plants in there to be doing what they're supposed to be doing. So there's a fine line between a sparse greenhouse [being] really easy to take care of because it's less time to water, plants are nice and far apart, you can pick leaves easy [*sic*] compared to, I'm going to space them together and just having that many in there respiring in the nighttime.

Discovering the ideal conditions for plant growth can be likened to creating the ideal conditions for a thriving school food environment in Newfoundland and Labrador. Taking small steps in its development, that is, through small grants and publicity campaigns, may be a safe starting point, but the conditions that developed at the St. Francis School in Harbour Grace give reason to believe that a bolder, long-term vision is necessary to address the complexity of the challenges faced by children and society today. The most important question has yet to be answered: How can the Newfoundland and Labrador school food environment be enhanced to optimize the education and health of children?

ACKNOWLEDGEMENT

This research project was funded under the Strategic Partnership — Harris Centre Student Research Fund.

REFERENCES

Ashe, L. M., and R. Sonnino. 2013. "At the crossroads: New paradigms of food security, public health nutrition and school food." *Public Health Nutrition* 16, 6: 1020. doi:10.1017/S1368980012004326.

Bagdonis, J., C. Hinrichs, and K. Schafft. 2009. "The emergence and framing of farm-to-school initiatives: Civic engagement, health and local agriculture." *Agriculture and Human Values* 26, 1: 107–19. doi:10.1007/s10460-008-9173-6.

Bassett-Gunter, R., J. Yessis, S. Manske, and D. Gleddie. 2015. "Healthy school communities in Canada." *Health Education Journal.* doi:10.1177/0017896915570397.

Black, J. L., C. E. Velazquez, N. Ahmadi, G. E. Chapman, S. Carten, J. Edward, . . . A. Rojas. 2015. "Sustainability and public health nutrition at school: Assessing the integration of healthy and environmentally sustainable food initiatives in Vancouver schools." *Public Health Nutrition* 18, 13: 2379–91.

Blay-Palmer, A. 2016. *Imagining Sustainable Food Systems: Theory and Practice.* London: Routledge.

Bowman, B. 1994. "St. Francis greenhouse." *The Compass,* 12.

———. n.d. "Second national award bestowed on St. Francis vice-principal." *The Compass.*

Card, A., and E. Doyle. 2008. "School health coordinators as change agents." *Health and Learning,* 3–11.

Coalition for School Nutrition. 2001. *School Survey of Food and Nutrition Policies and Services in Newfoundland and Labrador.* At: https://www.nlta.nl.ca/files/documents/archives/schnutrition_survey01.pdf.

Denzin, N. K., and Y. S. Lincoln. 2011. *The Sage Handbook of Qualitative Research.* Thousand Oaks, Calif.: Sage.

Deschesnes, M., C. Martin, and A. J. Hill. 2003. "Comprehensive approaches to school health promotion: How to achieve broader implementation?" *Health Promotion International* 18, 4: 387–96. doi:10.1093/heapro/dag410.

Flyvbjerg, B. 2006. "Five misunderstandings about case-study research." *Qualitative Inquiry* 12, 2: 219–45. doi:10.1177/1077800405284363.

Food First NL. 2015. "Everybody eats: A discussion paper on food security in Newfoundland and Labrador." At: http://www.foodfirstnl.ca/our-resources/everybodyeats.

Gaylie, V. 2011. *Roots and Research in Urban School Gardens.* New York: Peter Lang.

Goss Gilroy Inc. 2013. *Kids Eat Smart Foundation Newfoundland and Labrador Program Evaluation*. At: http://www.kidseatsmart.ca/wp-content/uploads/2014/10/KESF-NL-Program-Evaluation.pdf.

Hanrahan, M., and M. Ewtushik. 2001. *A Veritable Scoff: Sources on Foodways and Nutrition in Newfoundland and Labrador*. St. John's: Flanker Press.

Health Research Unit. 2012. "An evaluation of the delivery of the Kids Eat Smart (KES) Program." At: http://www.med.mun.ca/HRU/docs/KES_Report_Online.aspx.

Heart and Stroke Foundation. 2013. *School Food Gardens in Ontario: Educating for Health and Sustainability*. At: http://agardenineveryschool.ca/wp-content/uploads/2014/03/School-Food-Gardens-in-Ontario-Educating-for-Health-and-Sustainability-2013-Report.pdf.

Henderson, T. 2011. "Health Impact Assessment: HB 2800: Oregon farm to school and school garden policy." At: http://www.upstreampublichealth.org/resources/health-impact-assessment-hb-2800-farm-school.

International Union for Health Promotion and Education. 2009. "Achieving health promoting schools: Guidelines for promoting health in schools." At: http://www.iuhpe.org/images/PUBLICATIONS/THEMATIC/HPS/HPSGuidelines_ENG.pdf.

Joint Consortium for School Health. 2013. "Comprehensive School Health Framework." At: http://www.jcsh-cces.ca/index.php/about/comprehensive-school-health.

Kelly, W. 1997. "Decentralization of educational decision-making in the Newfoundland and Labrador education system reform process: Illusion or reality." M.Ed. thesis, Faculty of Education, Memorial University of Newfoundland. At: http://research.library.mun.ca/1510/.

Lang, T. 2009. "Reshaping the food system for ecological public health." *Journal of Hunger & Environmental Nutrition* 4, 3/4: 315–35. doi:10.1080/19320240903321227.

McIsaac, J.-L., K. Read, P. J. Veuglelers, and S. F. Kirk. 2013. "Culture matters: A case of school health promotion in Canada." *Health Promotion International*. doi:10.1093/heapro.dat055.

———, K. Storey, P. J. Veugelers, and S. F. L. Kirk. 2015. "Applying theoretical components to the implementation of health-promoting schools." *Health Education Journal* 74, 2: 131–43. doi:10.1177/0017896914530583.

McLeroy, K. R., D. Bibeau, A. Steckler, and K. Glanz. 1988. "An ecological perspective on health promotion programs." *Health Education & Behavior* 15, 4: 351–77. doi:10.1177/109019818801500401.

Morgan, K., and R. Sonnino. 2008. *The School Food Revolution: Public Food and the Challenge of Sustainable Development*. London: Earthscan.

Nesheim, M. C., M. Oria, and P. T. Yih, eds. 2015. *A Framework for Assessing Effects of the Food System*. Washington: National Academies Press.

Ogilvie, K. K., and A. Eggleton. 2016. "Obesity in Canada: A whole-of-society approach for a healthier Canada." At: http://www.senate-senat.ca/social.asp.

Olstad, D. L., K. D. Raine, and C. I. J. Nykiforuk. 2014. "Development of a report card on healthy food environments and nutrition for children in Canada." *Preventive Medicine* 69: 287–95. doi:10.1016/j.ypmed.2014.10.023.

Ozer, E. J. 2007. "The effects of school gardens on students and schools: Conceptualization and considerations for maximizing healthy development." *Health Education and Behavior* 34, 6: 846–63.

Pitt, R. D., and J. E. M. Pitt. 2015. "Harbour Grace." At: http://www.thecanadianencyclopedia.ca/en/article/harbour-grace/.

POWER UP! 2015. "Alberta's 2015 report card on children's healthy food environments and nutrition for children and youth." At: http://powerupforhealth.ca/report-card/.

Public Health Agency of Canada. 2009. "Newfoundland and Labrador: St. Francis School greenhouse — Local students have hands-on involvement in growing food and preparing healthy snacks." At: http://www.phac-aspc.gc.ca/publicat/2009/be-eb/newfoundland_labrador-terreneuve_labrador-eng.php.

Quinlan, A. J. 2012. "Building agricultural capacity in Newfoundland and Labrador." At: https://www.mun.ca/harriscentre/reports/arf/2011/11-SPHCSRF-Final-Quinlan.pdf.

Rayner, G., and T. Lang. 2012. *Ecological Public Health: Reshaping the Conditions for Good Health*. New York: Routledge.

Robert, S. A., and M. Weaver-Hightower. 2011. *School Food Politics: The Complex Ecology of Hunger and Feeding in Schools around the World*. New York: Peter Lang.

Robinson, A. 2015. "Greenhouse revived at St. Francis." *The Compass*, 11 May. At: http://www.cbncompass.ca/News/Local/2015-05-11/article-4142262/Greenhouse-revived-at-St.-Francis/1.

Rojas, A., W. Valley, B. Mansfield, E. Orrego, G. E. Chapman, and Y. Harlap. 2011. "Toward food system sustainability through school food system change: Think&EatGreen@School and the making of a community-university research alliance." *Sustainability* 3, 5: 763–88. doi:10.3390/su3050763.

Rowling, L., and V. Jeffreys. 2006. "Capturing complexity: Integrating health and education research to inform health-promoting schools policy and practice." *Health Education Research* 21, 5: 705–18. doi:10.1093/her/cyl089.

Samdal, O., and L. Rowling. 2013. *The Implementation of Health Promoting Schools: Exploring the Theories of What, Why and How*. London: Routledge.

Sheppard, T. 2003. "Means of survival: Youth unemployment and entrepreneurial training in Newfoundland." MA thesis, Faculty of Sociology, Memorial University of Newfoundland. At: http://research.library.mun.ca/7033/1/Sheppard_Tina.pdf.

Smith, G. A., and D. A. Gruenewald. 2008. *Place-based Education in the Global Age: Local Diversity*. New York: Lawrence Erlbaum Associates.

Statistics Canada. 2012. "Harbour Grace, Newfoundland and Labrador (Code 1001379) and Canada (Code 01) (table)." Census Profile. Catalogue no.98-316-XWE. At: http://www12.statcan.gc.ca/census-recensement/2011/dp-pd/prof/index.cfm?Lang=E.

Stuckler, D., and M. Nestle. 2012. "Big Food, food systems, and global health." *PLoS Medicine* 9, 6: e1001242. doi:10.1371/journal.pmed.1001242.

Sturmberg, J. P. 2013. *Handbook of Systems and Complexity in Health*. New York: Springer.

Sullivan, D. S. 2011. "From horticulture to healthy eating. *The Telegram*, 19 Mar. At: http://www.thetelegram.com/News/Local/2011-03-19/article-2346043/From-horticulture-to-healthy-eating/1.

Wallinga, D., M. Story, and M. Hamm. 2009. "Food systems and public health: Linkages to achieve healthier diets and healthier communities." *Journal of Hunger & Environmental Nutrition* 4, 3: 219–24. doi:10.1080/19320240903351463.

World Health Organization (WHO). 1986. "The Ottawa Charter for Health Promotion." At: http://www.who.int/healthpromotion/conferences/previous/ottawa/en/.

———. 2016. "School and youth health: What is a health promoting school?" At: http://www.who.int/school_youth_health/gshi/hps/en/.

4

Food as a Social Movement in Newfoundland and Labrador: The Role of Community Gardens

Kelly Vodden, Catherine Keske & Jannatul Islam

> Agricultural sustainability doesn't depend on agritechnology. To believe it does is to put the emphasis on the wrong bit of "agriculture." What sustainability depends on isn't agri- so much as *culture*. (Patel, 2010: 170)

INTRODUCTION

As noted in the book's introduction as well as in several chapters, La Vía Campesina food sovereignty movement reflects the right of all peoples to healthy and culturally appropriate food produced through ecological and sustainable methods. This includes the right of peoples to define their own food and agriculture (Desmarais, 2008).

In this chapter, we present an overview of the emergence of community gardening in Newfoundland and Labrador, followed by a qualitative, participatory action research case study documenting the process of establishing the Centreville-Wareham-Trinity (C-W-T) community garden in 2015 as an example of a community engaged in the early stages of redefining their own food and agriculture. Two of the authors were involved in the planning and implementation of the C-W-T garden, which yielded a harvest during the first two growing seasons (2015 and 2016). Yet, unless a community leader steps forward to organize the efforts, the sustainability of the C-W-T garden is uncertain. As we discuss in this chapter, nascent community gardens are rather typical, and presenting the chronology of a community garden's developmental stages and associated challenges fills a void in the literature where more work has been desired to guide other early-stage projects (Corrigan, 2011; Gottlieb and Fisher, 1996). Qualitative case studies also have potential to deliver valuable lessons in community food security and food sovereignty. These stories (often successful ones) may encourage others to engage in similar endeavours, which may in turn contribute to social movements around community gardens that are already in progress. To date, the majority of case studies written about community gardens have highlighted the value that the gardens bring to urban neighbourhoods (Teig et al., 2009; Hale et al., 2011; Hanna and Oh, 2000). We assert that the C-W-T community garden provides an interesting example of the resurgence of gardening in rural areas, where the once venerable household gardens all but disappeared a few decades ago, as discussed by Phillips (Chapter 1) and Roseman and Royal (Chapter 2). We posit that the renewed presence of community gardens in Newfoundland and Labrador reflects a social movement of increased community involvement in defining its food system, in the spirit of La Vía Campesina.

The chapter is organized as follows. First, we present an overview of the literature to identify key elements of social movements and lay the foundation for food as a social movement in Newfoundland and Labrador. We then present data collected from provincial and federal sources such as the Harris Centre and Statistics Canada to highlight historical trends in provincial food production and the recent establishment of community gardens. Finally, we present the case study of the development of the C-W-T community garden, asserting that the increase in community gardening in the Indian Bay and C-W-T communities and elsewhere reflects the emergence of a La Vía Campesina style social movement around food, with the potential to have an enduring influence on how food is produced, and consumed, in the province.

LITERATURE REVIEW

Defining Social Movements

Social movements involve "an organized set of constituents pursuing a common political agenda of change through collective action" (Batliwala, 2012: 3). Movements exhibit a degree of continuity over time and have a clear political agenda driven by some shared analysis of the social/structural conditions of concern, the changes being sought, and the targets involved in the change process (Batliwala, 2012). Social movements are dynamic, historical phenomena, and as such they "are shaped by circumstance; they are contingent things, which grow or shrink in response to factors that enable or constrain them" (Sogge and Dutting, 2010: 4). The development of social movements requires active and deliberate investment of labour, thought, and resources over time, while also having external environments conducive enough to enable them (Horn, 2013). Social movements are cultural productions that create spaces wherein people (re)construct social and political identities as they work collectively to pursue an agenda of social change (Doherty, 2006; Gottlieb, 2001; Hassanein, 2003).

Scholars such as Rao (2000), Sogge and Dutting (2010), Batliwala (2012), and Mondal (2015) identify the following four key features of social movements:

1. *Commitment to a cause or ideology:* The group's actions demonstrate a commitment to a specific cause (a political agenda of change). The existence of a cause or agenda is a distinguishing feature of any social movement. This cause is articulated and shared through framing processes that generate collective identities, meanings, and consciousness that create cohesion (although the group may be segmented based on personal, structural, and/or ideological ties, for example), and these characteristics are used to convince others to join the movement.
2. *Separation (real and/or perceived) from the established order:* The commitment to the cause is likely to involve significant separation from the established order, but with the aim of bringing a new set of values and a changed pattern of behaviour in individual(s) and society. This separation and the unification of actors within a social movement require real or perceived opposition from society at large or from the established order within which the movement has arisen and from the power-holders under this established order.
3. *Collective action over time:* Any social movement requires the involvement of a group of people engaged in collective actions. Such collective action emerges in response to particular

situations, often where inequality and/or socio-economic, cultural, and/or political oppression is experienced (leading to the opposition and commitment to change noted above). The collective actions should create interest among a relatively large group that can advance the cause and create changes in behaviour and ideology for the future. While the lifespans of social movements vary, movements tend to develop and sustain collective actions over time. Actions may range from confrontations to education and awareness strategies. These actions may be organized and the movements' constituency or membership mobilized in diverse ways, ranging from control by formal, elected organizations to relationships built on shared understanding or organization with limited governance.

4. *Social change:* Social movements seek to bring about (or resist) social change. Even if the change is only partially successful, the creation and growth of the movement involve some form of sustained change in patterns of behaviour, values, and/or goals. This change links back to the framing processes noted above.

We later reflect on the presence of each of these features in the C-W-T community garden and others, and more broadly in the food movement in Newfoundland and Labrador.

Decline in Farms, Increase in Direct Marketing, and the Re-emergence of Rural and Community-based Food Production Initiatives

As discussed in previous chapters, at one time there was a strong gardening and farming culture in outport communities like Indian Bay and C-W-T that eventually transitioned into larger-scale food and agriculture production after Confederation with Canada, as supplementary farming was discouraged in favour of larger, more concentrated commercial farms (Cadigan, 1998; Vodden, Hall, and Freshwater, 2013). However, since that time and to the present day, commercial farm operations have continued to decline, particularly with respect to the rest of Canada. By 2000, less than 600 farms remained in the province, and by 2011 there were only 510 farms, down from a peak of 4,226 farms in 1935 (Ramsey, 1998). The total acreage under farm production fell 13.5 per cent between 2006 and 2011 alone (Statistics Canada, 2010).

In the Census of Agriculture for 2016, Newfoundland and Labrador reported the fewest number of farms among the provinces, accounting for less than 1 per cent of all farms in Canada (Statistics Canada, 2017a). There were also one-fifth fewer farms in Newfoundland and Labrador between 2016 and 2011, the largest percentage drop in Canada during this five-year period. In comparison, the rest of Canada experienced a decrease of 5.9 per cent in the number of farms. In 2016 there were 25 per cent fewer farm operators in Newfoundland and Labrador compared to 2011. Although farming decreased during this five-year period, the proportion of Newfoundland and Labrador farms involved in direct marketing was the highest in the country. Approximately one-third of farms reported selling at least one commodity (typically fruits, vegetables, meat, poultry, and eggs) directly to consumers (Statistics Canada, 2017a). Despite the decline in the overall number of farms, the high proportion of direct sales from Newfoundland and Labrador farmers could ostensibly be viewed as evidence of a local food movement in alignment with La Vía Campesina.

Notwithstanding the overall shift in emphasis towards larger-scale commercial food production, community-based agricultural initiatives were also launched and/or continued during the 1960s and 1970s, typically fostered by Rural Development Associations (RDAs).[1] These initiatives became part of the rural development movement, which coalesced in response to declines in the fishery, resettlement, and other threats to rural livelihoods and communities (Vodden, Hall, and Freshwater, 2013). However, by the mid-1990s the RDAs that had fostered and supported many community and regional agriculture projects lost much of their government financial support. Many ceased to exist, as did their agricultural products. Subsequently, a new set of initiatives began to emerge, including community gardens, farmers' markets, "buy local" campaigns, community-led food assessments, community kitchens and freezers, and regional land-use planning efforts that consider multiple dimensions of agricultural production (Vodden, Hall, and Freshwater, 2013; Food First NL, 2015a, 2015b). Food First NL (formerly Food Security Network of NL) was founded in 1998 and incorporated in 1999. This was an important milestone for an emerging food movement in the province. The non-profit, membership-based group brought organizations and individuals together to undertake collective action with the explicit goal of enhancing food security in the province. Another important milestone was the 2007 Provincial Food Security Assembly, which attracted 117 delegates from varied backgrounds to work towards a food security agenda. One recommendation repeatedly made at the Assembly was the creation of, and support for, community gardens (FSN, 2007).

What Is a Community Garden?

A community garden has been described as "an organized, grassroots initiative whereby a section of land is used to produce food or flowers or both in an urban environment for the personal use or collective benefit of its members" (Glover, 2003: 265). Considering the context of the entire province, which includes many small towns and rural areas, Food First NL (FSN, 2011) offers a broadened conceptualization of a community garden as a shared space where people gather together and grow vegetables, fruits, flowers, native plants, and, potentially, livestock. A community garden is a piece of land (publicly or privately held) cultivated by a group of people rather than a single family or individual. Such a garden generally is managed and controlled by unpaid individuals or volunteers – usually the gardeners themselves (Corrigan, 2011; Gottlieb and Fisher, 1996) and may be organized in different ways: allotment-style, with individual plots allocated within a shared garden space, or communal-style, where individuals care for several garden beds together, sharing labour and produce (Hamilton-Nelson, 2011). Francis (2003) adds that community gardens are typically developed on vacant (often private) lands that may be developed and managed for and by local residents but are typically not viewed officially as part of a city's open space system. The community garden is therefore vulnerable to being displaced by housing and commercial development.

Brunetti (2010) states that the initial phase of community gardening in North America occurred in the nineteenth century, including early examples in Detroit in the 1890s. He suggests that community gardening was promoted by social and educational reformers along with other groups involved in the civic beautification movement. Community gardens (or "Victory Gardens") were encouraged during World War I to increase the domestic food supply

(Corrigan, 2011), but until the 1970s community gardening was generally considered a temporary solution to food shortages, economic depression, and civic crises.

Today, scholars and practitioners note that community gardens may have long-term functions to provide benefits to individuals, families, and communities (Brunetti, 2010) and are part of a growing global effort to build community food economies and provide alternatives to the dominant industrial food system (Cameron et al., 2014). As shown in Table 4.1, community gardens are places of production not only of food but also of a wide range of social, economic, physical, mental, and environmental benefits. In this table we summarize the results of 10 studies on community gardens and reorganize them into six categories of benefits: social, cultural, youth development, economic, health, and ecological. Social benefits include aspects that foster relationships, food knowledge, a sense of community, and social justice. Cultural benefits may encompass solidarity between generations or ethnic groups, by facilitating the growth of specialized crops and cultural exchange. Productive youth development opportunities may cultivate practical carpentry and food production skills and foster creativity by bridging gardening with more traditional subject matters like math, science, and art. Several of the economic benefits (such as reducing dependency on imported food and availability of fresh, high-quality food at relatively low cost, particularly for low-income families) are discussed throughout this book. However, community gardens may also reduce costs associated with packaging, waste, and fuel reliance. Several health benefits presented in Table 4.1 are addressed in Chapters 2, 3, and 5. Community gardens also present opportunities for collective work and social relationships, including reduced stress and isolation. Not surprisingly, community gardens may also provide ecological benefits arising from transitioning vacant spaces into green space, enhancing biodiversity, and facilitating preservation of heirloom varieties of flowers and vegetables.

Table 4.1. Benefits of community gardens.

Social	• Foster community engagement and relationships, including sharing of food and food-related knowledge. • Provide access to fresh food and space for social justice. • Foster a sense of community ownership, identity, and stewardship. • Provide training, leadership development, and opportunities for satisfying work (can lead to social empowerment, sites of grassroots political organizing). • Create a focal point for community events and programs. • Can reduce litter, vandalism, and crime by creating stronger neighbourhoods; challenge perceptions and negative stereotypes of a neighbourhood.
Cultural	• Offer unique opportunities for production of traditional crops and culturally preferred food by ethnic groups; can help protect and foster cultural identity. • Provide intergenerational exposure to cultural traditions. • Facilitate cultural exchange with other gardeners and volunteers. • Can create social solidity between generations and different ethnic groups.
Youth development	• Provide a place to teach environmental knowledge and awareness to youth; offer opportunities for learning about plant identification and life cycles, pollinators and insects and their roles in biodiversity. • Can be used to teach important skills and knowledge, including math, science, art, budgeting, history, nutrition, geography, home economics, and carpentry. • Develop a sense of community ownership and stewardship. • Enhance understanding about food systems, leadership, duties, and life skills. • Develop their creativity and share ideas in a socially meaningful and physically productive way.
Economic	• Help reduce dependency on imported food. • Reduce packaging, fuel inputs. • Offer low-cost, fresh, high-quality food. • Supplement budgets/reduce household budget pressures of low-income families. • Provide an opportunity to produce food to families and individuals without land. • May increase land value and security; less expensive to maintain and develop than parkland. • Can help create skills (e.g., leadership, organizing, and/or business skills) that may be used in the labour force or to develop entrepreneurs.

Health benefits	• Provide access to more nutritious food options than imported, commercially preserved foods, which encourages a more nutritious diet with increased consumption of fruits and vegetables. • Offer opportunities for relaxation and for exercise and physical activity. • Provide health benefits of collective work and social relationships, including reduced stress and isolation. • Create a sense of satisfaction and personal agency.
Ecological benefits	• Help reduce ecological footprint of food by reducing packaging and transport. • Develop increased awareness of and appreciation for living things through direct engagement with ecological processes necessary to maintain a garden plot; may contribute to a community sense of environmental stewardship. • Reduce air pollution through plant restoration of oxygen. • Provide a peaceful green place in urban environments (often from cleaned-up vacant space, resulting in improvements to the physical features of neighbourhood and public space). • Help to bring sustainable agriculture into the city. • Provide food and shelter for birds and insects, helping to enhance biodiversity. • Can help prevent the introduction of chemicals and preservatives. • Facilitate preservation of heirloom varieties of flowers and vegetables. • Encourage composting, helping to cycle outputs back into the system.

Sources: Brunetti (2010); Cameron et al. (2010); Nelson-Hamilton (2011); Corrigan (2011); Community Garden Alliance (2015); Doyle (2014b); Hale et al. (2011); Hanna and Oh (2000); Kortright and Wakefield (2011); Teig et al. (2009).

In summary, the benefits reviewed above illustrate that community gardens can play important roles for facilitating sustainable food systems and contribute to environmental health, economic and human health, and social equity. Community gardens may foster innovations for enhanced food security and waste diversion and create a shared space for communal work, where people gather together, celebrate their harvest, and enjoy a location for learning and recreation.

Community Gardens in Newfoundland and Labrador

A new wave of community gardening initiatives began in the province in the 2000s, with sufficient interest by 2008 to warrant the establishment of the Community Garden Alliance to connect community gardens throughout the province. As summarized in Table 4.2, Memorial University Botanical Garden (2014) listed a total of 26 community gardens in Newfoundland and Labrador: eight in the St. John's metro area, four in Avalon Peninsula and eastern Newfoundland, three in central Newfoundland, nine in western Newfoundland, and two in Labrador.

Table 4.2. Sample start dates for community gardens in Newfoundland and Labrador.

Name	Location	Year Started
St. Francis School greenhouse and gardens	Harbour Grace	1992
Brother Jim McSheffrey Community Garden	Mt. Scio Road, St. John's	mid-1990s
Community garden and greenhouse	Lamaline	2008
Rabbittown Community Garden	Graves Street, St. John's	2008
Placentia Area Community Garden	Placentia	2009
Blow Me Down Community Garden	Corner Brook	2010
Blue Crest Unity Garden	Grand Bank	2010
Exploits Community Garden	Botwood	2011
Peninsula Intergenerational Community Garden	Eastport	2012
St. Patrick's Organic Community Garden	Carbonear	2012
Cavell Park Community Garden	Cavell Road, St. John's	2012
Clarenville Age-Friendly Park Community Garden	Clarenville	2014
C-W-T Community Garden	Centreville-Wareham-Trinity	2015
Bay Roberts Community Garden	Bay Roberts	2015

Sources: www.communitygardensnl.ca; www.mun.ca/botgarden/gardening/comgarden/comgarden.php; and individual garden websites.

A review of provincial community garden websites shows a variety of community garden plot types and sizes, including 10' x 10' to 3' x 3' children's raised beds, ground plots, and greenhouse space. Factors such as location, demand, and local economic conditions appear to influence the variable prices charged for plots, with the rates ranging from "no charge" or "in kind" (e.g., donating a portion of the harvest to the Community Garden Committee for a celebratory dinner in Clarenville) to $60, depending on location and plot size.

Co-ordinating and Facilitating Community Gardening

Community gardens rely extensively on funding and volunteers. Omohundro (1982) suggests that local ecological and social conditions, as well as external influences, create challenges for gardening initiatives. Community gardens are no exception. He further explains that the project failure rate tends to be highest following the construction phase, when managerial skills are needed to establish a routine. Securing local initiative in the design and implementation of the project has also been shown to be important (Corrigan, 2011; Nelson-Hamilton, 2011).

Omohundro's research also identified a lack of co-ordination of gardening activities in the province that could promote gardening and help gardeners and entrepreneurs to extend their activities. Co-ordinated home gardening activities across the province were typically organized by neighbours, rather than through community–household partnerships, government, or intergovernmental efforts leading to the fulfillment of broader policy goals. The university and community home production educational programs from the twentieth century described by Lynne Phillips in Chapter 1 are being revisited today, but with a contemporary twist. Today, a growing number of organizations promote gardening activities and food security in the province. Food First NL, Memorial University Botanical Garden, and Community Garden Alliance are primary sources of community gardening knowledge and other resources. With some variability, these organizations hold workshops and seminars on gardening, food security, and native plants, and provide advice to new gardeners and food entrepreneurs.

In 2010 Dr. Anthony Brunetti prepared *The Community Garden Handbook for Newfoundland and Labrador* for the College of the North Atlantic and Avalon Regional Council (Rural Secretariat/Office of Public Engagement). The *Handbook* was compiled based on a review of European and North American experiences, together with surveys and interviews with community garden organizers throughout Newfoundland and Labrador. The goal was to provide assistance to citizens working to create community garden spaces (Brunetti, 2010). Also in 2010, the Food Security Network of NL (now Food First NL) developed the *Community Garden Best Practices Toolkit: A Guide for Community Organizations in Newfoundland and Labrador*, a tool to assist organizations wanting to start their own garden (FSN, 2011). The toolkit, which has since been adapted for use in other provinces, was created in partnership with two provincial agencies: the Poverty Reduction Division of the Department of Human Resources; and Labour and Employment and Health Promotion and Wellness within the Department of Health and Community Services.

Those interested in initiating a new garden can find information on possible funding supporters from these sources. While provincial and federal agricultural funding focuses on commercial operations, health and wellness agencies have provided funding to support community garden initiatives. Six wellness coalitions in the province, for example, bring together community, government and non-government agencies, and others with an interest in promoting wellness and enhancing the health and well-being of their regions. Each of these coalitions, through the provincial Health and Wellness grant program and the Community Healthy Living Fund, make community grants available at multiple times during the year. Many of the province's community gardens have benefited from this support. Federal funding is also available through the New Horizons for Seniors Program, which supports community-based projects that help seniors get involved in the lives of others and in their communities.

Numerous private-sector and non-government partners have also contributed financial and in-kind resources. Grand Bank's Blue Crest Unity Garden, for example, was launched with financial support from a Walmart-Evergreen Green Grant, the local Lions Club, and the involvement of the Burin Peninsula Heath Care Foundation and Burin Peninsula Environmental Reform Committee (Appleby et al., 2011). In Botwood, partners in the Exploits community garden include the Botwood Boys and Girls Club, as well as the Legion Action Committee, Central Regional Wellness Coalition, and Central Health (Food First NL, 2015b).

CASE STUDY: THE C-W-T COMMUNITY GARDEN EXPERIENCE

Figure 4.1 shows the Indian Bay and Centreville-Wareham-Trinity (C-W-T) region. The communities of Centreville, Wareham, and Trinity amalgamated in 1992 to become the town of C-W-T (population 1,160, according to 2016 NL

Statistics Agency data). Indian Bay is a neighbouring community located just five kilometres to the north, with a population of 175. The two communities share services such as firefighting, a school, clergy, an arena, and the Central Wellness Coalition, which promotes healthy eating and other aspects of healthy living. However, there are also independent institutions and organizations, such as the two town councils and several community groups. The communities' populations are aging, with an average age of 52 in C-W-T and 49 in Indian Bay (versus 44 for the entire province); 22 per cent of C-W-T's and 18 per cent of Indian Bay's population were 65 years of age or older in 2011, as compared to 16 per cent for the province as a whole (NL Statistics Agency, 2016).

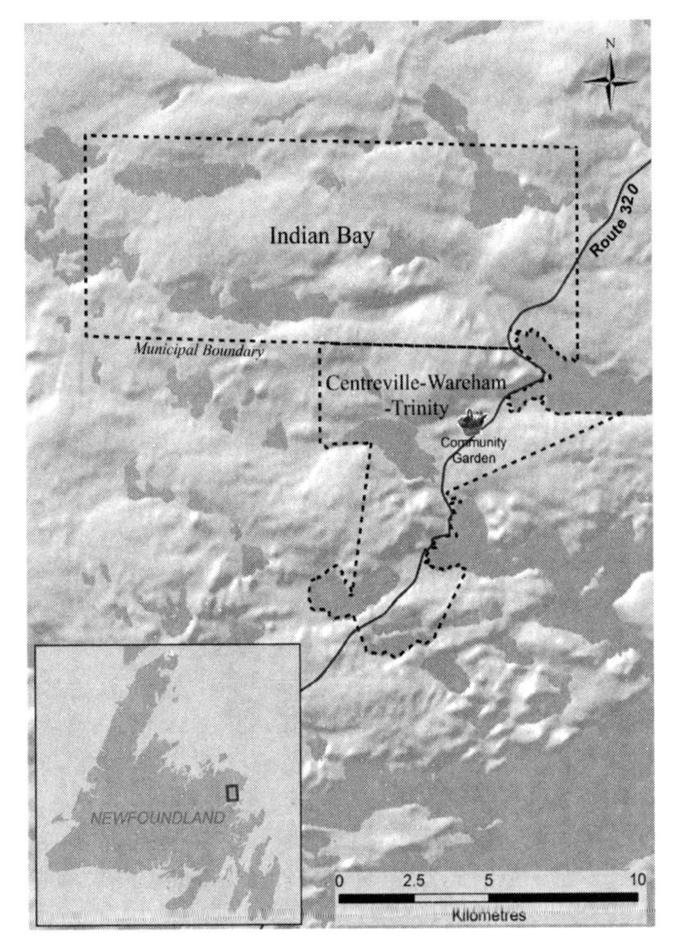

Figure 4.1. Map of Indian Bay and C-W-T study region.

Residents from the communities often drive nearly 100 kilometres in each direction (close to 200 km for a round trip) to buy their fresh food and vegetables in Gander, the regional shopping and service centre. Although there are several convenience and small grocery stores in the C-W-T/Indian Bay area and a grocery store in the neighbouring community of New-Wes-Valley (30 km away), these stores offer a more limited selection, particularly of fresh produce, than regional supermarkets. A former local grocery store operator and community garden volunteer explained that with the limited selection local stores are able to carry and the reduced freshness due to added transport, residents rarely purchase their fruits and vegetables locally. This leads to high levels of waste and costs to local small businesses already struggling from high levels of competition within the grocery sector, a finding validated by Lowitt and Neis in Chapter 8. Even in Gander, fruit and vegetable stocks are sometimes depleted due to problems with ferry transportation (Figure 4.2). One response to this geographic isolation has been the C-W-T community garden, which was implemented following methods similar to those presented in Hancock and Algozzine (2015).

Figure 4.2. Empty shelves in Gander grocery store. (Photo by Kelly Vodden)

Like other communities in the province, historically the area had a strong tradition of outport and supplementary small-scale farming to meet household demands. The communities that now make up the town of C-W-T are comprised mainly of families who were resettled from nearby Fair Islands during the late 1950s and early 1960s, where gardening and subsistence farming were essential parts of island life. A C-W-T town councillor and community garden representative recalled his father's farming culture in a community meeting, for example. He further suggested that the current problem of few people gardening or farming has been stimulated by the out-migration of young people for jobs, combined with the small size of families. He enjoys gardening but given their small family, they prefer shopping to the "hassle" of gardening and farming, which he further explains "people don't do nowadays" for the most part. That is, gardeners represent a minority of residents in the communities.

With the help of volunteers and various organizations, the C-W-T community garden was launched in 2015 as the first community garden initiative between the communities. In a 2013 C-W-T town hall meeting, attendees identified food security and the consistent availability of food as a concern in the community, while also noting the value of heightening community interest in gardening and/or small-scale household farming. Marjorie Horlick, an active volunteer and community leader, proposed the concept of a joint community garden for the communities. Members of the town council and the attending audience supported this idea. Marjorie and the mayor of C-W-T, Churence Rogers, then went to New-Wes-Valley to meet with representatives of Central Regional Wellness Coalition, a provincially funded organization that works to enact a vision of healthy people and healthy communities within the region through primary health-care facilitators.

They discussed how to establish the garden, and a regional primary health-care facilitator provided a copy of the *Community Garden Best Practices Toolkit* (FSN, 2011) to help them. A town staff member was appointed to work with Marjorie to write a grant proposal for a provincial Health and Wellness grant. With assistance from Food First NL and other information sources within the province, the budget and other aspects of the proposal were prepared. The town received a grant of $6,500 as a result of the proposal in 2014 and began to prepare for a 2015 community garden launch.

Dr. Kelly Vodden, Memorial University professor, local resident, and board member of the Indian Bay Ecosystem Corporation (IBEC), introduced the project and Marjorie to the project development officer of IBEC, and discussed the potential for the local non-profit organization dedicated to environmental stewardship to assist the town and volunteers to create the garden. As part of their Environment Canada-Ecoaction Community Stewardship Program, IBEC devoted both staff time and volunteer resources to assist with the initiative. The group held a meeting in February 2015 to plan how to proceed. In March 2015, they presented their plans for starting the community garden to the C-W-T town council, who offered their continuing support and appointed councillor and IBEC board member Ralph Ackerman as a council liaison.

Garden location was the first major issue up for discussion. After mulling several options, the council decided it would donate an empty plot adjacent to the council building with enough room for a 50-by-100-foot garden. A garden co-ordinator was selected. The land was cleared and prepared for the garden during warmer weather. Kelly turned to Memorial University's Environmental Policy Institute (EPI) to secure an intern to provide assistance. Jannatul Islam, a graduate candidate of EPI's Master of Arts in Environmental Policy, participated in a 12-week unpaid internship with IBEC, which provided accommodation and office space, while EPI provided travel and graduate assistance support.

An informal, six-person community garden committee was created that included Kelly and Jannatul. Other individuals, including IBEC staff and board members and other community residents, also lent a hand. An open meeting was advertised and held where work activities were listed and assigned. The major tasks ahead in the spring of 2015 were to design and agree on a layout, grow seedlings, clear the land and prepare the ground, construct boxes for the raised beds, prepare soil by mixing soil and compost, and purchase the tools and materials for gardening.

Local residents and businesses volunteered their help in different ways. Two elderly men in the community offered knowledge about how to grow vegetables in the area. A citizen and freelance artist designed a sign for the garden, while another friend of the garden from a neighbouring community provided a crab/lobster-based soil fertilizer. Other individuals and local businesses provided pallets and used wood to construct boxes. New Wood Manufacturers, a local firm, provided materials and donated equipment and manufacturing facilities.

By the middle of May 2015, the land had been cleared thanks to the town of Centreville and its municipal works staff and equipment. Marjorie discussed the community garden concept with many C-W-T citizens. A member of the Centreville Academy School Board suggested contacting the teachers at the school associated with science and gardening projects to invite students to participate. The school principal was contacted and agreed to provide students in Kindergarten to Grade 3 classes the opportunity to plant seedlings and then transplant them to the garden when it was ready. Committee members also grew different kinds of seedlings and supplied them to the community garden for planting, following the guidance of local gardeners and *The Old Farmer's Almanac Garden Planner* for the area (www.almanac.com).

The C-W-T community garden was officially opened on 23 June by C-W-T Mayor Churence Rogers. In addition to the team mentioned above, 50 K–3 students from Centreville Academy and other local residents attended the opening. Knowledge about native plants, gardening safety, composting, and habitat was conveyed to the schoolchildren and others in attendance. Central Wellness Coalition facilitator Shauna Humphries spoke with the children and provided educational games at the event. The Coalition also funded the nutrition break provided for the children. The students and gardeners planted their seeds and plants. The students painted their names and different pictures or symbols on wooden markers and placed them alongside their plants so that they could identify them as the plants grew. The teachers from the Academy expressed that the children genuinely enjoyed the planting experience, and one student even proclaimed that the opening was the most enjoyable day of his life! A Grade 2 student explained that she felt very proud when she was planting because her father works for IBEC, one of the most engaged organizations in the community garden project. (See also the Indian Bay Ecosystem Corporation [IBEC] website for more information on the garden and opening day.)

The cool weather of summer 2015 proved challenging, as did adhering to the agreed-upon maintenance schedule (with volunteers taking turns at watering and weeding). Throughout the summer, several students also visited the garden to water and look after their own plants, just as they had been encouraged to do on opening day. One Grade 3 student brought her grandmother and cousins, who travelled from Ontario, to show them her plants and excitedly explained gardening to them. While some plants such as pumpkins, peppers, and sunflowers did not fare well through the season, peas, beets, kale, radishes, and carrots flourished. Educational signage was also created to highlight native plants in and around the garden that provide food for people and for local wildlife. Examples included wild raspberries, blueberries, and strawberries, among others.

In October, the community gardening committee organized a feast event to celebrate the harvest and World Food Day. The event participants included local municipal officials, students from Centreville Academy, their grandparents and/or other family members, and local volunteers. The day started with a field trip to the garden where committee members hosted short gardening information sessions, then the students harvested some of the vegetables they had planted in the spring (for example, peas were still ready for picking and carrots for pulling). A lunch was served that had been prepared using primarily locally grown vegetables from LA Farms, along with wild Newfoundland moose and locally raised chicken. The meal was followed by videos and other educational activities, with assistance from the regional primary health-care facilitator and IBEC staff. The organizers deemed the event a success and predicted that it would help motivate volunteers and next year's youth gardeners.

In 2016, the garden entered into its second year of operation. IBEC obtained funding support from the provincial Multi-Materials Stewardship Board to enhance the environmental benefits of the garden, including the creation of a community compost station, workshops and demonstrations to increase waste diversion awareness, and promotion of local gardening to reduce the amount of commercial plastic packaging and other non-biodegradable consumer waste entering the regional landfill site. Approximately 60 Centreville Academy students (Kindergarten to Grade 3) once again participated in a spring planting day, with workshops on composting, planting, and garden care. The local Women's Institute became involved by renting two garden boxes to grow potatoes, carrots, and cabbage for a fall salt beef dinner. Plans also began for the construction of a greenhouse on the site. The nearby communities of New-Wes-Valley and Gander launched community gardens in 2016, indicating growth of a community garden movement within the region.

The summer of 2017 was the garden's third season. Without funding to support its involvement, IBEC became less engaged in the initiative. Garden volunteers were unable to organize the spring planting with Centreville Academy, further reducing the number of groups involved. Marjorie Horlick explains that efforts to seek volunteers from the local high school as part of students' community service requirements were also unsuccessful. The

Women's Institute continued to utilize two garden boxes to grow cabbage and carrots for their early October salt beef dinner. Other gardeners grew tomatoes, Swiss chard, and flowers. Strawberries and chives planted in 2016 grew successfully once again and were harvested by local gardeners. The garden continued to produce food but for a smaller number of local residents and with reduced support from organizations such as IBEC and EPI.

EXAMINING THE ROLE OF COMMUNITY GARDENS IN BUILDING A FOOD MOVEMENT

The stories of the C-W-T community garden and others in the province demonstrate multiple features of social movements generally, and specifically those aimed towards sustainable food systems. Comparing the C-W-T community garden case study to the characteristics of social movements illustrates the community garden's contribution to a social movement towards food security, sovereignty, and sustainability in the C-W-T region and in the province:

1. *Commitment to a cause or ideology:* By starting the C-W-T community garden, the volunteers and organizations involved in the project (including two of the chapter authors) demonstrated commitment to producing their preferred, locally grown, healthy, fresh food. The ideology reflects the community's historical tradition of gardening and small-scale farming and their desire to maintain and rejuvenate these traditions. This commitment is also born out of difficulty accessing fresh, affordable food at all times (essentially, food security concerns). Members further envision community benefits from the garden such as nurturing of intergenerational relationships and a reduced environmental footprint from waste and food production. These insights are being communicated with the hope of soliciting more involvement in the garden and growing the local food movement.
2. *Separation (real and/or perceived) from the established order:* Efforts by groups such as Food First NL raise awareness about the need for change and the inadequacies in our current food system. The C-W-T community garden joins Food First NL

and other community garden and food-related organizations in a call of action for a change in the status quo. This includes raising greater societal awareness of the importance of household and community production to address shortfalls in the commercial components of the food system. Although the majority of community members do not garden, the community garden aims to change this view. Volunteers are striving to increase gardening in an effort to reduce dependency on commercialized and industrialized food that is grown in and transported from other provinces and nations. The commitment to this cause involves separation from the established order and an attempt to create a new set of values and to change personal patterns of behaviour.

3. *Collective action over time:* The C-W-T community garden is a collective action of different individuals and organizations bringing together various skills, capacities, and resources. Examples include local businesses, academics, government agencies committed to health and social development, non-profit organizations, and individual citizens. The idea was first publicly suggested and garnered initial support in 2013, was launched in 2015, and it continued in 2016. In 2016, IBEC conducted a garden planting project with Kindergarten through Grade 3 students, who watered and weeded the garden boxes in July and August as time permitted. The 50+ Club of C-W-T also provided garden management during July and August as time permitted. However, like most nascent community gardens, the longevity of the garden remains uncertain, given the need for ongoing investment of time and resources from a limited number of already busy volunteers, although the project clearly extends beyond a one-time action.
4. *Social change:* The long-term goal of the community garden is to contribute to food sustainability in the communities of Centreville-Wareham-Trinity and Indian Bay. The hope is that the community garden may change patterns of behaviour to better align with the ideals of sustainable food systems and encourage food choices that enable food sovereignty, security,

> and sustainability. For now the changes have been much more modest. However, change has transpired. The garden site has been created, new relationships have been formed between the town of C-W-T, IBEC, and the Wellness Coalition, more than 70 students have learned gardening skills, gardening and food preparation knowledge has been shared across generations, and awareness of food security and sustainability issues has been heightened in the community.

There has clearly been community engagement with the project, including ties across generations and the public/private sectors, and we suggest that this nascent project reflects a step in the direction of developing a food sovereignty movement, where residents control their food system. However, given the small scale of its production, there are limitations to what a single community garden can contribute towards overarching food security issues. Further, significant challenges exist in ensuring the garden's long-term sustainability, including the need for increased community education and involvement. This must be balanced with the needs and preferences of different user groups, as well as the success in securing continued financial and institutional support.

Similar challenges have been experienced elsewhere (Nelson-Hamilton, 2011; Corrigan, 2011), but the C-W-T community garden can be viewed as a part of a larger, growing movement in the province that includes not only community gardens, but also farmers' markets, bulk-buying clubs, workshops, seed-saving efforts, and more (Brake, 2015). A Backyard Farming & Homesteading NL Facebook page has over 2,200 members, suggesting that the rejuvenation of a gardening and supplementary farming culture in Newfoundland and Labrador is well underway. Nelson-Hamilton (2011) notes how Rabbittown community garden organizers have described themselves as part of a local food movement in Newfoundland based on concerns about industrial agriculture practices and reliance on food imports. The garden itself became a space for organizing to establish other initiatives related to food security and food sovereignty in the city of St. John's, including new community gardens.

Community gardens can (and do) seed and support this growing interest in a food system that is more decentralized, diverse, self-reliant, co-operative in orientation, and harmonic with nature (Allen, 2004; Tieg et al., 2009). Community gardening emerges from a key concern about inadequate security, sovereignty, and sustainability within existing food systems. These efforts can be used to construct new discourses of identity and belonging, to generate new (or renewed) cultures, and to enact new forms of social relations through gardening and community activities and through changes in personal lifestyle choices. The examples given above demonstrate how social change may radiate from individuals and households through the community and provincial levels.

CONCLUSIONS AND FUTURE DIRECTIONS

In this chapter we presented a case study of the C-W-T community garden to illuminate how food can become a pivotal social movement within the province of Newfoundland and Labrador. In the spirit of La Vía Campesina, residents are coming together to take control of their food production through community gardens. In this situation, community gardening can facilitate food sovereignty at a small scale and serve as a tool to support a larger social movement towards food sovereignty and sustainability. For efforts to be successfully sustained, those who initiate community gardens require policy and financial supports as well as a commitment to inclusivity and innovative communication and marketing techniques in order to share their experiences and recruit volunteers and new members to the movement.

We suggest that Newfoundland and Labrador policy-makers reconsider the existing focus on commercialization and industrialization in favour of a multi-pronged approach to food system sustainability that aims to support and restore a tradition of supplementary household and community-level food production. Sharing of agricultural knowledge through initiatives such as community gardens and pastures can help escalate the social value placed on small-scale farming and encourage young people to consider food production as a part of their futures, whether at the household, community, or industrial scale. Much like the St. Francis greenhouse discussed by Doyle and Traverso-Yepez in Chapter 3, support for individual skill-building strategies such as cooking, composting, and gardening workshops, and promotion of community kitchens, farmers' markets, and buy-local campaigns can all help to facilitate this growing social movement. Despite challenges and recognized limitations, community gardens such as the C-W-T community garden have shown a potential to generate important socio-economic, cultural, environmental, health, and educational benefits for a sustainable community food system.

ACKNOWLEDGEMENTS

The authors would like to thank our key informants for their participation in this research, as well as Indian Bay Ecosystem Corp. and the Environmental Policy Institute, Grenfell Campus, MUN, for their support of the summer internship that led to this chapter, and all of the volunteers who have worked to create the C-W-T community garden. Finally, we are grateful to the reviewers for their very useful critiques and suggestions.

NOTES

1. Also referred to as Regional Development Associations in some sources.
2. The Census of Agriculture is implemented by Statistics Canada approximately once every five years. Data and reports are available at the Statistics Canada website. At the time of publication, preliminary results from the 2016 Census of Agriculture study are gradually becoming available. Interested readers may wish to peruse the Census of Agriculture website: http://www23.statcan.gc.ca/imdb/p2SV.pl?Function=getSurvey&Id=225699.

REFERENCES

Allen, P. 2004. *Together at the Table*. University Park: Pennsylvania State University Press.

Appleby, T., K. Armstrong, D. Nolan, and K. Jameson. 2011. *Burin Peninsula Community Led Food Assessment: 2010–2011*. Food Security Network NL. At: www.foodsecuritynews.com/Publications/Burin_Peninsula_CLFA_Final_Report.pdf.

Batliwala, S. 2012. *Changing Their World: Concepts and Practices of Women's Movements*, 2nd ed. Toronto: Association for Women's Rights in Development (AWID). At: https://www.awid.org/sites/default/files/atoms/files/changing_their_world_2ed_full_eng.pdf.

Brake, J. 2015. "Popularity of local food production on the rise in NL." *The Independent*, 22 May. At: http://theindependent.ca/2015/05/22/popularity-of-local-food-production-on-the-rise-in-nl/.

Brunetti, A. 2010. *The Community Garden Handbook for Newfoundland and Labrador*. St. John's: College of the North Atlantic and Provincial Rural Secretariat. At: http://www.ope.gov.nl.ca/pe/whatweredoing/cbr_reports/6.TheCommunityGardenHandbook.pdf.

Cadigan, S. 1998. "Agriculture." Heritage Newfoundland & Labrador. At: http://www.heritage.nf.ca/articles/economy/agriculture.php.

Cameron, J., K. Gibson, and A. Hill. 2014. "Cultivating hybrid collectives: Research methods for enacting community food economies in Australia and the Philippines." *Local Environment: The International Journal of Justice and Sustainability* 19, 1: 118–32. doi:http://dx.doi.org/10.1080/13549839.2013.855892.

Community Garden Alliance. 2015. "NL community gardens." At: http://communitygardensnl.ca/nl-community-gardens/.

Corrigan, M. P. 2011. "Growing what you eat: Developing community gardens in Baltimore, Maryland." *Applied Geography* 31, 4: 1232–41. doi:http://doi.org/10.1016/j.apgeog.2011.01.017.

Desmarais, A. 2008. "The power of peasants: Reflections on the meanings of La Vía Campesina." *Journal of Rural Studies* 24, 2: 138–49. doi:http://doi.org/10.1016/j.jrurstud.2007.12.002.

Doherty, K. E. 2006. "Mediating the critiques of the alternative agrifood movement: Growing power in Milwaukee." Master's thesis, Department of Geography, University of Wisconsin-Milwaukee. At: https://www.ssc.wisc.edu/~wright/Social%20Economy%20PDFs/Doherty%20--%20Growing%20Power.pdf.

Food First NL. 2015a. *Food First NL Annual Report 2014–2015*. St. John's: Food First NL. At: https://static1.squarespace.com/static/54d9128be4b0de7874ec9a82/t/5671c66bcbced6829d5ec316/1450296939769/2015-FFNLannualreport-web.pdf.

———. 2015b. *Everybody Eats: A Discussion Paper on Food Security in Newfoundland & Labrador*. St. John's: Food First NL. At: http://static1.squarespace.com/static/54d9128be4b0de7874ec9a82/t/56572239e4b0b807738c5ad0/1448550969272/Everybody+Eats_NL+Discussion+Paper+2015.pdf.

Food Security Network of Newfoundland & Labrador (FSN). 2007. *Healthy Food for All: Working toward Food Security in Newfoundland & Labrador*. Proceedings of the Food Security Assembly 2007. St. John's. At: http://www.foodsecuritynews.com/Publications/FSN_2007_Food_Security_Assembly_Conference_Proceedings.pdf.

———. 2011. *Community Garden Best Practices Toolkit: A Guide for Community Organizations in Newfoundland and Labrador*. St. John's: Food First NL. At: http://www.foodsecuritynews.com/Publications/Community_Garden_Best_Practices_Toolkit.pdf.

Francis, M. 2003. *Urban Open Space*. Washington: Island Press. At: http://depts.washington.edu/open2100/pdf/2_OpenSpaceTypes/Open_Space_Types/cgarden_typology.pdf.

Glover, T. D. 2003. "Community garden movement." In K. Christensen and D. Levinson, eds., *Encyclopedia of Community*, 264–66. Thousand Oaks, Calif.: Sage.

Gottlieb, R. 2001. *Environmentalism Unbound: Exploring New Pathways for Change*. Cambridge, Mass.: MIT Press.

——— and A. Fisher. 1996. "Community food security and environmental justice: Searching for common discourse." *Agriculture and Human Values* 3, 3: 23–32. Doi:10.1007/BF01538224.

Hancock, D. R., and B. Algozzine. 2015. *Doing Case Study Research: A Practical Guide for Beginning Researchers*. New York: Teachers College Press.

Hale, J., C. Knapp, L. Bardwell, M. Buchenau, J. Marshall, S. Fahriye, and J. S. Litt. 2011. "Connecting food environments and health through the relational nature of aesthetics: Gaining insight through the community gardening experience." *Social Science & Medicine* 72, 11: 1853–63.

Hanna, A. K., and P. Oh. 2000. "Rethinking urban poverty: A look at community gardens." *Bulletin of Science, Technology & Society* 20, 3: 207–16.

Hassanein, N. 2003. "Practicing food democracy: A pragmatic politics of transformation." *Journal of Rural Studies* 19: 77–86. doi:http://doi.org/10.1016/S0743-0167(02)00041-4.

Horn, J. 2013. *Gender and Social Movements: Overview Report*, Chapter 2: "Social movements: Evolution, definitions, debates and resources." Brighton, UK: BRIDGE, Institute of Development Studies. At: http://docs.bridge.ids.ac.uk/vfile/upload/4/document/1310/FULL%20REPORT.pdf.

Indian Bay Ecosystem Corporation. 2017. "C-W-T community garden." https://indianbayecosystem.com/pictures-3/bird-nesting-boxes/projects-2/c-w-t-community-garden/.

Memorial University Botanical Garden. 2017. "Community gardens of Newfoundland and Labrador." At: http://www.mun.ca/botgarden/gardening/comgarden/.

Mondal, P. 2015. "Social movements in India: Meaning, features and other details." At: http://www.yourarticlelibrary.com/india-2/social-movements-in-india-meaning-features-and-other-details/32941/.

Nelson-Hamilton, L. 2011. "Growing together: The intersections of food, identity, community and gardening in St. John's, Newfoundland." Master's thesis, Memorial University of Newfoundland.

NL Statistics Agency, Government of Newfoundland and Labrador. 2016. "Population by age and gender." At: http://nl.communityaccounts.ca.

Omohundro, J. T. 1982. "Subsistence gardens in Newfoundland." Tokyo: United Nations University. At: http://archive.unu.edu/unupress/food/8F073e/8F073E0a.htm.

Patel, R. 2010. *The Value of Nothing: How to Reshape Market Society and Redefine Democracy*. New York: Picador.

Ramsey, R. D. 1998. *Issues Affecting the Development of Agriculture in Newfoundland: A Case Study of the Lethbridge-Musgravetown Agricultural Development Area* (Research Paper No. 30). Truro, NS: Rural Research Centre, Nova Scotia Agricultural College.

Rao, M. S. 2000. *Social Movements in India: Studies in Peasant, Backward Classes, Sectarian, Tribal and Women's Movements*. New Delhi: Manohar.

Sogge, D., and G. Dutting. 2010. *Moving Targets: Notes on Social Movements* (Working Paper 2). The Hague: Hivos.

Statistics Canada. 2006. "Community profile." At: http://www12.statcan.gc.ca/census-recensement/2006/dp-pd/prof/92-591/index.cfm?Lang=E.

———. 2009. *2006 Census of Agriculture*. At: http://www.statcan.gc.ca/ca-ra2006/indexeng.htm.

———. 2010. "Farm cash receipts." Catalogue no. 21-011. At: http://www.statcan.gc.ca/pub/21-011-x/2011002/t031-eng.pdf.

———. 2011. "Farm and farm operator data." Catalogue no. 95-640-X. At: http://www.statcan.gc.ca/pub/95-640-x/2011001/p1/prov/prov-10-eng.htm#Farm_area.

———. 2017a. "Newfoundland and Labrador farms have the highest rate of direct marketing," 10 May. Catalogue no. 95-640X. At: http://www.statcan.gc.ca/pub/95-640-x/2016001/article/14800-eng.htm#b2.

———. 2017b. "Farm and farm operator data," 10 May. Catalogue no. 95-640X. At: http://www.statcan.gc.ca/pub/95-640-x/95-640-x2016001-eng.htm.

Teig, E., J. Amulya, L. Bardwell, M. Buchenau, J. Marshall, and J. Litt. 2009. "Collective efficacy in Denver, Colorado: Strengthening neighborhoods and health through community gardens." *Health & Place* 15: 1115–22. doi:http://doi.org/10.1016/j.healthplace.2009.06.003.

Town of C-W-T. 2017. "Community garden." At: http://www.townofcwt.com/community-garden-1.

Vodden, K., H. Hall, and D. Freshwater. 2013. *Understanding Regional Governance in Newfoundland and Labrador: A Survey of Regional Development Organizations*. St. John's: Harris Centre, Memorial University. At: http://www.mun.ca/harriscentre/reports/research/2013/1305_UnderstandingRegionalGovernance.pdf.

Part II

Lessons in Food Security and Food Sovereignty: Town, Bay, and Big Land

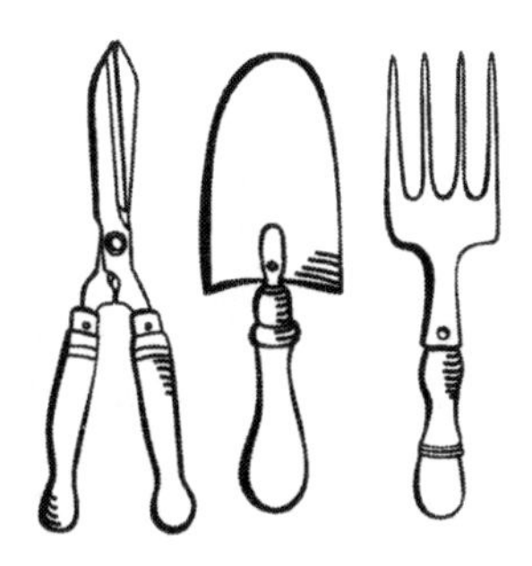

Pumpkin harvest © Dave Howells

5

The Lived Experience of Food (In)Security among Seniors and Single Parents in St. John's

Martha Traverso-Yepez, Atanu Sarkar, Veeresh Gadag & Kelly Hunter

INTRODUCTION

While food is physiologically essential for the body to function, what and how we eat is determined by a range of interrelated environmental, socio-economic, and cultural factors influencing the content and quality of our food consumption. By the same token, although we may be referring to a similar term, "food insecurity,"[1] in this study and throughout this edited volume, the ways food insecurity is expressed differ depending on the socio-cultural context (Pinstrup-Andersen, 2009).

Although the different concepts of food security and food insecurity are widely used in contemporary public policy and academic circles, food insecurity is not immediately recognized as an issue in developed economies, unless one is in close contact with low-income earners or with people who have difficulties in finding a job and are reliant on social assistance. During the past decades, food availability has exponentially increased — especially in the developed world — with the globalization of food trade and the commodification and increased supply of industrialized food (Carolan, 2011). However, the food industry has generated other concerns, with the abundance of processed, high-calorie, low-nutrient, but low-cost products that are made strategically appealing and palatable for mass consumption (Guthman, 2011; Wilson, 2004). As a result, a significant percentage of the global population consumes these calorie-dense, highly processed foods that are low in nutritional value because, among other reasons, they cannot afford the consumption of natural products (Guthman, 2011; Nestle et al., 1998; Winson, 2013).

Most of these people may not go hungry, but they are still "nutritionally insecure" (Pinstrup-Andersen, 2009). In Canada, for example, financial constraints in 12.5 per cent of households affect these individuals' access to quality or a sufficient quantity of food (Tarasuk, Mitchell, and Dachner, 2015). In accordance with the annual *Household Food Insecurity in Canada* reports, vulnerable populations face the most challenges in their ability to purchase and consume quality healthy food, taking a significant toll on their health and well-being (Tarasuk, 2009). However, a contextual understanding of how "food insecurity" is manifested among these populations is limited (Engler-Stringer, 2010; Keller, 2006; Sim, Glanville, and McIntyre, 2011).

FOOD SECURITY AMONG VULNERABLE POPULATIONS

Single parents and low-income elderly adults constitute two populations that are particularly vulnerable in confronting food security issues. Adult females represent the overwhelming majority of single-parent households in Canada, with 38 per cent of them experiencing food insecurity (Tarasuk et al., 2015). Families headed by single mothers are eight times more likely to report their children being food insecure compared to families with two parents, reflecting the significant economic constraints that female-led young families face (McIntyre and Rondeau, 2009; McIntyre, Bartoo, and Emery, 2014; Williams et al., 2012).

Food security issues within single-parent households can result in serious negative health impacts within such families, beyond the typical nutritional deficiencies. One American study based on longitudinal data links parental depression to food insecurity, due to their concern around poor infant health and anxieties over their inability to access adequate amounts of healthy food for their family (Bronte-Tinkew, Zaslow, Capps, Horowitz, and McNamara, 2007). Furthermore, there is an increased prevalence of behavioural, emotional, and academic problems among food insecure children compared to children from food secure families, as these food insecure parents may provide enough food to satisfy their children's hunger, but are unable to avoid nutritional deficiencies (Darmon and Drewnowski, 2008; Tarasuk, 2009; Veugelers and Fitzgerald, 2005).

Although they play a significant role, financial constraints are not the only barrier preventing lone parents from feeding their families healthy, nutritious meals. Various structural and socio-cultural factors challenge parents' ability to plan and prepare healthy meals, making "convenient," already-prepared food the more appealing option (Engler-Stringer, 2010). As Broughton, Janssen, Hertzman, Innis, and Frankish (2006: 215) highlight, lack of cooking skills, time constraints, and competing priorities often lead parents to "juggle taste, nutrition, cost and convenience in their food selections."

By the same token, elderly people, the fastest-growing age group in Canada, are also at risk for food security problems. By 2036, Newfoundland and Labrador is expected to have the highest proportion of seniors of all the provinces in Canada. Research shows that some seniors may lack the financial resources to provide for their own basic needs, as it is acknowledged that 6.3 per cent of elderly Canadians living alone encounter food insecurity (Tarasuk et al., 2015). A study conducted in Nova Scotia that compared the incomes and estimated expenses of four senior households, each reliant on different forms of fixed income (including the Guaranteed Income Supplement [GIS], Old Age Security [OAS], Canada Pension Plan [CPP], and personal pensions), found that seniors living alone had a much more limited grocery budget, as there is less protective financial support than when having two incomes (Green, Williams, Johnson, and Blum, 2008). Residing in an Atlantic province with comparable costs of living, seniors living in Newfoundland and Labrador may face similar circumstances as those in the study by Green et al.

That being said, it needs to be restated that the focus cannot solely be on finances when it comes to seniors' food insecurity. As Green-LaPierre et al. (2012: 2) point out: "Canada's current food security measurement tools, which focus on low-income as the most significant determinant of food insecurity, may not accurately capture the other major enablers and barriers seniors face in accessing sufficient, quality foods in socially acceptable ways." Limitations in functional abilities or disabilities, chronic diseases, cognitive decline, lack of social supports, and inability to access community programs clearly have a significant impact on personal food management strategies and food-related activities (e.g., grocery shopping, meal preparation), putting the elderly at higher risk of being food insecure (Green et al., 2008; Keller, 2006).

Social environments also have critical effects on experiences with food insecurity. In cases where seniors are reliant on others for food-related activities, social deprivation or lack of social networks and family supports are associated with increased household food insecurity (Carter, Dubois, Tremblay, and Taljaard, 2012).

Consequently, it becomes clear that in addition to income adequacy, other complex and interrelated factors are at play in food-related actions. This makes it necessary to explore the interconnected factors influencing the food practices of those who are considered food insecure, as food management strategies may contribute to differences in diet quality among these particular households (Sim et al., 2011). This exploration is especially important in the Newfoundland context, a region that has experienced increasing cost of living and a heavy reliance on high-cost, imported food (City of St. John's, 2015; Quinlan, 2012).

RESEARCH METHODOLOGY

Building upon a social justice framework and a systemic understanding of food insecurity as a social determinant of health, this chapter is comprised of a comparison analysis of data from a research study conducted in the St. John's metro area. The study explored the extent and experiential knowledge of food insecurity among seniors and single parents in St. John's. By examining the qualitative data collected in our face-to-face surveys and key informant interviews, we discuss experiences of food insecurity among these vulnerable populations and the ways that the socio-economic, cultural, and environmental contexts influence these populations' food practices. From a systemic understanding of this issue, we illustrate how potential solutions for food insecurity have been limited in their outreach at the individual, community, and macro-policy levels. We also suggest a comprehensive, multilevel, integrated food policy approach as a meaningful recommendation to guarantee healthy food environments for all.

For this project we developed a mixed-method research design, which included key informant interviews with four professionals either in the field of food security or working with these vulnerable populations, as well as face-to-face survey interviews with a sample of 50 single parents and 48 senior citizens above the age of 65 (approximately 1 per cent of these vulnerable populations in St. John's). The questionnaire used in the survey interviews was comprised of a selection of *standardized* questions adapted from the Canadian Community Health Survey (Statistics Canada, 2014). The topics explored included: (a) household practices related to meal preparation and meal consumption; (b) level of food awareness and food consumption practices; and (c) food purchasing practices, where food sufficiency, comprising of food quality and quantity, was explored. In addition, we also explored the level of engagement in community-based interventions to address food insecurity. The research was approved by the Health Research Ethics Authority (HREA).

There were also several open-ended questions for participants to express their opinions about the specific topics in the survey questionnaire. Socio-demographic variables included gender, age range, education, ethnicity, and major sources of income. As we were expecting lower literacy rates, the interviewer-administered surveys were conducted by an experienced research assistant. This provided opportunities for the participants to expand on particular issues or further explain their answers.

The survey results were entered into SPSS (Statistical Package for the Social Sciences) software for quantitative data analysis (Sarkar, Traverso-Yepez, Gadag, and Hunter, 2015), while key informant interviews, as well as the participants' additional comments and answers to the survey's open-ended questions, were logged and analyzed as qualitative data. Through this qualitative data, we are able to capture participants' experiences and perceptions of food-related practices, which are the focus of the present chapter.

SOCIO-DEMOGRAPHIC DIFFERENCES IN THE SAMPLE POPULATIONS

Although both seniors and single parents are considered vulnerable populations, our research shows that in our sample population, single parents' economic situations were more severe than were those of seniors. For example, 80 per cent of single parents had household incomes of less than $25,000 a year, compared to 48 per cent of seniors. When looking at the sources of income, 80 per cent of single parents were on income support, compared to 21 per cent of the senior population, with 48 per cent of seniors having job-related retirement pensions and 71 per cent receiving GIS or OAS. Receiving these additional benefits makes a considerable difference for this sector of the population, as such benefits provide supplementary financial resources.

Furthermore, it is important to recognize that single parents have to budget for a family, while many of the seniors in the sample only had to provide for themselves and, in some cases, their spouses. Thus, the financial constraints and budgets between the two groups are remarkably different.

Concerning the level of education, 29 per cent of the single parents in the sample had not completed high school, compared to 19 per cent of the seniors. In general, the seniors in our sample had higher educational levels, which may explain the higher income levels compared to single parents. The gaps in income between the two groups could be attributed to our recruitment strategies, but considering recent health inequities research these gaps could also be explained by the "gradient effect" (Marmot, 2005; Wilkinson and Pickett, 2010). This "gradient effect" infers that living with a lower income corresponds with poorer health and higher mortality rates in comparison to those with higher socio-economic statuses. Consequently, low-income seniors are more likely to have shorter lifespans; be significantly impaired by disease or mobility issues; and/or live in isolation at an earlier stage of their life. This could explain why we were more likely to find research participants among active seniors from affluent and better-educated populations, who have increased access to material and psychological resources (Marmot, 2005; Wilkinson and Pickett, 2010).

As we looked for a sample of these two populations living in more vulnerable conditions, most of the recruited single parents were clients of the Single Parent Association of Newfoundland (SPAN) or food bank users. In contrast, we had challenges in recruiting seniors living in vulnerable environments. Many seniors who participated were involved in community groups or activities, such as seniors' walking clubs, and already had strong community connections. Furthermore, it is worth noting that the average age of the senior participants fell between 65–70 years old. As one elderly participant pointed out, conducting this research with older seniors in their eighties and nineties may result in much different findings.

HOUSEHOLD MEAL PRACTICES

When asked how often participants ate at home for the main meal, 62 per cent of single parents and 48 per cent of seniors indicated that they typically eat at home every day. However, only 36 per cent of single parents, compared to 65 per cent of seniors, declared cooking mostly from whole, basic items when eating at home. A range of factors could be at play for this significant difference, including the time constraints that some single parents face in trying to raise children while being employed, which was a frequently mentioned issue among participants. Furthermore, discrepancies in what participants consider "whole, basic items" and "cooking from scratch" versus eating "already prepared foods" could have also affected their responses. Consequently, cooking from scratch and putting together meals mainly from canned products could be interpreted differently, as these terms are ambiguous and hard to differentiate (Engler-Stringer, 2010).

Regarding the time needed to prepare food from scratch, one single mother explained the dilemma she faces in choosing to work versus staying on social assistance. Since having her three kids, she has gone through periods of time where she is employed and periods where she relies on income support. While she is working, she has very little time to cook meals from scratch. So, although her income is higher than what it would be on social assistance, her family is not able to benefit from the increased grocery budget. She ultimately disclosed that her preference is to opt for social assistance because it gives her more time to cook healthy, homemade meals for herself and for her children, rather than buying frozen, prepared foods.

However, cooking from scratch was not the norm for many participants, even for those who suggested that time constraints were not an issue. This may be attributed to a decrease in investment in cooking skills, which was emphasized by our key informants who discussed different factors behind this shifting trend. Our findings confirmed other research evidence, showing how shifts in the social, economic, cultural, and physical environment are influenced by changing food-system environments (Story, Hamm, and Wallinga, 2009; Winson, 2013). These shifts include: advances in technology for industrial food storage, preparation, and cooking; increased availability of food commodities, shifting family time, and financial demands (labour-related market participation); and changing family priorities and values — all of which contribute to decreased opportunities for developing cooking and food preparation skills (Chenhall, 2010; Winson, 2013). As a result, during the past decades, trends in meal preparation had been deviating away from traditional cooking practices, both in the family and in the public education environments. Due to a combination of socio-cultural, economic, and environmental factors, people have become more reliant on available packaged or processed foods (Engler-Stringer, 2010), which may be reflected in the small percentage of the new generations of single parents who cook home-made meals (Chenhall, 2010; Groves, 2002).

Although some single parents discussed the importance of having children participate in purchasing and preparing food, most of these emphasized that this depends on their age and abilities. One single mother was emphatic in saying that it is important to involve her children as much as possible in food preparation and to teach them about healthy eating at an early age so that "they have the skills in case they have a better income when they are older."

Additionally, even if they had the option, getting children to eat healthy foods proved to be difficult for some of the parents. Some single parents emphasized the fact that their children are "picky eaters," limiting their ability to offer healthy meals to their kids. A couple of parents attributed picky eating to having raised their children on processed "junk" foods, which has conditioned them to preferring certain tastes; one of them emphasized the food industry's role in influencing children's preferences. Although some steps have been taken in Canada to limit food marketing towards children, evidence shows that even stricter regulations and policies need to be enforced to reduce the amount of advertising of "junk foods" to children and adolescents (Kelly et al., 2010; Nestle, 2006). Balancing food preference with economic viability was also mentioned by some participants, confirming what was found among low-income parents in other research (Dachner, Ricciuto, Kirkpatrick, and Tarasuk, 2010). Food systems researchers have studied how the "fast-food" culture has spread (Nestle, 2002), either out of necessity (lack of time to cook) or through marketing campaigns targeted at children. It is not difficult to promote fast foods, junk food, and heavily processed, packaged food, especially because these are usually the cheapest options available and are loaded with sugar, salt, and artificial flavours and colours in order to be more appealing to children (Nestle, 2002; Winson, 2013).

Nonetheless, as living conditions differ, one parent made the distinction between her children's eating habits and her own, describing her children's as being "excellent," while hers were depicted as being only "fair." This may reflect parents' willingness to sacrifice their own nutrition in order to be able to provide healthier foods to their children in the face of food insecurity (Hamelin, Beaudry, and Habicht, 2002; McIntyre et al., 2003). A couple of other single parents shared this sentiment, mentioning that they have skipped meals in periods of extreme hardship so their children are able to eat.

As expected, seniors with chronic illnesses expressed the difficulty of maintaining a proper diet while living on a fixed income. However, they also elaborated on other obstacles impeding their ability to prepare healthy meals, such as issues of mobility and fine motor skills, which hinder their independent cooking skills. Reliance on others for food-related practices has been associated with poorer diets among seniors with mobility issues (Keller, 2006). A poor-quality diet can have a serious impact on dealing with chronic diseases, worsening these individuals' health and well-being (Green et al., 2008). This was emphasized by one participant who was diagnosed with diabetes. Although her diet should be significantly more constrictive for health purposes, her budget limits her ability to eat as well as her doctor recommends. This notion is confirmed in the research of Green et al. (2008), which shows the cyclical relationship between food insecurity and chronic diseases that require specialized diets that are often much more expensive.

Some seniors also voiced the lack of motivation to prepare meals if living alone, which often led them to be dependent on processed food, something emphasized by one key informant: "I know there's this sort of common experience among seniors where, because you're typically living alone, you do not have the desire or drive to be preparing healthy meals." This lack of motivation may stem from difficulty in preparing small servings of food or the inability to consume larger amounts of homemade food before it spoils. A couple of widowed seniors emphasized the difficulty they had in adjusting their meal preparation after the death of their spouses. This was especially true for one participant whose wife had passed away a couple of years ago. Adhering to gender roles, she had done the majority of the cooking in the household, so the participant had to begin to prepare food himself after her death. He expressed that his lack of food preparation skills, together with his lack of motivation to cook, has resulted in his reliance on frozen foods and pre-prepared meals.

LEVEL OF AWARENESS WITH REGARD TO HEALTHY FOOD CONSUMPTION

We also found that level of food and nutrition awareness is higher among seniors, as only 64 per cent of single parents expressed being concerned about what they eat, compared to 81 per cent of seniors. In contrast, 36 per cent of single parents and 19 per cent of seniors declared being "little" or "not at all" concerned about their eating practices. Out of the 19 per cent of seniors who did not indicate concern around eating practices, a few participants attributed their indifference to having "given up" due to old age. Another identified reason was the role that traditional food habits have on many in the elderly population. For example, a senior discussed how her husband grew up eating bologna and that – despite health issues – "he was stuck in his ways" and was not open to eating healthier options.

Most seniors, however, said that adhering to healthy diets had been influenced by the advice from their health-care providers. They primarily described their eating habits as being either "excellent" (10 per cent), "very good" (39 per cent), or "good" (29 per cent), while a minority of seniors indicated it as being either "fair" (8 per cent) or "poor" (12.5 per cent). In contrast, only two of the single parents described their eating habits as "excellent." The remainder either considered them "very good" (20 per cent), "good" (34 per cent), or even "fair" (28 per cent) or "poor" (14 per cent). Many single parents discussed their strategies in trying to get the healthiest diet possible by "matching good food with bad food." In other words, they will supplement some items with more affordable options, while trying to keep meals as healthy as possible, depending on the weekly budget: "If I have chicken I may use frozen vegetables for a stir-fry. If I don't have frozen vegetables, I will use canned stuff. Our ideal stir-fry is what we had last night: chicken with a wide variety of vegetables out of the fridge. But I know by this time next week, the fridge won't have those options."

While understandings of food and nutrition are developed within the complexity of an individual's specific environmental context, perceptions around what is considered a "good" versus a "bad" diet vary significantly from person to person. As a subjective kind of knowledge influenced by people's life experiences, potential discrepancies about this kind of understanding, as with any health-related self-evaluation, can vary significantly among individuals and between cultures (Keller, 2006).

In the question regarding participants' sources of diet and nutrition information, Canada's Food Guide was one of the more frequently referred to sources, with 39 per cent of single parents and 60 per cent of seniors having cited this document as being one of their main resources. Additionally, almost all participants said they were "very familiar" or "somewhat familiar" with the four main food groups. This knowledge can likely be attributed to the wide distribution of Canada's Food Guide among the population, often through schools, health-care centres, and community service centres (Health Canada, 2011). The guide, a document under periodic revisions, provides Canadians with information on food's nutritional values and suggestions for the amount of servings of particular food groups that they should consume every day, depending on their age and sex. However, despite the large uptake of Canada's Food Guide as a guiding source for healthy eating information, it has been criticized as being too focused on individual behaviours without consideration of the larger socio-economic and cultural context, while also oversimplifying the complexity of nutrition (Andresen, 2007; Kondro, 2006).

Moreover, our research indicated discrepancies between participants' self-perceived levels of nutrition awareness and their personal assessments of their diets, which were often described as being "fair" or "poor." This suggests that many participants are unable to eat healthy diets due to a range of limitations and constraining factors, despite having indicated that they are well aware of nutrition and healthy eating. This infers that health promotion campaigns around healthy eating with a focus on education and awareness, such as Canada's Food Guide, will not be effective without considering people's socio-economic, environmental, and cultural circumstances (Cook, 2008).

FOOD PURCHASING PRACTICES AND THE LIMITATION OF LIVING WITH FIXED LOW INCOME

The section of the survey that focused on food purchasing allowed us to grasp the critical financial situations of some seniors and (more often) single parents. Relying on a fixed income was a common concern for single parents and seniors, as the cost of living has increased dramatically in the province over the past several years. In our survey, 69 per cent of single parents and 44 per cent of seniors declared not always having sufficient money when shopping for groceries. By the same token, when questioned whether they had to cut out some foods because of budget, 96 per cent of single parents and 57 per cent of seniors responded that they either often or sometimes have to do so. One senior participant discussed how she relies on eating carbs (i.e., Kraft Dinner) because it is what she can afford; this has resulted in her substantial weight gain over the past 10 years.

Interestingly, the difference in the vulnerabilities that single parents face compared to seniors was a matter acknowledged by key informants, as well as by some of the elderly participants. At least five of the elderly participants mentioned how single parents face more severe vulnerabilities in their daily lives that may hinder their ability to provide healthy, nutritious foods to their children. Two of the participants who discussed these issues had worked in schools or in early childhood education and understood the negative effects that poor nutrition can have on children's health and development, having witnessed it first-hand within their classrooms. One participant in particular emphasized how single parents and young parents may face more difficulty in "trying to raise a family on low incomes, not only for being able to afford healthy foods but also because society has become so materialistic and competitive, so parents want to be able to provide their kids with everything to be able to keep up with other families."

Transportation proved to be another barrier in many individuals' ability to access nutritious and affordable foods, as we expected from reviewing the literature on this topic (Cook, 2008; Green-LaPierre et al., 2012; Williams et al., 2012). Several seniors who do not own vehicles complained about the lack of transportation available to go to the supermarket. Some parents indicated similar problems, but not to the same extent as seniors. While public transportation in urban centres greatly reduces transportation costs (Green et al., 2008), participants who rely on public transport criticized it as being a somewhat unreliable and inconvenient method of transportation, particularly for the single parents who have to bring their young children with them while shopping for groceries. As transportation can be problematic, some participants indicated that this is a reason why they rely on non-perishable foods, "rather than [on] produce and items that expire quickly."

In attempting to make healthy choices on a limited budget, some participants discussed their personal strategies in purchasing more nutritious food options for the cheapest price possible. This included shopping for sales, buying generic rather than brand-name items, or buying certain foods in bulk from stores such as Costco or Bulk Barn. Other studies around food purchasing practices of low-income individuals found that participants saw the value in devoting extra time and energy to procuring food at the best possible price, whether through shopping at more than one store, using coupons, or locating sales and specials (Dachner et al., 2010; Engler-Stringer, 2010).

A couple of participants discussed the potential positive outcome that bulk-buying clubs or programs could have for those who have busy schedules, lack their own transportation, or have limited grocery budgets. Additionally, bulk-buying clubs can allow single parents to save on child-care costs or the stress of bringing young children with them to the supermarket. As food is typically cheaper when purchased in bulk, those with strict grocery budgets could benefit from "going in" with others on items from places like Costco, without having to spend a large amount of money at one time or risking wasting food from it spoiling (Dachner et al., 2010). That being said, some of the participants recognized the difficulty in organizing these types of systems. Rather, they suggested that it is perhaps a program that community centres or support groups could consider implementing.

The food management behaviours discussed above fall under the category of "planning," where deliberate actions are undertaken with regard to food procurement and preparation in order to make the grocery budget last as long as possible (Sim et al., 2011). Despite these planning strategies, many participants indicated that they still feel they are limited to choosing cheaper alternatives, such as processed options. Again, many of the single parents emphasized lack of resources as the main reason why they are not always able to address their concerns around nutrition. Overwhelmingly, the main barrier to a proper diet was affordability, with more than half of the single parents and one-third of seniors complaining about the cost of food, expressed both in answering a specific question concerning barriers and throughout the interviews.

As an isolated island with limited local food production, the vast majority of foods are imported to Newfoundland from the mainland, last estimated at 90 per cent (Quinlan, 2012). This has contributed to the high price of food in the province — a problem that many participants noted has significantly increased over the past several years. The price of food substantially limits their ability to make food-purchasing choices based on nutritional quality and diversity. As one single father explained, he is able to afford enough food for him and his daughter so that they are not hungry, but he does not try new foods or healthier alternatives as he finds it "difficult to add variety while staying within the budget." These notions were also disclosed by the key informants, who emphasized how the lack of cooking skills may be influencing the inability to think about low-budget food alternatives. A key informant, in addition to discussing high food prices, commented about the poor quality of imported, perishable produce, especially when distributed to rural areas.

Other budget priorities, such as rent, heat, medications, and school supplies, may come before grocery budgets, reducing these individuals' ability to spend more on healthy foods, which is reflected in other studies on this issue (Hamelin et al., 2002). In fact, 88 per cent of single parents and 52 per cent of seniors responded affirmatively to the question about whether other budget priorities impact their food choices. Additionally, 88 per cent of single parents and 39 per cent of seniors affirmed that in the past 12 months (either "often" or "sometimes") the food they purchased did not last until the end of the month and they could not afford to buy more, reflecting their critical financial constraints.

That being said, the situation seems harder for those reliant on social assistance as their main source of income. A common complaint was the extremely small budget that this provides: "[t]he cost of food is rising, so is the cost of living, but yet social assistance remains the same." This was a sentiment shared by seniors living on a fixed income, such as Old Age Security: "I've been retired for 15 years and up to this point I'm doing ok but I do worry about my finances in another 10–15 years. Will I be able to maintain my present standard of living? I doubt it, and food will be one of the few things where I'll be able to try to save money." Some participants asserted that it is the government's responsibility to reduce the price of foods and cost of living, as well as to increase social assistance rates to allow for a more realistic budget. One participant in particular framed the inability for seniors to afford food as being a social rights concern: "everyone deserves the right to 'eat healthy' but many seniors have difficulty with this because of the cost of healthy choices."

In the discussion about cost of living and budgets, a fair number of participants brought up the increasing price of housing in St. John's, which has been apparent over the past decade (City of St. John's, 2015). One senior who owns her house knows how fortunate she is that she does not have a mortgage. As she is living on a fixed income, if she had a mortgage or was renting she would be in a much more vulnerable situation than she is now. She also expressed her sympathy for seniors who struggle in being able to afford the basic necessities when the cost of living goes up and their incomes stay the same.

LEVELS OF ENGAGEMENT WITH EXISTING FOOD (IN)SECURITY INTERVENTIONS

To fully illustrate the barriers and opportunities of current food security options in St. John's, we now explore participants' engagement in interventions concerned with addressing food insecurity. To frame our discussion, we have adopted the continuum suggested by Cook (2008) that breaks interventions down into three categories: efficiency interventions or "emergency solutions," transition strategies, and redesign interventions.

Efficiency Interventions: Food Bank Consumption

Used to address immediate food insufficiencies, food banks act as an emergency solution to provide supplementary food for those with financial constraints. While research shows that food bank use in the province has decreased from 2003 to 2013 (Food Banks Canada, 2013), our findings suggest the use of food banks was much higher among single parents: 82 per cent of single parents (53 per cent often and 29 per cent sometimes) compared to 26 per cent of seniors (17 per cent often and 9 per cent sometimes) indicated that they use food banks. There were mixed opinions about food banks, with some participants expressing their appreciation for them to supplement extra food within their family's diets.

While in our study many people considered food banks as necessary, there were also comments of criticism. Since most food bank donations are non-perishable, in the form of canned and packaged goods, some participants indicated their frustration with the lack of variety in available foods. Although food banks receive some perishable foods and have space in their budgets to occasionally purchase items such as dairy products, meats, and produce, donations tend to vary so their ability to offer quality products fluctuates. As expressed by a key informant working at a food bank, "the food bank does not have control over the quality of food, other than making sure we don't give out outdated foods and poison people. But we can only give out what we get from Community Food Sharing or anyone who does a food drive. So again, it tends to be canned goods, boxed goods, and tinned goods." As this informant explained, because food banks rely primarily on provisions from corporate sponsors, a lot of the donations are what cannot be sold in supermarkets, meaning it is close to the expiration date or has defective packaging, for example. This also means that a big portion of this food is not of high quality or high nutritious standards.

One single parent in particular raised concerns about food banks, saying that the products they provide are often defective or spoiled. She stressed that there should be more regulations to ensure better quality in food that is given "charitably" to those suffering from the increased cost of food in Newfoundland. She also commented that it "shouldn't be acceptable to give people subpar food simply because they're poor"; rather, people living on social assistance should also have the ability to maintain a healthy standard of living. The instability of food banks was discussed at length by a couple of key informants, who also highlighted how ineffective food banks are at addressing the structural problems behind food insecurity.

That being said, food banks have become the normalized way for governments to address food insecurity, acting as "secondary extensions of weakened social safety nets" (Riches, 2002: 648). Rather than confronting the structural issues behind food insecurity, such as poverty reduction or redesigning an unsustainable agricultural and food production system, governments rely on non-profit food banks to avoid taking responsibility for these issues. This reflects neo-liberal, laissez-faire policies that promote individualism and minimal governmental involvement in the social sector, reducing their accountability and their obligation to protect vulnerable populations (Riches, 2002; Tarasuk, 2009). Moreover, quite a bit of stigma tends to be attached to food banks, often making those who use them feel shame or embarrassment that they are unable to access food in the same manner as most people, as exemplified in research with low-income individuals using food banks (Williams et al., 2012). Therefore, food banks are not considered "socially acceptable" and are only "Band-Aid" solutions to the problem of food insecurity (Riches, 2002).

Transition Strategies: Community-Based Solutions

Community-based, "transitional" solutions comprise the second category on the food security intervention continuum, attempting to diminish the impact of food issues with a focus on capacity-building within communities (Cook, 2008). These strategies consist of knowledge-dissemination programs, skills-building initiatives (e.g., budgeting, shopping, cooking), and opportunities to access homemade meals or locally grown food in the form of community kitchens or community gardens.

It is undeniable that some necessary, "transitional" solutions have been implemented due to an increased concern about food-related issues among certain sectors of society. This is reflected in the work developed by the community organizations Food First NL and the Single Parent Association of Newfoundland, as well as Food Advocacy Research at Memorial University. While these groups are committed to raising awareness and providing community-based programs to address food-related issues, they are also critical in putting pressure on key stakeholders to adopt a more holistic approach in policy-making.

However, our research shows that the uptake of such programs may face challenges, as they provide necessary, but not sufficient, conditions for food security. For example, in terms of their interest in growing their own food, 85 per cent of single parents expressed interest in having a vegetable garden, although none of them already do. Despite there being a remarkably high percentage of single parents who expressed interest in gardening, many of these participants admitted that it is unlikely that they ever will have a garden due to the lack of time to commit to maintaining it or not having the yard space to devote to a vegetable garden. Additionally, 75 per cent of single parents responded that they were at least "somewhat" interested in participating in a community garden. The reasons impeding their ability to do so are similar to the previous question, although the limited amount of community gardening sites within the city may also prevent their involvement. Also, considering soil and weather conditions for gardening in Newfoundland, growing food may be as expensive as buying quality food products. Consequently, unless there is some institutional support available, this is not affordable for people with limited budgets.

With regard to seniors' interest in gardening, only 43 per cent indicated their interest in having a garden, while 21 per cent already do and another 34 per cent of seniors expressed interest in having a plot at a community garden. Due to age, it is more likely that physical impediments would prevent them from being able to garden, as some seniors discussed back pain as a hindrance. Further, seniors living in apartment buildings lack the space required to plant a garden.

Despite the expressed limitations towards gardening, many participants recognized the benefits of growing food locally and lamented the high price of locally grown and organic produce. Discussion of local foods sometimes brought about suggestions for community agricultural share programs, where local farms offer produce boxes for members for affordable rates. A couple of participants also recommended expanding the city's farmers' market.

Community kitchens are another example of a "transitional" solution, offering opportunities for culinary skill-building, as well as a chance to socialize with others within the community. In our study, 77 per cent of single parents indicated that they would be interested in participating in a community kitchen, with only four single parents having had the experience of participating in one of these programs in the past. These individuals emphasized how these programs offered not only an opportunity to improve cooking and food preparation skills, but were also valuable occasions for social interaction with others within the community and provided opportunities for them to connect with and support other single parents while sharing a meal. However, a couple of the participants explained that the community kitchens they attended struggled to finance these programs on limited budgets. A smaller amount of seniors (43 per cent) indicated their interest in community kitchens, while only one of them has participated in the past. It is likely that these individuals live alone.

The key informants for this study explained that community kitchens often depend on small grants and funding from community centres, limiting their ability to offer consistent programming within communities. Additionally, one of the key informants discussed the particular barriers that single parents may face in trying to participate in community kitchens or cooking classes. Not only is it difficult for parents to make time to spare a couple of hours a week to participate in these programs, especially if they are working, but finding child care can be very challenging. One single parent echoed these concerns, saying she is interested in participating in cooking classes, but is waiting until her children are old enough to look after themselves so that she is able to leave them for an evening. Even when child care is provided at the community centre, hectic lives can get in the way. One key informant gave the example of parents having to cancel their plans when their children are sick. So even if these programs are offered, the uptake may not be as high as expected for a variety of reasons.

School meal programs that offer breakfasts or lunches for students for either no cost or a low cost are another example of a "transitional" strategy mentioned several times by participants, especially by parents of younger children. These participants emphasized school meal programs' relevance in alleviating food insecurity. However, not all schools in the city offer these programs, as some parents regretfully explained.

At 85 per cent, the overwhelming majority of single parents wished to learn more about nutrition and food budgeting. Another 6 per cent of single parents felt that they are already quite familiar with this topic. In contrast, 50 per cent of seniors indicated that they would like to learn more about nutrition and food budgeting, while 23 per cent felt they already are knowledgeable in this area. The difference between how many seniors felt they are knowledgeable in nutrition, cooking skills, and food budgeting compared to single parents corroborates the fact that these skills are more common among the elderly population.

The great proportion of single parents (91 per cent) said they would like to learn fast, easy, and low-cost recipes and meal ideas. Comparatively, 78 per cent of seniors indicated the same. That being said, similar to the issue with community kitchens, it is difficult to determine how to offer skill-building opportunities for cooking nutritious meals, as community programming is typically not well funded or there is often not much uptake due to people's busy lives or issues in mobility, as discussed above. However, the mixed messages from participants illustrate that "transitional" strategies are worth considering together with redesign interventions, to maintain a multi-level approach.

Redesign Interventions: A More Holistic Approach

The third and final category, "redesign strategies," consists of solutions that take a holistic approach to food security to include structural changes, such as poverty reduction and addressing food policies (Cook, 2008). It is based on a multi-level approach, where transition interventions (including food knowledge, education, and awareness strategies, community kitchens and community gardens, etc.) are expected to complement more structural measures.

For example, discussions with a key informant on the food industry and its relation to the province's lack of assertive food policies brought about comments and suggestions for government intervention. Similarly, a couple of participants, when talking about locally produced food, indicated that they would like to be able to access fresh, locally caught fish, and emphasized the need for better distribution strategies as the prices are much too high. These participants explained that in order to get seafood for an affordable price, one would need to have connections with a friend or family member who fishes; for instance, one senior emphasized her luck in being able to get fresh fish regularly from her brother.

Newfoundland's food system was a frequently discussed topic, from the lack of locally grown food to the quality of food imported to the province, which one participant described as being "second class." Some participants discussed their reservations about the quality of produce being shipped in: "even when you eat healthy food from the grocery store, you still don't know where it came from as it's shipped from across the world and probably contains chemicals." Agriculture is not a major industry in the province, due to perceptions of a challenging climate and poor soil. Although there are farms across Newfoundland, the reality, as noted above, is that 90 per cent of food in the province is imported (Quinlan, 2012). Emphasis should be given to production of local foods (particularly green vegetables) and promotion of their consumption. This initiative may bring down the cost of nutritious food through increased access (Doyle and Sarkar, 2015).

Nonetheless, a participant expressed her skepticism about systemic change and the government's likelihood of putting any pressure on the food industry, going as far as to doubt the necessity for more research in this area, as "the government probably is already aware of the issues but isn't willing to do anything about them." Unfortunately, food policies in the province have not been given the attention they deserve. For instance, seafood, a traditional staple in our economy and in our homes, has not been duly protected for domestic consumption and is instead prioritized for the export market (Song and Chuenpagdee, 2015; see also Foley and Mather, Chapter 9). Nonetheless, considering food from a more holistic perspective is becoming more widespread within the province, which is likely, ultimately, to have a beneficial influence on the broader sectors of our population.

Observing the work of Food First NL, as well as through our key informant interviews, we see a push for recovering the culture of local food production in Newfoundland and Labrador, which has been partially dismantled due to the prevalence of imported, industrial agriculture in the province (Food First NL, 2015). This relates to the increasing debate about systemic changes to the current unsustainable global, industrialized food system, manifested in the example of the food sovereignty approach, which has been gaining momentum globally, including in Canada.

FINAL CONSIDERATIONS

Our research shows that the single parents in our sample are in more stressful financial situations than the senior participants. Through participants' responses, we can see how challenging it is to make ends meet while raising children with a single income that often comes from social assistance. We observed that time constraints, lack of knowledge about how to get the best nutritional value for the food dollar, lack of cooking skills, and having grown up in a "fast-food culture" are also influencing people's food consumption practices. As a result, people tend to rely heavily on processed, canned, or takeout foods.

Meanwhile, most seniors are able to maintain more comfortable lifestyles with their retirement pensions or Old Age Security. However, the sample also included some seniors with very limited resources, for whom maintaining a healthy diet while living with chronic conditions was not affordable. Seniors also reported having a range of other challenges, such as mobility issues, lack of fine motor skills, and lack of motivation to prepare food for themselves if living alone.

One of the main challenges in tackling the complexity of food-related issues is to see beyond the existing fragmentation of the food system. Food insecurity is depoliticized and slips under the radar the moment that lack of income is addressed through social or patronizing measures (i.e., food banks). Furthermore, despite their close interrelations, food production, nutrition, and the health and social sectors and their policies continue to operate in silos (Burchi, Fanzo, and Frison, 2011). It is also important to realize that while socio-economic conditions frame food consumption patterns and preferences, these are difficult to challenge without addressing their underlying causes (Stead, McDermott, MacKintosh, and Adamson, 2011). The restrictive vision of the food security/anti-poverty explanation, focused on individuals and on the consumer level, although relevant, neglects a more comprehensive understanding of the food system and how food-related practices become culturally embedded in populations. The linkages to depoliticization and structural problems of the neo-liberal food system are further discussed in Chapter 9. Adopting a food sovereignty perspective would incorporate a wider lens to include the varying yet interconnected levels of food production, distribution, and consumption in line with social justice concerns (Story et al., 2009; Suschnigg, 2012).

The main contribution of our research is to show evidence of this complexity, recognizing the multiple levels of influence, from economic constraints to cultural lifestyles that are framed by living conditions, the shifting pace of life, and the food industry's responses to these changes. These factors interconnect to shape current food environments in the province, rendering new generations more exposed and more accustomed than ever to cheap, highly dense, processed foods, which are affordable, accessible, and gratifying, despite conditioning unhealthy food consumption patterns (Nestle, 2002; Winson, 2013). This is especially relevant to consider in Newfoundland and Labrador, one of the provinces with the highest rates of diabetes, heart attacks, strokes, and certain types of cancer compared to the Canadian average; all of these illnesses are very much related to food practices (Ewtushik, 2003; Government of Newfoundland and Labrador, 2006).

Our research also shows the need to work with a multi-level, multi-faceted approach, to limit efficiency interventions (e.g., food banks) to emergency situations, and to put more consistent funding in transitional solutions (e.g., including food-related activities in the educational system, enhancing cooking skills, and encouraging gardening activities). As some of the participants in this study suggested, educational, hands-on programs in schools are necessary to allow for an ecological and more sustainable approach to food issues. Consequently, a very important investment resides on raising the level of awareness about healthy eating among the new generations in times of industrialized food production.

More importantly, there is a need to address vulnerabilities and to guarantee a fairer distribution of economic and cultural resources, which can only be done by adopting an upstream approach that looks critically at neo-liberal forms of governance and macro-economic policies that generate an increased concentration of wealth in the hands of a few (Solar and Irwin, 2010; Winson, 2004). Part of the mandate of a neo-liberal profit-driven economy is to neglect the complex interconnections among the food industry, food consumption, the socio-economic environment, and people's health, and to ignore the need for more inclusive "health in all policy" approaches. Consequently, addressing the current fragmentation of food production, nutrition, health, and social policy systems will help to reduce the existence of significant health inequities among the most vulnerable populations.

ACKNOWLEDGEMENTS

The investigators would like to thank the Harris Centre at Memorial University for funding this research and the members of our Advisory Committee: Ivette Walton (Single Parents Association of Newfoundland), Kristie Jameson (Food First NL), and Kelly Heiz (Seniors' Resource Centre). We also extend much gratitude to our key informants and to all the research participants in this project.

NOTE

1. "Food insecurity" is described as "the inability to acquire or consume an adequate diet quality or sufficient quantity of food in socially acceptable ways, or the uncertainty that one will be able to do so" (McIntyre and Rondeau, 2009: 188). The term "food security" originated in 1974, when the World Food Conference was concerned about the global food supply. Since then, there have been many definitional versions of "food security," the latest having emerged from the Food and Agriculture Organization (FAO) in 2001, when it was considered as "the situation that exists when all people, at all times, have physical, social and economic access to sufficient, safe and nutritious food that meets their dietary needs and food preferences for an active and healthy life" (Clay, 2002: 28).

REFERENCES

Andresen, M. 2007. "Mixed reviews for Canada's new food guide." *Canadian Medical Association Journal (CMAJ)* 176, 6: 752–53. doi:10.1503/cmaj.070240.

Bronte-Tinkew, J., M. Zaslow, R. Capps, A. Horowitz, and M. McNamara. 2007. "Food insecurity works through depression, parenting, and infant feeding to influence overweight and health in toddlers." *Journal of Nutrition* 137, 9: 2160–65. At: http://jn.nutrition.org/.

Broughton, M. A., P. S. Janssen, C. Hertzman, S. M. Innis, and C. J. Frankish. 2006. "Predictors and outcomes of household food insecurity among inner city families with preschool children in Vancouver." *Canadian Journal of Public Health* 97, 3: 214–16. At: http://journal.cpha.ca/index.php/cjph.

Burchi, F., J. Fanzo, and E. Frison. 2011. "The role of food and nutrition system approaches in tackling hidden hunger." *International Journal of Environmental Research and Public Health* 8, 2: 358–73. doi:10.3390/ijerph8020358.

Carolan, M. 2011. *The Real Cost of Cheap Food*. London: Earthscan.

Carter, M. A., L. Dubois, M. S. Tremblay, and M. Taljaard. 2012. "Local social environmental factors are associated with household food insecurity in a longitudinal study of children." *BMC Public Health* 12. doi:10.1186/1471-2458-12-1038.

Chenhall, C. 2010. *Improving Cooking and Food Preparation Skills: A Synthesis of the Evidence to Inform Program and Policy Development* (Report No. 1). Government of Canada. At: http://www.hc-sc.gc.ca/fn-an/nutrition/child-enfant/cfps-acc-synthes-eng.php.

City of St. John's. 2015. "Living in St. John's: Cost of living." At: http://www.stjohns.ca/living-st-johns/newcomers/about-st-johns/cost-living.

Clay, E. 2002. "Food security: Concepts and measurement." In *Trade Reforms and Food Security: Conceptualising the Linkages*, 25–34. Rome: Food and Agriculture Organization of the United Nations.

Cook, B. 2008. *Food Security Issues in a Public Health Context*. Antigonish, NS: National Collaborating Centre for the Determinants of Health.

Dachner, N., L. Ricciuto, S. I. Kirkpatrick, and V. Tarasuk. 2010. "Food purchasing and food insecurity: Among low-income families in Toronto." *Canadian Journal of Dietetic Practice and Research* 71, 3: e50–e56. doi:10.3148/71.3.2010.e50.

Darmon, N., and A. Drewnowski. 2008. "Does social class predict diet quality?" *American Journal of Clinical Nutrition* 87, 5: 1107–17. At: http://ajcn.nutrition.org/.

Doyle, E., and A. Sarkar. 2015. "New terroir de terre-neuve: Contemplating this province's food culture." *Newfoundland Quarterly* 108, 2: 39–42. At: http://www.mun.ca/nq/.

Engler-Stringer, R. 2010. "The domestic foodscapes of young low-income women in Montreal: Cooking practices in the context of an increasingly processed food supply." *Health Education & Behavior* 37, 2: 211–26. doi:10.1177/1090198109339453.

Ewtushik, M. 2003. *The Cost of Eating in Newfoundland and Labrador*. St. John's: Dieticians of Newfoundland and Labrador.

Food Banks Canada. 2013. *HungerCount 2013*. (Report No. 1). Toronto: Food Banks Canada. At: http://www.foodbankscanada.ca/FoodBanks/MediaLibrary/HungerCount/HungerCount2013.pdf.

Food First NL. 2015. *Everybody Eats: A Discussion Paper on Food Security in Newfoundland and Labrador*. St. John's: Food First NL.

Government of Newfoundland and Labrador. 2006. *Eating Healthier in Newfoundland and Labrador: Provincial Food and Nutrition Framework and Action Plan* (Report No. 1). St. John's: Government of Newfoundland and Labrador.

Green, R. J., P. L. Williams, C. S. Johnson, and I. Blum. 2008. "Can Canadian seniors on public pensions afford a nutritious diet?" *Canadian Journal on Aging* 27, 1: 69–79. doi:10.3138/cja.27.1.069.

Green-LaPierre, R. J., P. L. Williams, N. T. Glanville, D. Norris, H. C. Hunter, and C. G. Watt. 2012. "Learning from 'knocks in life': Food insecurity among low-income lone senior women." *Journal of Aging Research*. doi:10.1155/2012/450630.

Groves, A. 2002. "Children's food: Market forces and industry responses." *Nutrition Bulletin* 27, 3: 187–90. doi:10.1046/j.1467-3010.2002.00228.x.

Guthman, J. 2011. *Weighing In: Obesity, Food Justice, and the Limits of Capitalism*. Berkeley: University of California Press.

Hamelin, A., M. Beaudry, and J. Habicht. 2002. "Characterization of household food insecurity in Québec: Food and feelings." *Social Science & Medicine* 54, 1: 119–32. doi:10.1016/s0277-9536(01)00013-2.

Health Canada. 2011. "Eating well with Canada's food guide." At: http://www.hc-sc.gc.ca/fn-an/food-guide-aliment/index-eng.php.

Keller, H. H. 2006. "Reliance on others for food-related activities of daily living." *Journal of Nutrition for the Elderly* 25, 1: 43–59. doi:10.1300/j052v25n01_05.

Kelly, B., J. C. Halford, E. J. Boyland, K. Chapman, I. Bautista-Castaño, C. Berg, . . . C. Summerbell. 2010. "Television food advertising to children: A global perspective." *American Journal of Public Health* 100, 9: 1730–36. doi:10.2105/ajph.2009.179267.

Kondro, W. 2006. "Proposed Canada food guide called 'obesogenic'." *CMAJ* 174, 5: 605–06. doi:10.1503/cmaj.060039.

Marmot, M. 2005. *The Status Syndrome: How Social Standing Affects Our Health and Longevity*. New York: Holt.

McIntyre, L., A. C. Bartoo, and J. Emery. 2014. "When working is not enough: Food insecurity in the Canadian labour force." *Public Health Nutrition* 17, 1: 49–57. doi:10.1017/s1368980012004053.

———, N. T. Glanville, K. D. Raine, J. B. Dayle, B. Anderson, and N. Battaglia. 2003. "Do low-income lone mothers compromise their nutrition to feed their children?" *CMAJ* 168, 6: 686–91. At: http://www.cmaj.ca/.

——— and K. Rondeau. 2009. "Food insecurity." In D. Raphael, ed., *Social Determinants of Health: Canadian Perspectives*, 188–204. Toronto: Canadian Scholars' Press.

Nestle, M. 2002. *Food Politics: How the Food Industry Influences Nutrition and Health*. Berkeley: University of California Press.

———. 2006. "Food marketing and childhood obesity—A matter of policy." *New England Journal of Medicine* 354, 24: 2527–29. doi:10.1056/nejmp068014.

———, R. Wing, L. Birch, L. DiSogra, A. Drewnowski, S. Middleton, . . . C. Economos. 1998. "Behavioral and social influences on food choice." *Nutrition Reviews* 56, 5: 50–64. doi:10.1111/j.1753-4887.1998.tb01732.x.

Pinstrup-Andersen, P. 2009. "Food security: Definition and measurement." *Food Security* 1, 1: 5–7. doi:10.1007/s12571-008-0002-y.

Quinlan, A. J. 2012. *Building Agricultural Capacity in Newfoundland and Labrador* (Report No. 1). St. John's: Harris Centre, Memorial University. At: https://www.mun.ca/harriscentre/reports/arf/2011/11-SPHCSRF-Final-Quinlan.pdf.

Riches, G. 2002. "Food banks and food security: Welfare reform, human rights and social policy. Lessons from Canada?" *Social Policy & Administration* 36, 6: 648–63. doi:10.1111/1467-9515.00309.

Sarkar, A., M. Traverso-Yepez, V. Gadag, and K. Hunter. 2015. *A Study on Food Security among Single Parents and Elderly Population in St. John's* (Final Report). St. John's: Harris Centre, Memorial University. At: http://www.mun.ca/harriscentre/reports/SARKAR_ARF_13-14_FINAL_REPORT.pdf.

Sim, S. M., N. T. Glanville, and L. McIntyre. 2011. "Food management behaviours: In food-insecure, lone mother-led families." *Canadian Journal of Dietetic Practice and Research* 72, 3: 123–29. doi:10.3148/72.3.2011.123.

Solar, O., and A. Irwin. 2010. *A Conceptual Framework for Action on the Social Determinants of Health*. Social Determinants of Health Discussion Paper 2. Geneva: World Health Organization.

Song, A. M., and R. Chuenpagdee. 2015. "A principle-based analysis of multilevel policy areas on inshore fisheries in Newfoundland and Labrador, Canada." In S. Jentoft and R. Chuenpagdee, eds., *Interactive Governance for Small-scale Fisheries: Global Reflections*, 435–56. Dordrecht, Netherlands: Springer.

Statistics Canada. 2014. *Canadian Community Health Survey: Annual Component – 2014 Questionnaire*. At: http://www23.statcan.gc.ca/imdb-bmdi/instrument/3226_Q1_V11-eng.pdf.

Stead, M., L. McDermott, A. M. MacKintosh, and A. Adamson. 2011. "Why healthy eating is bad for young people's health: Identity, belonging and food." *Social Science & Medicine* 72, 7: 1131–39. doi:10.1016/j.socscimed.2010.12.029.

Story, M., M. W. Hamm, and D. Wallinga. 2009. "Food systems and public health: Linkages to achieve healthier diets and healthier communities." *Journal of Hunger & Environmental Nutrition* 4, 3/4: 219–24. doi:10.1080/19320240903351463.

Suschnigg, C. 2012. "Food security? Some contradictions associated with corporate donations to Canada's food banks." In A. Winson, J. Sumner, and M. Koc, eds., *Critical Perspectives in Food Studies*. Toronto: Oxford University Press.

Tarasuk, V. 2009. "Health implications of food insecurity." In D. Raphael, ed., *Social Determinants of Health: Canadian Perspectives*, 205–20. Toronto: Canadian Scholars' Press.

———, A. Mitchell, and N. Dachner. 2015. *Household Food Insecurity in Canada, 2013* (Report No. 3). Toronto: Research to identify policy options to reduce food insecurity (PROOF). At: http://proof.utoronto.ca/wp-content/uploads/2015/10/foodinsecurity2013.pdf.

Veugelers, P. J., and A. L. Fitzgerald. 2005. "Prevalence of and risk factors for childhood overweight and obesity." *CMAJ* 173, 6: 607–13. doi:10.1503/cmaj.050445.

Wilkinson, R., and K. Pickett. 2010. *The Spirit Level: Why Equality Is Better for Everyone*. London: Penguin.

Williams, P. L., R. B. MacAulay, B. J. Anderson, K. Barro, D. E. Gillis, C. P. Johnson, . . . D. E. Reimer. 2012. "'I would have never thought that I would be in such a predicament': Voices from women experiencing food insecurity in Nova Scotia, Canada." *Journal of Hunger & Environmental Nutrition* 7, 2/3: 253–70. doi:10.1080/19320248.2012.704740.

Winson, A. 2004. "Bringing political economy into the debate on the obesity epidemic." *Agriculture and Human Values* 21, 4: 299–312. doi:10.1007/s10460-003-1206-6.

———. 2013. *Industrial Diet: The Degradation of Food and the Struggle for Healthy Eating*. Vancouver: University of British Columbia Press.

6

"Just About Self-Sufficient": Cases in the History of Self-Provisioning in Newfoundland and Labrador

Adrian Tanner

INTRODUCTION

Although today the food supply system in Newfoundland and Labrador is largely market-oriented, in the recent past a significant contribution came from household subsistence production of food. In rural areas hunting, gardening, fishing, and berry picking were important supplements to purchased food, particularly, for example, during the Depression of the 1930s. In this chapter I refer to the production of food that is destined to be consumed by those who produce it as "self-provisioning."[1] This source of food has been of special importance throughout the history of Newfoundland and Labrador. Moreover, as is shown from the examples in the chapters in this volume, self-provisioning continues to play a role under modern conditions.

This chapter examines four examples of particular times and specific places in which self-provisioning played a significant part in the domestic economy. While three of these involve Indigenous people, I have not included all such cases. For example, I do not include coverage of the self-provisioning practices of the historic Inuit of the north coast of Labrador, studied by Woolett (2004) and Taylor (1977), nor of the historic Mi'kmaq of the island of Newfoundland, as described, for example, in books by Jackson (1993) and Jeddore (2015).

Before European contact, the Labrador Innu, like other Indigenous people of northern Canada, depended on hunting and foraging, primarily for their own food and other needs, although some food was shared outside the household that produced it, through gifts to others in the community and at communal feasts. Following the arrival of European traders the Innu began to exchange furs for tools and supplies, while continuing to spend most of their time directly harvesting animals for their food. Rather than adopting the economic motives and values of the traders, they integrated the production of furs destined for the market within those traditional Innu values and practices associated with self-provisioning.

With the arrival of traders to southern Labrador, a second group, people of mixed Inuit and European descent, gained prominence in the Lake Melville region as trappers and fishers. They created a new economic adaptation, based on a combination of Inuit, European, and Innu influences. In their salmon fishing and fur trapping they exhibited market motives and values, while their gardening, hunting, and gathering activities were governed by seasonal opportunities to satisfy known needs, mainly for food.

When European fishing people, who initially came to catch and make salt fish, began to overwinter, they found that they needed to supplement whatever store supplies they could afford by producing their own food, particularly in times of low world prices for salt fish. They did so by hunting, fishing, foraging, gardening, and animal husbandry, all of which were aimed to respond, at least in part, to "known needs." Over time they established a tradition in this province of self-provisioning, alongside a market-oriented economy.

Finally, I add what is essentially a footnote to Chapter 7 by Schiff and Bernard. In the mid-twentieth century the settlement of the Labrador Indigenous peoples into villages, to provide them with health, education, and welfare services, unintentionally helped to set the stage for a food security crisis. People found that hunting from the new centres required them to travel to and from the hunting grounds, using snowmobiles, aircraft, or more recently, trucks and ATVs, all of which require more cash than could be obtained from the sale of furs or sealskins alone. Settlement was part of a general government policy to have the Innu and Inuit become workers in the market economy, despite the lack of jobs. In these circumstances many Indigenous people found themselves effectively confined to their villages, their ability to hunt severely diminished, making them dependent on high-cost imported food.

These cases illustrate differing kinds of relationship between self-provisioning and the market, two kinds of economies that I argue entail different motivations and values. The historic Innu kept the market at arm's length by extending the same economic attitudes to fur trapping as they did for hunting for food. Both the nineteenth-century and early twentieth-century Southern Inuit of Lake Melville and the nineteenth-century Newfoundland salt cod fishers engaged in market production at particular seasons, while at other times of the year they hunted and gardened, mainly for their family's food needs. Finally, participation in the self-provisioning sector by the twentieth-century Indigenous people of Labrador became dependent on an inadequate market sector of the economy.

THE HISTORIC LABRADOR INNU

What we know of how the Innu lived in the nineteenth and early twentieth centuries comes partly from contemporary writings of traders and missionaries, although few of them actually accompanied Innu families as they moved from place to place in the interior of Labrador, hunting and trapping. In 1975 and 1976 I conducted research with Innu who had been settled in the village of Sheshatshiu since the 1960s (Tanner, 1976).[2] This included recording elders' recollections of their way of life before settlement. I also made use of documentary sources, the most important of which included the material gathered by P. T. McGrath for the Labrador boundary dispute (McGrath, 1926). This source included affidavits collected from many Labrador land users. Even though no Innu were interviewed, several trappers who regularly travelled far into the interior gave McGrath details of where both the part-Inuit trappers and the Innu hunters resided and travelled. In the 1940s, V. Tanner led a large Finnish research team in making a general study of Labrador. They visited many parts of Labrador, including to some Innu living in the interior, and documented the land-use pattern of both the Innu and the part-Inuit trappers (Tanner, 1947: 584–85).

These sources, along with the detailed studies by Mailhot (1997) and Zimmerly (1975), provide an understanding of the historic Innu, as well as, in Zimmerly's case, of the historic Lake Melville part-Inuit group, the subject of the subsequent section of this chapter. In my analysis of these data, I also drew on a previous study I had made the economic system of Iiyuu (Cree) hunters in northern Quebec, in which I came to some general conclusion about the household economics of northern Canadian Indigenous hunters and trappers living within a classic fur trade monopoly (Tanner, 2014).

My 1975–76 Innu research showed that the present occupants of Sheshatshiu were actually made up of several subgroups, each formerly based in one or another general area in the interior, each in different directions from Northwest River. The boundary between Labrador and Quebec, which was not clearly established until 1927, did not affect historic Innu land-use patterns. In fact, there were many kin connections between the Innu of Labrador and those based in the adjacent parts of Quebec. Groups moved freely back and forth, with the north–south rivers draining into the St. Lawrence providing travel routes. A visit to a Hudson's Bay Company post was not only undertaken for trade, but also to meet with a missionary — for children to be baptized, for marriages, and for funerals. In the nineteenth century Catholic missionaries were not always at Northwest River, so in some years the Innu might travel to one of the trading posts on the north shore of the St. Lawrence rather than to a post in Labrador (Tanner, 1976: 23–24).

Before being settled into permanent villages over the past half-century, the Labrador Innu depended primarily on hunting for their food security. For most of the year they lived in loosely organized multi-family hunting groups that travelled over large parts of the Quebec–Labrador interior, utilizing land-based resources, and only came to the coast for short periods in the summer, while some remained inland all year (Tanner, 1947: 629–32). Their hunting and gathering activities were organized on an annual basis, rather than on the need to hunt every day to feed themselves, given that stores of food were kept for months at a time. During substantial parts of the year storage of meat or other food was achieved by simply leaving it outside on a cache platform to freeze; at other times they employed drying, smoking, or pickling (Stopp, 2002). Besides the various forms of meat and berries consumed as food, the skins of the animals provided material for clothing and were used to make equipment such as snowshoes. Meat from the hunt was also regularly shared outside the family, within a system of generalized exchange. The sharing of meat, particularly of large animals like caribou and bear, was considered a sacred obligation to the animal spirits, and occurred through gifts and at communal feasts (Speck, 1977; Henricksen, 2009).

With the appearance of European traders, the Innu began to exchange furs for factory-made tools such as axes, guns, traps, cooking pots, and blankets.

Imported food did not become a significant trade item for the Innu until the twentieth century.[3] While products from the European industrial economy were new for the Innu, trade itself was familiar, as it was to other northern Indigenous peoples. For example, archaeologists have shown that the kind of stone that was very suitable for making tools, such as chert from Ramah Bay, Labrador, was traded over long distances (e.g., Anstey and Renouf, 2011). We can assume that, by the same token, the pre-contact Indigenous peoples of Newfoundland and Labrador also would have had access to exotic items traded in from distant groups.

In dealing with European fur traders the Innu applied their existing self-provisioning economic attitudes to the new activity of trapping for furs. One indication of this is that the Innu had insisted on maintaining a debt relationship with traders. Each year tools and supplies for a hunting season were advanced to hunters, to be repaid at the season's end. The fur-trade historian Arthur Ray writes:

> Over the long course of the [Hudson's Bay] company's history senior officials had repeatedly attempted without success to eliminate or curtail sharply the debt system. Not only did Indians depend on receiving it, but they believed the company had an obligation to provide it.... Credit represented a kind of reciprocal obligation... as long as competitors extended credit, the Hudson's Bay Company had no choice but to do likewise. (Ray, 1990: 85)

The debt system provided a necessary interface between two otherwise incompatible economic systems. It allowed the Innu to continue to produce on the basis of self-provisioning values and motivations, and still be able to acquire goods from traders, whose goals in the transaction were as agents of the mercantile commodity market. For the Innu the production of furs, as for meat, was planned to fulfill specific needs known in advance. Just as hunting for meat ended once they had enough to meet the known food needs of each particular season of the year, so fur trapping ended once they had sufficient to satisfy known needs, as set by the debt system. While these plans could be disrupted by weather or animal shortage, by setting goals for fur production, the debt system established known needs for the imported supplies required for hunting and trapping.

Northern Indigenous hunters did not organize fur production using the logic of the market, as is shown by their relative unresponsiveness to pressures from traders for them to maximize high-value fur production. Ensuring production of meat from hunting was more important than was the cash value of trapping. Rather than focus on trapping for high-priced fine fur, like marten, found in Labrador in major river valleys, the Innu hunted and trapped on the more open environment of the plateau areas to the south, west, and northwest of Northwest River, beyond the major river valleys. This was in part so they would be more likely to encounter caribou, and also because the region includes beaver habitat, an animal especially important to the Innu, as much for its meat[4] as for the market value of its fur.

This tendency was noticed by HBC traders as early as 1749. As the fur-trade historian E. E. Rich observes:

> English economic rules did not apply to the Indian trade. On the contrary, all who had any knowledge of the trade were convinced that a rise in prices would lead [to] the Indians bringing down fewer furs. This was because, according to one view at the time, the Indians wanted only a given quantity of trade goods, and so higher prices would bring in fewer furs. (Rich, 1960: 47)

This approach to production is in line with what Max Weber, writing in the first quarter of the twentieth century, observed when he defined "traditional labour" as "work expended until reaching an accustomed level of livelihood" (Weber, 1984 [1923]). The discipline of economics took note of this apparent "irrationality" of self-provisioning economies through studies of the Russian peasantry. The agrarian economist Alexander Chayanov noted that small-scale independent peasant farmers did not act according to the rules of supply and demand. If the price they were paid for a particular commodity rose, their production did not increase, but, perversely, was actually reduced. For these peasant farmers, who also directly produced a lot of their own food, their needs for cash were specific and limited, so that once these needs were met, they exercised a preference for leisure (Chayanov, 1966).

An additional characteristic of the Innu self-provisioning economy was their disinterest in the accumulation of property. The Jesuit missionary Paul

Le Jeune made the following observation in 1634 about the Innu in the area of what is now Labrador and the adjacent parts of Quebec:

> Moreover, if it is a great blessing to be free from a great evil, our [Montagnais] Savages are happy; for the two tyrants who provide hell and torture for many of our Europeans, do not reign in their great forests, — I mean ambition and avarice . . . as they are contented with a mere living, not one of them gives himself to the Devil to acquire wealth. (cited by Sahlins, 1972: 14)

This disinterest in material accumulation was particularly appropriate for nomadic hunters, as their way of life involved frequent movement that would have been hampered by excessive material possessions.

THE HISTORIC SOUTHERN LABRADOR INUIT OF NUNATUKAVUT

The first traders who came to Labrador either from Quebec or the island of Newfoundland were initially drawn to trade for fur with the Innu. However, around the same time another group of trapper/fishers, composed of European men, most of them with Inuit wives, along with their descendants, became established around Lake Melville. For detailed histories of two such families, see Way (2014) and Stopp (2014). They developed a different economic adaptation to the Labrador environment from those of either the Innu or the Inuit, in that they did not feel limited to either a coastal or an interior adaptation. As noted by Zimmerly (1975: 70), "The technology available to the settlers came partially from their old European skills, but included borrowed traits from both the Indians and the Eskimos." In their fur trapping they responded to European kinds of market attitudes, values, and motivations, as they did in their commercial salmon fishing. The Inuit wives brought with them the skills to supply warm clothing, sleds, and other specialized items of hunting technology, while from the Innu the men learned trapping techniques and how to travel and survive in the Labrador interior.

After 1836, when the Hudson's Bay Company established a virtual monopoly in the region, with posts at Rigolet and Northwest River, this new group expanded, as more men arrived in the area. Many of them were former HBC employees who had finished their five-year contracts and remained, several of them also marrying Inuit women. They trapped and fished salmon commercially, each at specific seasons, while at other times they engaged in hunting, gardening, and gathering berries and firewood for their own direct needs (Zimmerly, 1975; Tanner, 1947; Goudie, 1975; Montague and Dawson, 2013). During the trapping season the women remained at the coast, while each man focused his efforts on as quickly as possible trapping as much of whatever species of fine fur commanded the best price at the time. By bringing to the trapline most of the food they needed for the trapping season, and by sleeping each night in permanent tilts (cabins) they built along their traplines, returning to the coast after about three months, they were able to focus on maximizing their trapping returns, spending as little time as possible in hunting for their own food. At other seasons, while at the coast, they hunted, gardened, and gathered, motivated by specific needs at each season. In summary, unlike the Innu, these Lake Melville part-Inuit had an economy that employed a combination of both market and self-provisioning practices, separated by season.[6]

The new group also introduced European-influenced land tenure practices, quite different from those used at the same time by either the Innu or the Inuit. They had permanent coastal dwellings where many of them asserted European-type exclusive rights to particular salmon rivers. In time they also developed a system of individual traplines along one of the main rivers from the interior emptying into Lake Melville. These trappers began to use the valleys of the Churchill and Naskapi rivers for their traplines, which at first did not cause problems for the Innu. These were not government-administered registered traplines, but a system developed by the trappers themselves. During this period there are several accounts of friendship and assistance between the part-Inuit trappers and Innu hunters (McGrath, 1926). Around 1900, when all the valley trapping areas were in use, some trappers moved beyond the valleys and started to establish traplines on the plateau region, an area the Innu considered their territory. When, as a result, conflicts arose between the trappers and Innu hunters, the courts effectively legitimized the traplines ex post facto (Tanner, 1976: 30, 34).

THE NINETEENTH-CENTURY SALT COD FISHERY

As with the Lake Melville part-Inuit group, among those engaged in the nineteenth-century Newfoundland and Labrador salt cod fishery self-provisioning took place within the same household economic context as the market production of fish. In this case, as Ommer et al. point out: "the two [subsistence and the cash economy] are interdependent — subsistence activities stretch scarce dollars, and cash is needed to fuel the subsistence pump. Some subsistence activities also have a cultural and recreational value and are not just engaged in for livelihood purposes" (Ommer, Turner, MacDonald, and Sinclair, 2007: 121).

People who were otherwise engaged in the fishery or wage labour came to depend on hunting, gardening, and gathering for some of their food. In addition to food, many Newfoundland and Labrador fisher-people also built their own houses, made their own furniture, and built their own boats and fishing stages. These are all aspects of self-provisioning in what was otherwise a market-based economy. Fishing families had some access to food from the fish they caught, fish that would have otherwise been traded. According to Rosemary Ommer (1989b: 11, n. 39), the self-provisioning of fish "allowed people to continue to survive when they were cut off, and to keep store purchases low when they were poorly off."

In time both the fish merchants and the government came to promote the idea that fishing families should cultivate gardens to supplement their food supply when salt fish prices were low (Ryan, 1971: 4). But these were not the first fishing people to do so; for example, at an earlier period the settlers at Ferryland and Cupids also grew some of their own food (Tourigny, 2009; Gilbert, 2008). After 1832, faced with declining prices for salt cod, fishing families were encouraged to engage in subsistence gardening. However, growing conditions were generally poor, and their crops were usually insufficient for the food needs of families, particularly during periods of low prices for salt cod. The promotion of gardening was unable to prevent a food security crisis — in the form of desperate petitions to the government for relief, incidents of breaking into stores, and, in 1816–17, 1832, and 1833, food riots (Cadigan, 1992: 60).

By the twentieth century a pattern of fishermen regularly supplementing whatever food they could afford from the trader, by means of hunting, gardening, and animal husbandry, had become established. It was a mixed economy of commercial fishing combined with hunting, berry picking, and homegrown food. As is noted in Chapters 1 and 2 in this volume, the 1930s Depression led to a new emphasis on household gardening. Edgar House recalled the central role of women in self-provisioning:

> I admired these women because when the men were fishing, they had to build the kitchen gardens and tend them. They grew a lot of potatoes and turnips, carrots and beets — the usual things. Sometimes they'd have to make hay. Cut hay and make hay. They had to do all that. They worked hard, these women. . . . My mother's brother, in Champneys East, and her sister in Champneys West were just about self-sufficient. They fished, so they had lots of fish for the winter. They grew vegetables, they had hens. They had cows and sheep and goats. . . . There were quite a few sheep, all the way from Bonaventure down to English Harbour. I suppose every five or six homes had a loom and they made their own wool. They knitted sweaters and socks and so on. There weren't many horses, but they kept dog teams. (House, 2015: 161–62)

As noted by Phillips in Chapter 1, this pattern of self-provisioning declined in the period of rapid modernization following Confederation. However, in some areas it continues to be part of contemporary household economies; in one region of Newfoundland and Labrador self-provisioning has been documented with a field study on the Great Northern Peninsula (Omohundro, 1994, 1995). Yet relatively little is known of its contribution to a rural household's total economy, either now or in the past. Few province-wide statistics on these activities are available, particularly given that this kind of production tends to escape government documentation, and because it does not tend to involve any government intervention or the collection of statistics, self-provisioning is sometimes stigmatized with terms like the "informal" or, even more negatively, the "underground" economy.

THE NORTHERN INDIGENOUS FOOD CRISIS

Self-provisioning in the context of the market continues to play an important part of the economy of many Indigenous people, especially in the North, and women are especially important in this economy (Kuokkanen, 2011). "Country food" continues to be an important part of the Labrador diet (Felt et al., 2012; Hanrahan, 2012; Mackay and Orr, 1987, 1988; Mitchell, 2014). In the case of contemporary Labrador Indigenous people, whose serious food insecurity is documented by Schiff and Bernard in Chapter 7, the harvesting of these traditional foods has been under threat for some years (Ames, 1977). One factor in this is that travel between the urban centres, where the Indigenous people have been settled, and their hunting grounds requires some kind of mechanized transport, which in turn requires cash. Before the oil crisis of the mid-1970s the sale of pelts would cover a substantial part of these costs of hunting, but more recently the expenses have risen, while at the same time, the European ban on seal products and anti-trapping campaigns have undermined furs and sealskins as sources of cash. Hunting for game meat has thus become dependent on inputs from other parts of the market economy.[7]

There has been some response by government, including financing of these up-front costs of going hunting (Castro, 2016). Because the cash value to food is not accounted for in the hunting support programs, they are generally seen as a subsidy. During the period when the provincial government was administering the Indigenous people in the province (1947-2002) such subsidies were generally inadequate. Per capita spending on Indigenous people in Newfoundland and Labrador was much lower, compared to those living in similar conditions in the rest of Canada (Tanner et al., 1994). The subsidies were justified, politically, as a healing initiative, intended to get people away from the acute social problems in the settlements. If the value of the meat was taken into account, the costs of the capitalization of hunting might well be justified differently, as a rational form of food production.

For both the Innu and the Inuit there are also important non-market cultural values connected to hunting, in addition to the value of food obtained. For example, Castro (2016) has documented the contemporary social importance for the Labrador Innu of the sharing of game meat, especially caribou. The social value of caribou meat, outside its nutritional role, is also expressed as a complex set of symbolic acts celebrating the spiritual relationship hunters have with the animals, such as in rituals enacted as part of the harvesting and eating of animal meat. These expressions also include the forms of respect directed towards to the animals' inedible remains, like the bones and antlers, which are decorated and displayed (Armitage, 1992).

THE RELATION OF MARKET PRODUCTION TO SELF-PROVISIONING

This chapter confirms what some others have observed — that we lack an adequate theoretical understanding of the economics of self-subsistence, particularly where it exists in the context of an industrial economy. According to Durrenberger, this lack of theory is explained in part because as yet there are no agreed-upon measures of the values of what is produced — no statistics, no testable laws. He cites the case of Mississippi shrimpers, who have an approach to their own household economies that is similar to self-provisioning. They are motivated to fulfill known needs, but from the perspective of government economic policy they should behave like capitalist firms. In contrast to the capitalist tendency towards the over-exploitation of open access, common property resources, the shrimpers limit their production, showing a preference for leisure when their harvests are adequate for their needs. This kind of behaviour leads fishery administrators to conclude that, in their actions, the shrimpers are crazy (Durrenberger, 1994). As I have shown in this chapter, the issue is that there is a necessary relationship between market economics and self-provisioning, even though each embodies different values and motivations.

The self-provisioning of food from fishing, hunting, garden crops, and wild berries, as well as for non-food requirements such as the wood harvested for building supplies and for firewood, were, and still are, an aspect of the household economies of some rural people across Canada (Teitelbaum and Beckley, 2007). As I have shown, at different times and places in Newfoundland and Labrador self-provisioning has existed, always with some kind of relationship to the market economy. The historic Innu kept the market at arm's length by means of the debt system. In hunting for food they were motivated by their family's "known needs," and similarly for trapping, another "known need" — the debt they owed for their imported supplies — influenced fur production. Both the southern Labrador Inuit around Lake Melville and the Newfoundland salt cod fishers effectively had a dual economy, directly engaging in both kinds. They produced for the market, using market motivations, while in other seasons they also gardened, hunted, and gathered to supplement the food they could obtain on the basis of the market production alone. For the contemporary Indigenous people of Labrador, the two economies of the market and self-provisioning have become inextricably linked, such that hunting for food requires inputs from the cash economy, and at times when the market economy is insufficient to supply these inputs, a food insecurity crisis can result.

NOTES

1. I use this term in preference to "subsistence," which is defined as "the minimum (as of food and shelter) necessary to support life" (Merriam-Webster, 2017). By contrast, I assert that self-provisioning economies may also support forms of luxury.
2. The purpose of this report was to document the land use and occupancy of the Sheshatshiu Innu, one of the requirements for making a claim through the federal government's Aboriginal land claims process at the time. This report was to demonstrate that the group making the claim had used and occupied that area in question since time immemorial, as well as to show over time their pattern of land use.
3. Imported foods have now become part of the necessary supplies for an Innu hunt (Castro, 2016).
4. The average animal supplies 13 kg of high-quality food (Ashley, 2002: 46).
5. Writing of the late 1800s, Margaret Baikie noted that for her Inuit grandmother's generation "[n]early all the Eskimo girls were getting married to the Englishmen" (Baikie, n.d.: 51).
6. Today, many of the descendants of this group, the Southern Inuit, belong to the NunatuKavut Indigenous organization.
7. Chapter 2, by Roseman and Royal, also makes the point that today gardening for food requires input from the market economy.

REFERENCES

Ames, Randy. 1977. *Social, Economic and Legal Problems of Hunting in Northern Labrador*. Report for the Labrador Inuit Association. Nain, NL: Labrador Inuit Association.

Anstey, Robert J., and M. A. P. Renouf. 2011. "Down the Labrador: Ramah chert use at Phillip's Garden, Port au Choix." In M. A. P. Renouf, ed., *The Cultural Landscape of Port au Choix*, 189–207. New York: Springer.

Armitage, Peter. 1992. "Religious ideology among the Innu of eastern Quebec and Labrador." *Religiologiques* 6: 64–110.

Ashley, Bruce. 2002. *Edible Weights of Wildlife Species Used for Country Food in the Northwest Territories and Nunavut*. Yellowknife, NWT: Government of the Northwest Territories.

Baikie, Margaret. n.d. "Labrador memories. Reflections at Mulligan." *Them Days*.

Cadigan, Sean T. 1992. "The staple model reconsidered: The case of agricultural policy in northeast Newfoundland, 1785–1855." *Acadiensis* 21, 2: 48–71.

Castro, Damian. 2016. "Meating the social: Sharing atiku euiash in Sheshatshiu, Labrador." Doctoral dissertation, Department of Anthropology, Memorial University of Newfoundland.

Chayanov, Alexander V. 1966. *The Theory of Peasant Economy*. Homewood, Ill.: Richard D. Irwin (for the American Economic Association).

Durrenberger, E. 1994. "Shrimpers, processors and common property in Mississippi." *Human Organization* 53, 1: 74–82. doi:10.17730/humo.53.1.2736220061197n06.

Felt, Lawrence, David C. Natcher, Andrea Procter, Nancy Sillitt, Katie Winters, Tristan Gear . . . Roland Kemuksigak. 2012. "The more things change: Patterns of country food harvesting by the Labrador Inuit on the North Labrador Coast." In David Natcher, Lawrence Felt, and Andrea Procter, eds., *Settlement, Subsistence, and Change among the Labrador Inuit: The Nunatsiavummiut Experience*, Vol. 2, 139. Winnipeg: University of Manitoba Press.

Gilbert, William. 2008. "Archaeology at Cupids 2003–2005." Newfoundland Baccalieu Trail Heritage Corporation. At: http://www.cupidslegacyentre.ca/digs/archaeology.php.

Goudie, Elizabeth. 1973. *Woman of Labrador*. Toronto: Peter Martin Associates.

Hanrahan, Maura. 2012. "Tracing social change among the Labrador Inuit: What does the nutrition tell us?" In David Natcher, Lawrence Felt, and Andrea Procter, eds., *Settlement, Subsistence, and Change among the Labrador Inuit: The Nunatsiavummiut Experience*, vol. 2, 121. Winnipeg: University of Manitoba Press.

Henriksen, Georg. 2009. *I Dreamed the Animals. Kaniukutat: The Life of an Innu Hunter*. New York: Berghahn Books.

House, Doug, and Adrian House. 2015. *An Extraordinary Ordinary Man: The Life Story of Edgar House*. St. John's: ISER Books.

Jackson, Doug. 1993. *"On the Country": The Micmac of Newfoundland*. St. John's: Harry Cuff.

Jeddore, John Nick. 2015. *Moccasin Tracks: A Memoir of Mi'kmaw Life in Newfoundland*. St. John's: ISER Books.

Kuokkanen, Rauna. 2011. "Indigenous economies, theories of subsistence, and women: Exploring the social economy model for Indigenous governance." *American Indian Quarterly* 35, 2: 215–40.

Mackey, M. G. Alton, and R. D. Moore Orr. 1987. "An evaluation of household country food use in Makkovik, Labrador, July 1980–June 1981." *Arctic* 40, 1: 60–65.

——— and ———. 1988. "The seasonal nutrient density of country food harvested in Makkovik, Labrador." *Arctic* 41, 2: 105–08.

Mailhot, José. 1997. *The People of Sheshatshit: In the Land of the Innu*. St. John's: ISER Books.

McGrath, P. T. 1926. McGrath Papers, Centre for Newfoundland Studies, Memorial University of Newfoundland Archives. St. John's, coll-175.

Merriam-Webster.com. 2017. "Subsistence." At: https://www.merriam-webster.com/dictionary.

Mitchell, Gregory. 2014. "'We don't have any Klik or Spam in the house — How about a piece of boiled salmon for lunch?' Country food in NunatuKavut." In John Kennedy, ed., *History and Renewal of Labrador's Inuit-Métis*. St. John's: ISER Books.

Montague, Louie. 2013. *I Never Knowed It Was Hard: Memoirs of a Labrador Trapper*. Elizabeth Dawson, ed. St. John's: ISER Books.

Ommer, Rosemary E. 1989. "Merchant credit and the informal economy: Newfoundland 1919–1929." *Historical Papers / Communications Historiques* 24, 1: 167–89. doi:10.7202/031001ar.

———, Nancy J. Turner, Martha MacDonald, and Peter Sinclair. 2007. "Food security and the informal economy." In Christopher C. Parish, Nancy J. Turner, and Shirley M. Solberg, eds., *Resetting the Kitchen Table: Food Security, Culture, Health and Resilience in Coastal Communities*, 115–28. New York: Nova Science.

Omohundro, J. 1994. *Rough Food: The Seasons of Subsistence in Northern Newfoundland*. St. John's: ISER Books.

———. 1995. "Living off the land." In Lawrence Felt and Peter R. Sinclair, eds., *Living on the Edge: The Great Northern Peninsula of Newfoundland*, 103–27. St. John's: Institute of Social and Economic Research, Memorial University of Newfoundland.

Ray, Arthur J. 1990. *The Canadian Fur Trade in the Industrial Age*. Toronto: University of Toronto Press.

Rich, E. E. 1960. "Trade habits and economic motivation among the Indians of North America." *Canadian Journal of Economics and Political Science* 26, 1: 35–53. doi:10.2307/138817.

Ryan, Shannon. 1971. "The Newfoundland cod fishery in the 19th century." MA thesis, Faculty of Humanities and Social Science, Memorial University of Newfoundland.

Sahlins, Marshall. 1972. *Stone Age Economics*. Chicago: Aldine Atherton.

Speck, Frank G. 1977 [1935]. *Naskapi: The Savage Hunters of the Labrador Peninsula*. Norman: University of Oklahoma Press.

Stopp, Marianne. 2002. "Ethnohistoric analogues for storage as an adaptive strategy in northeastern subarctic prehistory." *Journal of Anthropological Archaeology* 21, 3: 301–28. doi:10.1016/S0278-4165(02)00004-1.

———. 2014. "'I, old Lydia Campbell': A Labrador woman of national historic significance." In John Kennedy, ed., *History and Renewal of Labrador's Inuit-Métis*. St. John's: ISER Books.

Tanner, Adrian. 1976. "Land use and land tenure in the southern part of Labrador." Unpublished report for the Naskapi Montagnais Innu Association. Copy in the Newfoundland Room, Queen Elizabeth II Library, Memorial University.

———. 2014. *Bringing Home Animals*, 2nd ed. St. John's: ISER Books.

———, John C. Kennedy, Susan McCorquodale, and Gordon Inglis. 1994. *Aboriginal Peoples and Governance in Newfoundland and Labrador*. A Report for Governance Project, Royal Commission on Aboriginal Peoples, Oct. St. John's. At: http://publications.gc.ca/site/archivee-archived.html?url=http://publications.gc.ca/collections/collection_2016/bcp-pco/Z1-1991-1-41-79-eng.pdf.

Tanner, Vino. 1947. *Outlines of the Geography, Life and Customs of Newfoundland-Labrador*. Cambridge: Cambridge University Press.

Taylor, J. Garth. 1977. "Moravian Mission influence on Labrador Inuit subsistence: 1776–1830." In D. A. Muise, ed., *Approaches to Native History: Papers of a Conference at the National Museum of Man, October 1975*, 16–29. Ottawa: National Museums of Canada.

Teitelbaum, Sara, and Thomas Beckley. 2006. "Harvested, hunted and home grown: The prevalence of self-provisioning in rural Canada." *Journal of Rural and Community Development* 1, 2: 114–30.

Tourigny, Eric. 2009. "What ladies and gentlemen eat for dinner: The analysis of faunal materials recovered from a seventeenth-century high-status English household. Ferryland, Newfoundland." MA thesis, Department of Archaeology, Memorial University of Newfoundland.

Way, Patricia. 2014. "The story of William Phippard." In John Kennedy, ed., *History and Renewal of Labrador's Inuit-Métis*. St. John's: ISER Books.

Weber, Max. 1981 [1927]. *General Economic History*. Ira J. Cohen, trans. New Brunswick, NJ: Transaction.

Woolett, James. 2004. "Labrador Inuit subsistence in the context of environmental change: An initial landscape history perspective." *American Anthropologist* 109, 1: 69–84.

Zimmerly, David William. 1975. *Cain's Land Revisited: Culture Change in Central Labrador, 1775–1972*. St. John's: Institute of Social and Economic Research.

7

Food Systems and Indigenous Peoples in Labrador: Issues and New Directions

Rebecca Schiff & Karine Bernard

INTRODUCTION

Food insecurity in northern Canadian Indigenous communities is a serious and complex public health problem.[1] The lack of fresh, healthy, high-quality, and affordable food in northern Indigenous communities has been linked to rising rates of chronic illnesses, such as cardiovascular disease, diabetes, and some cancers. The alarming findings are that food insecurity has been shown to be tightly entangled with impaired well-being among Canadians. Numerous studies have documented the relationship between food insecurity among Indigenous people and health behaviours and outcomes, such as obesity, poor general health, high stress, poor diet, smoking, chronic diseases, and growth retardation. Moreover, studies point out that severe food insecurity experienced during childhood could have lasting effects on health outcomes later in life. Food insecurity has been clearly documented as leading to poor nutritional health and related physical health risks in northern Indigenous communities.

Food security is also documented as a significant proximal determinant of health for Indigenous people in Canada. There is also clear documentation that food insecurity negatively impacts other determinants of health, including economic, social, and mental health.

In general, Canada's northern and remote regions experience high rates of food insecurity, exceptionally high food costs, environmental concerns related to contamination and climate change, and a diversity of other uniquely northern challenges related to food production, acquisition, and consumption. Indigenous communities in Labrador are no exception. They experience significant food-related challenges attributable to factors specific to northern and remote regions. Indigenous communities in Labrador utilize a combination of store-bought foods, foods grown within or near communities, and "country foods" to meet nutritional needs.[2] Limiting factors on food security are related to all of these food sources. This chapter provides a review of the literature on issues related to food security and sustainable food systems for Indigenous communities in Labrador. It concludes with some directions for planning, policy, and future research to support food security in Labrador.

DEMOGRAPHIC AND GEOGRAPHIC CONTEXT OF LABRADOR

Labrador is located on the eastern coast of the Canadian mainland. It is northwest of the island of Newfoundland and borders the province of Quebec to the east and north. The population of the region is approximately 27,000 (Statistics Canada, 2011). Residents are of mixed descent and include Innu, Inuit, Southern Inuit (Inuit-Métis), and non-Aboriginal people. Within Labrador, there are multiple political, social, and geographic centres. These centres include the Indigenous political organizations (NunatuKavut, Nunatsiavut, and Labrador Innu Nation) as well as the various non-Indigenous politically defined communities. The Nunatsiavut government represents approximately 7,000 Inuit of the Labrador Inuit land claim area. The Inuit primarily live in the northern coastal Labrador communities of Nain, Hopedale, Postville, Makkovik, and Rigolet (see Figure 7.1). Many Nunatsiavut beneficiaries also live in Happy Valley–Goose Bay (HVGB) and North West River. The NunatuKavut Community Council (formerly Labrador Métis Nation) represents the 6,000 Southern Inuit people of Labrador. NunatuKavut members live primarily in communities along the southeast coast of Labrador from Cartwright to Forteau. As shown in Figure 7.1, many members also live in the central Labrador area of HVGB and Mud Lake. The Innu Nation in Labrador is comprised of two communities: Sheshatshiu Innu First Nation (SIFN) and Mushuau Innu First Nation (MIFN). The Sheshatshiu population is close to 2,000 while the MIFN population consists of approximately 900 people located in the community of Natuashish. Sheshatshiu is located close to HVGB and Natuashish close to

Davis Inlet. These Indigenous political organizations (which are also cultural and social entities) represent individuals dispersed over multiple, often geographically remote, municipal communities.

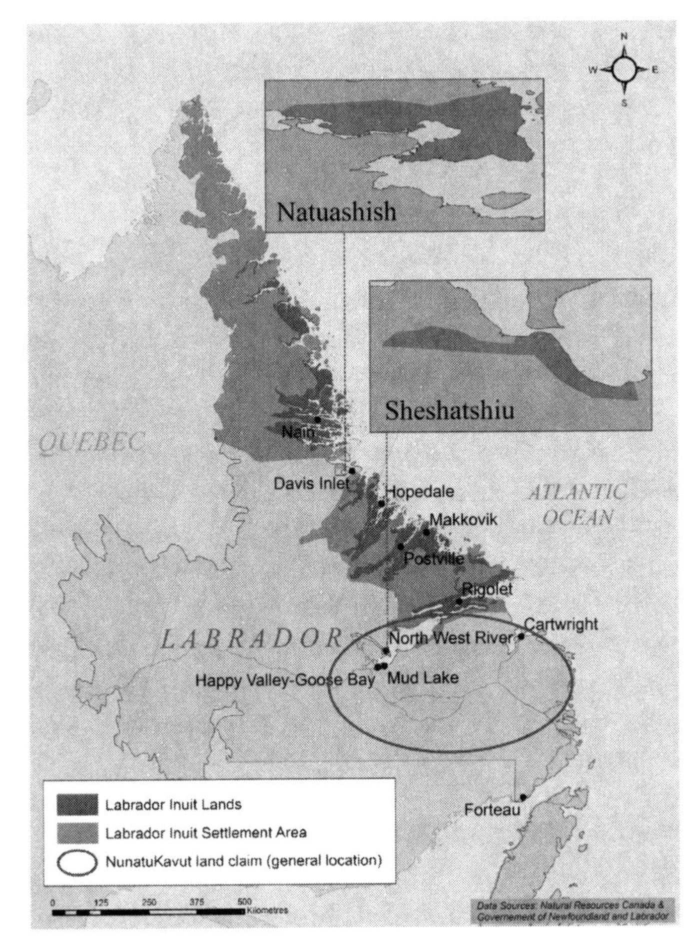

Figure 7.1. Map of Inuit lands and settlement areas in Labrador. (Cartography by Myron King)

Labrador was one of the first locations known to European explorers, first visited by Norse explorers in the eleventh century. Processes of European colonization are thought to have begun when British, Portuguese, and French explorers and fishermen came to the region in the late fifteenth and early sixteenth centuries. These early visitors initially came to the region on a seasonal basis, with more permanent contact and settlement established in the seventeenth and eighteenth centuries. Colonization in Labrador has been a complex process, experienced in diverse ways in different regions of Labrador, by different ethnocultural groups, over several centuries. As with other Indigenous and northern communities, these processes have had a significant impact on lifestyle, traditional food systems, and food security. While it is outside the scope of this chapter to provide a detailed discussion of these processes, several other works specifically focus on the impact of colonization on the food system in Labrador, including Felt et al. (2012), Hanrahan (2008), Nudell (2006), and Martin (2011).

PREVALENCE OF FOOD (IN)SECURITY AMONG NORTHERN INDIGENOUS COMMUNITIES IN CANADA

Across northern Canada, Indigenous communities experience extremely high rates of food insecurity. In Kugaaruk (Nunavut), Lawn and Harvey (2003) found that 83 per cent of households experienced food insecurity. The Inuit Health Survey conducted in 2007 and 2008, which included 36 communities from the Inuvialuit Settlement Region in the western Arctic, Nunavut, and Nunatsiavut revealed similar findings: 62.6 per cent of households were food insecure (Egeland, Pacey, Cao, and Sobol, 2011). Among them, 33.6 per cent were moderately food insecure and 29.1 per cent were severely food insecure. Research examining food insecurity among First Nations households in northern Manitoba and northern Ontario identified extremely high rates: 75 per cent and 70 per cent of households, respectively, were found to be food insecure (Skinner, Hanning, Desjardins, and Tsuji, 2013; Thompson et al., 2011).[3]

PREVALENCE OF FOOD (IN)SECURITY AMONG INDIGENOUS COMMUNITIES IN LABRADOR

Across the province of Newfoundland and Labrador, the rate of household-level food insecurity decreased from 15.7 per cent in 2007 to 13.4 per cent in 2012 (Tarasuk, Mitchell, and Dachner, 2014). However, these rates mask the uneven distribution of food insecurity across communities, with much higher rates seen in northern and Indigenous communities. Martin et al. (2012) found that 22.2 per cent of NunatuKavut[4] households were food insecure and 5.8 per cent were severely insecure. Though these rates are not as high as elsewhere in Canada, the food insecurity prevalence was much higher in NunatuKavut compared to province-wide rates reported by

Tarasuk et al. (2014). Nunatsiavut communities[5] report even higher rates of food insecurity, where 46 per cent of households with children were found to be food insecure, with 16 per cent being severely food insecure (Allard and Lemay, 2012).[6] Although there are no reports on food insecurity rates for the Innu of Sheshatshiu and Natuashish, high levels of poverty in those communities suggests that food issues may be significant. Some reports, such as that of the Gathering Voices Project Team (2011) in regard to Sheshatshiu, provide qualitative evidence of significant food security concerns.

FACTORS LIMITING FOOD SECURITY FOR INDIGENOUS COMMUNITIES IN LABRADOR

The reasons behind higher rates of food security for Indigenous communities in Labrador are complex. Northern Indigenous communities experience food security issues that are unique and specific to northern and remote locations. Indigenous communities in Labrador utilize a combination of store-bought foods, foods grown within or near communities, and "country foods" to meet nutritional needs. In the following section, we examine current knowledge about these limiting factors, with particular attention to the context of Labrador, in order to understand potential avenues for alleviating food insecurity in those communities.

Issues in Access to and Availability of Store-Bought Food in Labrador

Numerous factors affect access to safe, healthy, and adequate food for Indigenous communities in Labrador. Long-distance transportation to remote areas has a significant impact on the availability, quality, and cost of store-bought foods. Fuel and other costs associated with food transportation contribute to food costs that are significantly higher than those found in Canada's urban centres (Aboriginal and Northern Affairs Canada, 2008; Boult, 2004; Myers, Powell, and Duhaime, 2004). While food costs are higher in northern communities when compared to their southern counterparts, costs in less accessible northern communities are disproportionately higher than in northern service centres.[7] Food costs are inconsistent across the North such that more remote communities (such as Nain and Black Tickle) experience much higher costs than those found in remote service centres such as HVGB.[8] Issues related to the high costs of retail foods are further complicated for the significant number of low-income households who face additional economic pressures in accessing healthy foods in the North (Boult, 2004; Myers et al., 2004).

The Nutritious Food Basket (NFB) provides basis for evaluating and comparing food prices across communities for about 60 different foods representing a nutritious diet (Tarasuk et al., 2014). Similarly to elsewhere in Canada, the NFB is more expensive in Labrador (north), than in Newfoundland (south), as shown in Table 7.1. As an indicator, the NFB for Nunatsiavut communities, located on the northern coast of Labrador, exceeds $300 per week for a family of four compared to an average of less than $180 on the island of Newfoundland (Government of Canada, 2013; Food Security Network, 2011). Though the NFB does not report prices specifically for NunatuKavut communities or the Innu First Nations, other studies document significant food costs compared to southern counterparts (Martin et al., 2012; Gathering Voices Project Team, 2011).

Table 7.1. Nutritious Food Basket (NFB) cost in Newfoundland and Labrador.

Geographical Area	Cost (Canadian dollars)
Newfoundland	$178.84
Labrador	$211.49
Urban	$201.89
Rural	$229.37
Nunatsiavut	
Hopedale	$362.95
Makkovik	$318.85
Nain	$321.82

Note: NFB cost per week for a family of four, consisting of a man and woman (25–49 years), a boy (13–15 years), and a girl (7–9 years). Source: Department of Health and Community Services, Government of Newfoundland and Labrador, June 2011.

In addition to the high cost of food, quality and availability are also concerns (Boult, 2004; Myers et al., 2004). Poor availability of fresh foods are reported among communities, mainly due to long-distance transportation and weather (Martin et al., 2012). Many remote Indigenous communities in Labrador depend entirely on ferry and plane services to transport food. As a result, the quality and availability of food is frequently compromised (Martin et al., 2012). During summer, availability is more reliable in some communities where food can be transported via seasonal roads. However, residents report that the quality of food is poorer since transport by road leads to lengthier transportation times (Martin et al., 2012).

A survey of food quality and availability conducted in Labrador in 2001 revealed substantial concerns related to the quality of perishable foods found in local stores (Ladouceur and Hill, 2001). Ladouceur and Hill (2002) state that 80 per cent of respondents also reported poor availability, indicating that there was never or only sometimes enough variety of fresh fruits and vegetables available for purchase. Additional cultural and socio-economic factors complicate issues of cost and availability. These include knowledge about how to prepare different foods, a limited range of choices for different foods, and the ability to prepare and consume healthy foods (Beaumier and Ford, 2010; Myers et al., 2004).

Issues Impacting Community-based Food Production

The geographical location of Labrador communities creates challenges for community-based food production. A variety of factors limit the ability to produce or acquire food through gardening, farming, and fishing in Indigenous communities of Labrador. Short growing seasons, light levels, permafrost, and poor soil quality impact the capacity to grow food (Jóhannesson, 2012; Juday et al., 2010; Leahey, 1954). Despite an abundance of water, some communities (such as Black Tickle) face irrigation challenges due to water quality issues caused by lack of water treatment infrastructure (Sarkar, Hanrahan, and Hudson, 2015). Other communities experience difficulty in accessing safe water for irrigation due to various other issues, such as the impact of industrial development, mining, and hydroelectric projects on water quality (Airhart, Janes, and Jamieson, 2011; Jóhannesson, 2012; Myers et al., 2004; Thompson, 2005). Access to agricultural and fishing supplies is also limited (in terms of cost and selection) due to transportation issues (Airhart et al., 2011; Jóhannesson, 2012).

Traditional Food and Food Security among Indigenous Peoples in Labrador

For many Indigenous people in Labrador, country foods play critical nutritional and cultural roles. Access, consumption, and sharing of traditional food, such as caribou, moose, fish, and wild berries, have been shown to improve food security, develop diet quality, and provide social benefits (Egeland, Johnson-Down, Cao, Seikh, and Weiler, 2011; Boult, 2004; Chan et al., 2006; Beaumier and Ford, 2010; Myers, Powell, and Duhaime, 2004). Activities associated with traditional food acquisition and preparation also provide spiritual connection to the land, the community, and the past (Pufall et al., 2011). Research has shown that these activities have important effects on emotional, mental, and spiritual health (Alton-Mackey, 1984; Martin, 2009, 2011; Martin et al., 2012). Moreover, activities related to the procurement, processing, and consumption of traditional food reinforce cultural expressions of identity and pride (Martin, 2009; Hanrahan, Sarkar, and Hudson, 2014).

While there has been some research on community food production and store-bought food issues, much of the food-related research with Indigenous communities in Labrador has focused on country foods. Research has examined historical evidence and changes over time to country food systems, the contemporary significance of country foods, and human impacts on availability of and access to country foods.

Changes in the Traditional Food System of Indigenous Communities in Newfoundland and Labrador over the Past Century

In Newfoundland and Labrador, as elsewhere in Canada, Indigenous communities used to rely exclusively on traditional food systems. Indigenous peoples of Labrador obtained their food through hunting, fishing, trapping, harvesting, and sharing. No parts of animals were wasted, as they used to eat all animal parts including blood, liver, bones, and intestines (Hanrahan, 2008). Labrador Inuit practised seasonal hunting and fishing. In the summer, fish were caught and dry-cured. In late summer, Inuit moved inland to hunt caribou, drying and storing much of the meat for the winter. In the fall, they used to hunt harp seals and in the winter they caught and ate ringed seal, walrus, and seabirds (Hanrahan, 2008). Innu used to follow caribou and catch fish during the summer (Indian and Northern Affairs Canada, 1996). Some of these practices remain intact, but over the past century many have come to rely increasingly on store-bought foods.

However, the arrival of Europeans in the late fifteenth century impacted the daily life of Indigenous people in Labrador as well as traditional food systems. The most dramatic changes in traditional lifestyles may have occurred in the past century through increased exposure to colonial policy, including the residential school system and resettlement policies (Hanrahan, 2008; Martin, 2009). Under the Newfoundland government's resettlement policy, several Indigenous communities were relocated to larger and more permanent settlements in the 1950s and 1960s (Indian and Northern Affairs Canada, 1996; Maritime History Archive, 2010). The residential school system also had a significant impact on traditional food systems in Labrador. Among the numerous adverse impacts, resettlement and residential schools impaired the transmission of knowledge about traditional food systems and traditional ways of living.

Over the past century, government legislation imposed restrictions on hunting and fishing practices and diminished communities' control over their lands. Changes in the abundance and distribution of food species due to human-induced environmental changes affected availability of and access to country foods. Through industrialization and globalization, store-bought foods began to appear, creating an increased reliance on imported market foods alongside a decrease in the consumption of traditional foods (Adelson, 2005; Hanrahan, 2008; Martin, 2009). From a traditional food diet rich in nutrients, Indigenous people in Newfoundland and Labrador have undergone a shift to a diet composed heavily of store-bought foods high in carbohydrates such as starches and sugar. This has led to dietary issues due to the poor quality, variety, and high cost of market foods (Egeland, 2010; Ladouceur and Hill, 2002; Martin et al., 2012). Along with this shift, several diet-related diseases appeared. Poor dental health, chronic constipation, and vitamin deficiencies emerged as early as the middle of the nineteenth century (Hanrahan, 2008). More recently, chronic diseases such as type 2 diabetes, high blood pressure, and dyslipidemia have significantly affected health for Indigenous communities in the region (Egeland, 2010; Hanrahan, 2008; Martin et al., 2012).

Continued Importance of the Traditional Food System

Nonetheless, amid these tremendous changes, the traditional food system has continued to play a crucial role in the life of Indigenous people in the province. The harvest of country foods is still vital to community and individual well-being (Nain Research Centre Kaujisapvinga, 2015; Martin, 2009). It strengthens social networks; fosters cultural pride and continuity; connects people to the land and with the past; promotes emotional, mental, and spiritual health; enables the transmission of cultural values, skills, and spirituality; and contributes to the economies of communities (Egeland et al., 2011; Martin, 2009; NunatuKavut, 2013; Power, 2008; Natcher, Felt, McDonald, and Ford, 2012).

Hunting, fishing, trapping, gathering plants, and harvesting eggs are part of the everyday life for a majority of Indigenous households across Labrador, although there are differences between and within communities. Among Nunatsiavut communities, about 85 per cent of households participate in harvesting activities (Felt et al., 2012). Among communities on the southeast coast, 63 per cent of NunatuKavut members supplement their food through hunting and 78 per cent supplement their food through fishing and gathering berries (Martin et al., 2012). The traditional food system contributes significantly to food intake among Indigenous people in Labrador. For 56 per cent of Inuit households in Nunatsiavut, wild meat and fish make up more than half of the meat or fish eaten (Tait, 2007). Nearly 45 per cent of Inuit-Métis supplement their food several times a week with traditional food. The most popular wild foods are large game (42.1 per cent), followed by fish (42.4 per cent), whereas gull eggs are consumed several times a week by only 5.3 per cent of people (Martin et al., 2012). Atikesse, Bouche de Grosbois, St. Jean, Penashue, and Benuen (2010) and others (Gathering Voices Project Team, 2011) have reported a similar significance of country foods for Innu communities.

Impact of Human Activities on the Traditional Food System

Human activities have had significant impact on the traditional food systems of Indigenous peoples in Labrador. The loss and contamination of Indigenous land and waters have profoundly disrupted traditional food systems, with a subsequent impact on the well-being of Indigenous individuals and communities. Numerous human activities, related to military developments, mining, hydroelectric projects, and climate change have affected the distribution and abundance of species as well as the safety of county food consumption.

Military developments have impacted all communities in Labrador. The military development at Happy Valley–Goose Bay has created ongoing challenges in terms of contamination (Canadian Environmental Assessment Agency, 2012). Contamination of water, soil, and biota has significantly impacted surrounding areas (Canadian Environmental Assessment Agency, 2012) that historically served as hunting grounds for Indigenous peoples in the region. Low-level flying out of the Goose Bay air base was also the subject of significant protest by the Labrador Innu, who were concerned over the impact on the health and distribution of caribou and other species (Belanger and Lackenbauer, 2015). The military radar station at Hopedale has also created environmental issues in that region, impacting traditional food sources. Contamination by polychlorinated biphenyls (PCBs) from the site exceeded the maximum allowable amount specified in the Canadian Environmental Protection Act PCB material storage regulations. PCBs have been discovered in marine sediments, ringed seals, shorthorn sculpins (bottom-feeding fish), and black guillemots (seabirds) (Brown, Sheldon, Burgess, and Reimer, 2009; Brown, Fisk, Helbing, and Reimer, 2014; Kuzyk, Stow, Burgess, Solomon, and Reimer, 2005; Nain Research Centre Kaujisapvinga, 2015).

Indigenous communities in Labrador have also expressed concern over the impact of mining on traditional food sources. Uranium exploration, with improperly disposed waste at the Kitts Pond site near Makkovik, has been implicated by some as having impacted soil, water, plants, fish, and animals in the region (Schiff, Sarkar, Choi, and Anstey, 2014). The Voisey's Bay nickel mine located near Nain and Natuashish has also created concern among the Innu and Nunatsiavut Inuit. In particular, Inuit and Innu individuals and groups have expressed concern over potential effects on the environment, employment, social impacts, and, most importantly, the potential impact of resource development on traditional activities such as hunting, fishing, and trapping (Voisey's Bay Nickel Company, 1997).

Major hydroelectric plants, including the Upper Churchill project of the 1970s and the more recent damming of the Lower Churchill River at Muskrat Falls have been implicated as having a significant impact on traditional food systems. The Upper Churchill project is commonly accepted as having had serious and major effects on the Labrador Innu (Higgins and Shalev, 2007). More recently, concern has grown over the potential impacts of flooding that may result from the new hydroelectric development underway at Muskrat Falls. Flooding associated with the creation of new reservoirs for hydroelectric development has been shown to increase bioaccumulation of mercury in fish and seals through a combination of factors that enhance methylmercury production in aquatic environments. The area of inundated land for the Muskrat Falls reservoir is 41 km^2 and it is estimated that peak fish mercury concentrations may increase by about 1.5 to 4.5 times from estimated baseline concentrations (Nalcor Energy, 2009). Research involving dietary surveys and hair sampling with residents of all ages is underway in order to assess the level of exposure to methylmercury among communities around the Churchill River (Nalcor Energy, 2014).

Northern regions, including Labrador, are also severely impacted by accelerated warming (Allard and Lemay, 2012; Lévesque, Hermanutz, and Gérin-Lajoie, 2012). In Labrador, abrupt warming began around 1993 (Allard and Lemay, 2012). Communities and scientists have reported impacts on the distribution and abundance of species, which affect traditional food availability across the Arctic (Beaumier and Ford, 2010; Ford, Pearce, Duerden, Furgal, and Smit, 2010; Meakin and Kurvits, 2009). Although snowfalls are expected to increase in other Arctic areas, decrease over much of Nunatsiavut is projected. Resultant changes in ice formation and composition make travel to access country food more difficult and dangerous. The travelling seasons and, consequently, the harvesting seasons are getting shorter and pose challenges to country food access (Natcher and Davis, 2007).

STRATEGIES TO ADDRESS FOOD SECURITY IN NORTHERN CANADIAN INDIGENOUS COMMUNITIES

A number of strategies to address issues related to food security and food systems have been implemented in Labrador. These include programs focused primarily on food prices and country food, while a few strategies have also taken a more comprehensive approach to food systems issues for Indigenous communities in Labrador.

Store-Bought Food: The NL Air Foodlift Subsidy Program

In addition to the federal Nutrition North (formerly Food Mail) program, the provincial Air Foodlift Subsidy program is a key component to Newfoundland and Labrador's food security strategy. It was established in 1997 to ensure that nutritious, perishable items such as fruit, vegetables, and dairy products are available along the coast of Labrador. The program serves all communities in Nunatsiavut, Natuashish Innu First Nation, and two communities in the NunatuKavut region. Although the most isolated and remote communities can benefit from the program, other communities in Labrador also experience challenges related to high food prices and could benefit from programming to address their needs. There are also a number of administrative and conceptual issues with the program, similar to those highlighted by the Auditor General of Canada with respect to Nutrition North (Auditor General, 2014).

Community Food Production: Agriculture and Gardening Programs

Several communities in Labrador have implemented community gardening programs as part of broader food security strategies. Happy Valley–Goose Bay has demonstrated particular success in supporting gardening and agricultural production. Besides development of a children's vegetable garden, the community food security network developed a food garden in a social housing area. In 2012, Newfoundland and Labrador Housing Corporation provided the Food Security Network of NL (now called Food First NL) with a vacant block of land in an area of town primarily dedicated to social housing. Funding given through a Wellness grant from the provincial health ministry provided support for involvement of low-income seniors in the garden project. Two community kitchens were also established in the same neighbourhood. The kitchens attracted several hundred participants in the first year. They were run collaboratively by the Community-Led Food Assessment (CLFA) co-ordinator and a Health Canada nutritionist. Operation of the kitchens was also supported through food donations from local food retailers and provincial funding from the Newfoundland and Labrador Housing Corporation.

HVGB also developed a community farmers' market in 2011, which met with initial success. The market provided an additional outlet for the distribution of food produced by several farms in the region, helping to boost sales and visibility of local food. Farmers in the region work continuously to design innovative strategies to address the relatively harsh growing conditions found in Labrador.

Communities in Nunatsiavut have also met with success in implementing community and home-based gardening programs. In Hopedale, a pilot home-based vegetable gardening program started in 2012. In 2013, its expansion to other communities was set as a top priority to improve food security in Nunatsiavut (Food Security Network, 2013).

Country Foods: Promoting Hunting, Fishing, and Trapping Activities

In addition to community-based food production and fishing, country foods continue to play a crucial role in the life of Indigenous people in Labrador. Due to the challenges faced in regard to the access and distribution of country food, there has been a need to develop programs that can address these issues. Ensuring that people have access to community-based resources, such as equipment and space that can be used to produce or prepare foods, has been an effective strategy in increasing access to traditional food, encouraging food sharing, and strengthening skills (Thompson et al., 2011). To this aim, both community freezer and harvest support programs have been shown to foster food security and food sovereignty (Council of Canadian Academies, 2014).

Over the past decade, increased involvement of Indigenous people in sustainable forest management in Labrador has had an impact in influencing forestry planning and food security (Wyatt, Merrill, and Natcher, 2011). Ecological reserves and protected areas have been implemented to conserve the cultural and environmental features of the landscape. Information related to culture and social values; traditional uses by Indigenous people; sites of importance for cultural heritage; and hunting, fishing, and travel routes all are integrated in public planning. However, according to Roberts (2006), public planning can risk being too focused on forest management; thus, it needs to consider inclusion of practices focused on integrated ecosystem impacts, including the effects of those elements on food security and country foods.

With respect to harvesting, sharing, and consuming traditional food, community freezer programs are widespread among several Indigenous communities. Community freezer programs aim to preserve traditional food, such as moose, caribou, and partridge provided by community hunters, in order to provide access for individuals who would otherwise be unable to obtain traditional food. They play an important role in providing storage facilities. In Labrador, Food First NL collaborated with several Nunatsiavut communities, including in HVGB, Hopedale, and Nain, to develop community freezers (Food Security Network, 2012).

In collaboration with academic partners at Memorial and Trent universities, Nunatsiavut established a modified community freezer, hunting, and skill-sharing program, named "Going Off, Growing Strong," where both youth and adults harvest in teams to replenish freezers. The program demonstrated positive outcomes for the community (Organ, Castleden, Furgal, Sheldon, and Hart, 2014). It was successful in increasing the availability of traditional food, fostering sharing, promoting intergenerational transfer of traditional knowledge, and empowering youth with skills to provide for their family and community (Organ et al., 2014; Council of Canadian Academies, 2014; Food Security Network, 2013). Despite a positive impact, the program still faces challenges with regard to ongoing environmental, economic, and socio-cultural issues that affect food access, supply, and utilization (Organ et al., 2014).

Country food acquisition has also been negatively influenced by a number of natural resource, hunting, and fishing regulations. NunatuKavut members and other Indigenous communities in Labrador have expressed concerns about the increasing restrictions that affect when, how much, and how often traditional foods can be collected (Martin et al., 2012). Methods for decolonizing policy-making processes need to be further investigated so that regulatory environments recognize the inherent rights and needs of Indigenous communities in Labrador with respect to food security and resource management.

NEW DIRECTIONS: DECOLONIZATION AND FOOD SYSTEMS APPROACHES

The programs described above have successfully addressed some issues related to store-bought foods, country foods, and community food production. However, Indigenous communities in Labrador still face numerous challenges related to food security. Existing programs have been unable to address gaps such as the interconnected roles of different food acquisition methods and other structural factors that limit food security. A significant issue with programs such as the Air Foodlift Subsidy and some country food programs is that they focus primarily on market activities or traditional food acquisition and ignore other structural factors that limit food security in these communities. Many existing strategies have not been attentive to issues of colonialism and the experiences of the people "on the ground" — experiences that are integral to truly addressing and understanding northern food insecurity.

Community-Led Food Assessment and Food Networks

Lack of capacity within many existing programs is a limiting factor to facilitating the interrelated roles of community food production, food retail, and traditional food activities that support food security and community wellness. However, some positive developments have been supported through partnerships between community and government organizations, Food First NL, and university partners. Community-led food assessments in HVGB and Nunatsiavut have played an important role in identifying gaps through comprehensive examination of community food systems (Airhart et al., 2011; Flowers, Nochasak, and Jameson, 2010). These strategies have fostered integrated development of programs, which can attend to the interrelated roles of country foods, store-bought foods, community-produced foods, and food education in meeting community food security needs. In 2012, McTavish, Furgal, Popp, and McCarney produced a guide for implementing community-led food assessments, which may be useful for Inuit and other Indigenous communities (see Food First NL, n.d.). Support for assessments, and for community-led implementation of their recommendations, could be beneficial for other Indigenous communities in the region.

Food networks, councils, and coalitions have shown that they can help to foster the development of comprehensive food systems strategies and to address food policy issues for Indigenous communities (Council of Canadian Academies, 2014; Schiff and Brunger, 2013). Food networks provide an ongoing venue for cross-sectoral collaboration to examine local and regional food-related concerns and interests on an ongoing basis. Further work should also focus on examining such strategies and new approaches that can evaluate programmatic, policy, and regulatory issues related to store-bought foods, country foods, and community food production.

Decolonizing Food

While understanding of specific, successful programmatic and policy approaches is critical to building sustainable food systems for Indigenous communities in Labrador, there is also a need to address underlying structural issues. Addressing food insecurity among Indigenous communities requires an in-depth understanding of the current and historical contexts of food security and food systems for Indigenous communities. Colonialism has had a significant impact in shaping northern Indigenous communities through continual exclusion in resource and policy discussions. These exclusionary practices have impacted knowledge of traditional methods, including the ability to produce, acquire, and utilize food resources in culturally relevant ways. Traditional community food production methods and access to country foods have been impaired by colonialism through limited autonomy, dispossession from traditional lands, forced resettlement of communities, and residential schools. Colonialism continues to influence communities, perpetuating inequalities through structural violence and ongoing support for historical and new colonial legislation (Adelson, 2005; Agriculture and Agri-Food Canada, 1998; Frohlich, Ross, and Richmond, 2006; King, Smith, and Gracey, 2009; Raphael, Curry-Stevens, and Bryant, 2008; Richmond and Ross, 2009). The ongoing impacts of colonialism continue to impair the right to food for Indigenous communities (United Nations Committee on Economic, Social and Cultural Rights, 1999), as is reflected in the overwhelmingly high rates of food insecurity among Indigenous communities in Canada (Tarasuk et al., 2014; DeSchutter, 2013). Decolonizing Indigenous food systems is key to food security. Embracing the right to food and food sovereignty discourse may support further work towards decolonizing Indigenous food systems.

Canada has a long-standing international commitment related to food security as a signatory of the 1948 Universal Declaration of Human Rights and the 1966 International Covenant on Economic, Social and Cultural Rights, which enshrine the right to food. In 1998, following the World Food Summit in Rome, Canada's Action Plan for Food Security reiterated the right to food and acknowledged the importance of traditional food acquisition methods by Indigenous communities and the important role of hunting, fishing, gathering, and trading towards food security in Canada (Agriculture and Agri-Food Canada, 1998). More recently, the Canadian government supported the United Nations Declaration on the Rights of Indigenous Peoples, which specifies the rights of Indigenous peoples over the resources and territories they have traditionally occupied, and the right to develop these lands by reason of their traditional ownership or use (United Nations, 2008).

Future programs and policy development must recognize that food security for Indigenous peoples depends on decolonization and self-determination. The ability of Indigenous communities and individuals to take control of their own food system, as well as to obtain the necessary education, knowledge, and skills, is crucial to achieve food security (World Forum on Food Sovereignty, 2001). As discussed in the Introduction to this volume, principles of food sovereignty can provide some guidance for future efforts towards the development of policy and processes that protect Indigenous rights (Council of Canadian Academies, 2014; Hanrahan et al., 2014). In conclusion, although northern Indigenous communities clearly have obstacles to overcome, the opportunity to define their own food system presents Labradorians with a new and positive direction.

NOTES

1. For the purposes of this chapter we follow the definition of "food security" offered by Hamm and Bellows (2003: 37), where "community food security" is "a situation in which all community residents obtain a safe, culturally acceptable, nutritionally adequate diet through a sustainable food system that maximizes community self-reliance and social justice."
2. The "traditional food system" is defined as: "all food within a particular culture available from local natural resources and culturally accepted. It also includes the sociocultural meanings, acquisition, processing techniques, use, composition, and nutritional consequences for the people using the food" (Kuhnlein and Receveur, 1996: 418). In Labrador, many individuals also refer to such foods as "country foods." For the purposes of this discussion, and to reflect contemporary usage, we use the terms "country foods" and "traditional foods" interchangeably.
3. It is important to note that the studies we report on here use different methodologies for measuring levels of food security. The challenge of comparing food security issues and rates across communities, due to differing methodologies, is an issue highlighted in a comprehensive report (Council of Canadian Academies, 2014) on the state of food security in northern Indigenous communities. Accounting for differences in methodologies, food insecurity rates are still much higher in isolated and northern communities than elsewhere.
4. NunatuKavut members identify as Southern Inuit and primarily reside in 10 communities located on the southeast coast of Labrador, from north to south: Cartwright, Domino/Black Tickle, Paradise River, Norman Bay, Charlottetown, Pinsent's Arm, William's Harbour, Port Hope Simpson, St. Lewis, Mary's Harbour, and Lodge Bay. Many NunatuKavut members also reside in the central Labrador region in and near Happy Valley–Goose Bay and Mud Lake.

5. Nunatsiavut includes five communities located on the northeast coast of Labrador. These communities are (from north to south): Nain, Hopedale, Makkovik, Postville, and Rigolet.
6. This raises the issue that methodologies for determining levels of food security may be inconsistent across studies and may lead to different interpretations as to whether rates are actually higher or lower in different communities.
7. Service centres in the North are those communities where important government, health, and other services are located and to which residents of smaller communities travel to access those services. HVGB is often considered to be a service centre for central and coastal Labrador, since it houses the majority of government, health, and other services for the region.
8. Examination of the data provided by Indigenous and Northern Affairs Canada indicates a trend when comparing service centres with smaller and more isolated communities (see Aboriginal and Northern Affairs Canada, 2008).

REFERENCES

Aboriginal and Northern Affairs Canada. 2008. "Regional results of price surveys" [tables]. At: http://www.statcan.gc.ca/.

Adelson, N. 2005. "The embodiment of inequity: Health disparities in Aboriginal Canada." *Canadian Journal of Public Health* 96: S45–S61. At: http://journal.cpha.ca/index.php/cjph/index.

Agriculture and Agri-Food Canada. 1998. *Canada's Action Plan for Food Security: In Response to the World Food Summit Action Plan.* At: http://www.agr.gc.ca/misb/fsec-seca/pdf/action_e.pdf.

Airhart, J., K. Janes, and K. Jamieson. 2011. *Food Security Upper Lake Melville: Community-led Food Assessment.* St. John's: Food Security Network of Newfoundland and Labrador.

Allard, M., and M. Lemay. 2012. *Nunavik and Nunatsiavut: From Science to Policy. An Integrated Regional Impact Study (IRIS) of Climate Change and Modernization.* At: http://www.arcticnet.ulaval.ca/pdf/media/iris_report_complete.pdf.

Alton-Mackey, A. 1984. *An Evaluation of Country Food Use, St. Lewis, Labrador.* St. John's: Faculty of Medicine, Memorial University of Newfoundland.

Atikesse, L., S. Bouche de Grosbois, M. St. Jean, B. Penashue, and M. Benuen. 2010. "Innu food consumption patterns: Traditional food and body mass index." *Canadian Journal of Dietetic Practice* 71, 3: e41–e49. doi:10.3148/71.3.2010.125.

Auditor General of Canada. 2014. "Nutrition North Canada — Aboriginal Affairs and Northern Development Canada." Chapter 6 of *2014 Fall Report of the Auditor General of Canada*. At: http://www.oag-bvg.gc.ca/internet/English/parl_oag_201411_06_e_39964.html.

Beaumier, M., and J. D. Ford. 2010. "Food insecurity among Inuit women exacerbated by socio-economic stresses and climate change." *Canadian Journal of Public Health* 101, 3: 196–201. At: http://journal.cpha.ca/index.php/cjph/index.

Belanger, Y., and P. W. Lackenbauer. 2015. *Blockades or Breakthroughs? Aboriginal Peoples Confront the Canadian State*. Montreal and Kingston: McGill-Queen's University Press.

Boult, D. A. 2004. *Hunger in the Arctic: Food (In)security in Inuit Communities*. At: http://www.naho.ca/documents/it/2004_Inuit_Food_Security.pdf.

Brown, T. M., T. A. Sheldon, N. M. Burgess, and K. J. Reimer. 2009. "Reduction of PCB contamination in an Arctic coastal environment: A first step in assessing ecosystem recovery after the removal of a point source." *Environmental Science & Technology* 43, 20: 7635–42. doi:10.1021/es900941w.

———, A. T. Fisk, C. C. Helbing, and K. J. Reimer. 2014. "Polychlorinated biphenyl profiles in ringed seals (*Pusa hispida*) reveal historical contamination by a military radar station in Labrador, Canada." *Environmental Toxicology and Chemistry* 33, 3: 592–601. doi:10.1002/etc.2468.

Canadian Environmental Assessment Agency. 2012. "5 Wing Goose Bay remediation project" (reference no. 07-01-26393). At: http://www.ceaa.gc.ca/052/details-eng.cfm?pid=26393.

Chan, H. M., K. Fediuk, S. Hamilton, L. Rostas, A. Caughey, H. Kuhnlein, G. Egeland, and E. Loring. 2006. "Food security in Nunavut, Canada: Barriers and recommendations." *International Journal of Circumpolar Health* 65, 5. doi:10.3402/ijch.v65i5.18132.

Council of Canadian Academies. 2014. *Aboriginal Food Security in Northern Canada: An Assessment of the State of Knowledge*. Ottawa: Expert Panel on the State of Knowledge of Food Security in Northern Canada.

DeSchutter, O. 2013. *Mission to Canada*. Report presented to the 22nd Session of the United Nations Human Rights Council. Geneva: UN.

Egeland, G. M. 2010. *Inuit Health Survey 2007–2008: Nunatsiavut*. At: https://www.mcgill.ca/cine/files/cine/adult_report_-_nunatsiavut.pdf.

———, L. Johnson-Down, Z. R. Cao, N. Seikh, and H. Weiler. 2011. "Food insecurity and nutrition transition combine to affect nutrient intakes in Canadian Arctic communities." *Journal of Nutrition* 141, 9: 1746–53. doi:10.3945/jn.111.139006.

———, A. Pacey, Z. Cao, and I. Sobol. 2010. "Food insecurity among Inuit preschoolers: Nunavut Inuit Child Health Survey 2007–2008." *Canadian Medical Association Journal* 182, 3: 243–48. doi:10.1503/cmaj.091297.

Felt, L., D. C. Natcher, A. Procter, N. Sillitt, K. Winters, T. Gear, . . . R. Kemuksigak. 2012. "The more things change: Patterns of country food harvesting by the Labrador Inuit on the North Labrador Coast." In D. C. Natcher, L. Felt, and A. Procter, eds., *Settlement, Subsistence, and Change among the Labrador Inuit: The Nunatsiavummiut Experience*, 139–70. Winnipeg: University of Manitoba Press.

Flowers, J., S. Nochasak, and K. Jameson. 2010. *NiKigijavut Hopedalimi: "Our Food in Hopedale"*. At: http://www.foodsecuritynews.com/Publications/NiKigijavutHopedalimiReportFINAL.pdf.

Food First NL. n.d. "Community-led food assessment." At: http://www.foodfirstnl.ca/our-resources/community-led-food-assessment.

Food Security Network. 2011. "Newfoundland and Labrador nutritious food basket." At: http://www.foodsecuritynews.com/nutritious-food-basket.html.

———. 2012. *2012 Community Report*. At: http://www.foodsecuritynews.com/Publications/FSN_2012_Community_Report.pdf.

———. 2013. *Engaging Communities Project Update*. At: http://www.foodsecuritynews.com/Publications/Engaging%20Communities_Project%20Update_November%202013.pdf.

Ford, J. D., T. Pearce, F. Duerden, C. Furgal, and B. Smit. 2010. "Climate change policy responses for Canada's Inuit population: The importance of and opportunities for adaptation." *Global Environmental Change* 20, 1:, 177–91. doi:10.1016/j.gloenvcha.2009.10.008.

Frohlich, K. L., N. Ross, and C. Richmond. 2006. "Health disparities in Canada today: Some evidence and a theoretical framework." *Health Policy* 79, 2/3: 132–43. doi:10.1016/j.healthpol.2005.12.010.

Gathering Voices Project Team et al. 2011. "Crying out for help: Brief presented to the Panel for the Environmental Assessment of the Lower Churchill Hydroelectric Development." 23 Mar. At: http://www.ceaa.gc.ca/050/documents/49197/49197E.pdf.

Government of Canada, Nutritious North Canada. 2013. "Cost of the revised northern food basket in 2012–2013." At: http://www.nutritionnorthcanada.gc.ca/eng/1369313792863/1369313809684.

Hamm, M. W., and A. C. Bellows. 2003. "Community food security: Background and future directions." *Journal of Nutrition Education and Behavior* 35, 1: 37–43. doi:10.1016/s1499-4046(06)60325-4.

Hanrahan, M. 2008. "Tracking social change among the Labrador Inuit and Inuit-Métis: What does the nutrition literature tell us?" *Food, Culture and Society* 11, 3: 315–33. doi:10.2752/175174408x347883. At: http://www.nunatukavut.ca/home/files/pg/exploring_water_insecurity-hanrahan_sarkar_hudson.pdf.

———, A. Sarkar, and A. Hudson. 2014. "Exploring water insecurity in a northern Indigenous community in Canada: The 'never-ending job' of the Southern Inuit of Black Tickle, Labrador." *Arctic Anthropology* 51, 2: 9–22. doi:10.3368/aa.51.2.9.

Higgins, J., and G. Shalev. 2007. "The 1969 contract." Heritage Newfoundland & Labrador. At: http://www.heritage.nf.ca/articles/politics/churchill-falls-impacts.php.

Jóhannesson, T. 2012. "Arctic-quality certification." At: http://www.bioforsk.no/ikbViewer/Content/75386/Torfi%20Proposal_english_revised.pdf.

Juday, G. P., V. Barber, P. Duffy, H. Linderholm, S. Rupp, S. Sparrow, . . . M. Wilmking. 2010. "Forests, land management, agriculture." In International Arctic Science Committee, *Arctic Climate Impact Assessment*. At: http://www.acia.uaf.edu/PDFs/ACIA_Science_Chapters_Final/ACIA_Ch14_Final.pdf.

King, M., A. Smith, and M. Gracey. 2009. "Indigenous health part 2: The underlying causes of the health gap." *The Lancet* 374, 9683: 76–85. doi:10.1016/s0140-6736(09)60827-8.

Kuhnlein, H. V., and O. Receveur. 1996. "Dietary change and traditional food systems of Indigenous peoples." *Annual Review of Nutrition* 16, 1: 417–42. doi:10.1146/annurev.nu.16.070196.002221.

Kuzyk, Z. A., J. P. Stow, N. M. Burgess, S. M. Solomon, and K. J. Reimer. 2005. "PCBs in sediments and the coastal food web near a local contaminant source in Saglek Bay, Labrador." *Science of the Total Environment* 351 & 352: 264–84. doi:10.1016/j.scitotenv.2005.04.050.

Ladouceur, L. L., and F. Hill. 2002. *Results of the Survey on Food Quality in Six Isolated Communities in Labrador, March 2001.* At: http://www.aadnc-aandc.gc.ca/DAM/DAM-INTER-HQ/STAGING/texte-text/survfoo2001_1100100035899_eng.pdf.

Lawn, J., and D. Harvey. 2003. *Nutrition and Food Security in Kugaaruk, Nunavut: Baseline Survey for the Food Mail Pilot Project.* At: https://www.aadnc-aandc.gc.ca/DAM/DAM-INTER-HQ/STAGING/texte-text/kg03_1100100035822_eng.pdf.

Leahey, A. 1954. "Soil and agricultural problems in Subarctic and Arctic Canada." *Arctic* 7, 3/4: 249–54. doi:10.14430/arctic3852.

Lévesque, E., L. Hermanutz, and J. Gérin-Lajoie. 2012. "Trends in vegetation dynamics and impacts on berry productivity." In M. Allard and M. Lemay, eds., *Arctic Nunavik and Nunatsiavut: From Science to Policy. An Integrated Regional Impact Study (IRIS) of Climate Change and Modernization*, 223–47. At: http://www.arcticnet.ulaval.ca/pdf/media/iris_report_complete.pdf.

Maritime History Archive. 2010. "'No great future': Government sponsored resettlement in Newfoundland and Labrador since Confederation." Maritime History Archive, St. John's. At: https://www.mun.ca/mha/resettlement/rs_intro.php.

Martin, D. 2009. "Food stories: A Labrador Inuit-Métis community speaks about global change." Doctoral dissertation, Dalhousie University.

———. 2011. "'Now we got lots to eat and they're telling us not to eat it': Understanding changes to south-east Labrador Inuit relationships to food." *International Journal of Circumpolar Health* 70, 4: 384–95. doi:10.3402/ijch.v70i4.17842.

———, J. Valcour, J. Bull, M. Paul, J. Graham, and D. Wall. 2012. "NunatuKavut community health needs assessment: A community-based research report." Happy Valley–Goose Bay, NL: NunatuKavut Community Council.

Meakin, S., and T. Kurvits. 2009. *Assessing the Impacts of Climate Change on Food Security in the Canadian Arctic.* At: http://www.grida.no/files/publications/foodsec_updt_LA_lo.pdf.

Myers, H., S. Powell, and G. Duhaime. 2004. "Setting the table for food security: Policy impacts in Nunavut." *Canadian Journal of Native Studies* 24, 2: 425–45. At: https://www.brandonu.ca/native-studies/cjns/.

Nain Research Centre Kaujisapvinga. 2015. "Hopedale: Hopedale radar site clean-up." At: http://nainresearchcentre.com/hopedale/.

Nalcor Energy. 2009. "Part A. Biophysical Assessment." In *Lower Churchill Hydroelectric Generation Project: Environmental Impact Statement*, vol. 2. At: http://www.nalcorenergy.com/uploads/file/Volume%20IIA.pdf.

———. 2014. "Muskrat Falls Project: Dietary survey & hair sampling program." At: http://muskratfalls.nalcorenergy.com/wp-content/uploads/2013/03/Muskrat-Falls-Dietary-Survey-Hair-Sampling-Info-Sheet-Nov2014_Final-2.pdf.

Natcher, D. C., and S. Davis. 2007. "Rethinking devolution: Challenges for Aboriginal resource management in the Yukon Territory." *Society & Natural Resources* 20, 3: 271–79. doi:10.1080/08941920601117405.

———, L. Felt, J. McDonald, and R. Ford. 2012. "The social organization of wildfood production in Postville, Nunatsiavut." In D. C. Natcher, L. Felt, and A. Procter, eds., *Settlement, Subsistence, and Change among the Labrador Inuit: The Nunatsiavummiut Experience*, 171–88. Winnipeg: University of Manitoba Press.

Nudell, Z. 2006. "Colonization embodied: Diabetes in Sheshatshiu." Master's thesis, St. Mary's University, Halifax.

Organ, J., H. Castleden, C. Furgal, T. Sheldon, and C. Hart. 2014. "Contemporary programs in support of traditional ways: Inuit perspectives on community freezers as a mechanism to alleviate pressures of wild food access in Nain, Nunatsiavut." *Health & Place* 30: 251–59. doi:10.1016/j.healthplace.2014.09.012.

Pufall, E. L., A. Q. Jones, S. A. McEwen, C. Lyall, A. S. Peregrine, and V. L. Edge. 2011. "Perception of the importance of traditional country foods to the physical, mental, and spiritual health of Labrador Inuit." *Arctic* 64, 2: 242–50. doi:10.14430/arctic4103.

Raphael, D., A. Curry-Stevens, and T. Bryant. 2008. "Barriers to addressing the social determinants of health: Insights from the Canadian experience." *Health Policy* 88, 2/3: 222–35. doi:10.1016/j.healthpol.2008.03.015.

Richmond, C. A. M., and N. A. Ross. 2009. "The determinants of First Nation and Inuit health: A critical population health approach." *Health & Place* 15, 2: 403–11. doi:10.1016/j.healthplace.2008.07.004.

Roberts, B. A., N. P. P. Simon, and K. W. Deering. 2006. "The forests and woodlands of Labrador, Canada: Ecology, distribution and future management." *Ecological Research* 21, 6: 868–80. doi:10.1007/s11284-006-0051-7.

Royal Commission on Aboriginal Peoples. 1996. *Final Report*, vol. 1, *Look Forward, Looking Back*. At: https://qspace.library.queensu.ca/handle/1974/6874.

Sarkar, A., M. Hanrahan, and A. Hudson. 2015. "Water insecurity in Canadian Indigenous communities: Some inconvenient truths." *Rural and Remote Health* 15. At: http://www.rrh.org.au/.

Schiff, R., and F. Brunger. 2013. "Northern food networks: Building collaborative efforts for food security in remote Canadian aboriginal communities." *Journal of Agriculture, Food Systems, and Community Development* 3, 3: 121–38. doi:10.5304/jafscd.2013.033.012.

———, A. Sarkar, M. Choi, and Z. Anstey. 2014. *Community Perspectives on Uranium Exploration, Mining, and Health Impacts in Makkovik, Labrador*. Happy Valley-Goose Bay, NL: Nunatsiavut Government.

Skinner, K., R. M. Hanning, E. Desjardins, and L. J. Tsuji. 2013. "Giving voice to food insecurity in a remote Indigenous community in subarctic Ontario, Canada: Traditional ways, ways to cope, ways forward." *BMC Public Health* 13, 1: 427. doi:10.1186/1471-2458-13-427.

Statistics Canada. 2011. "Census of Population program (2011)." At: http://www12.statcan.gc.ca/.

Tait, H. 2007. "Harvesting and country food: Fact sheet." At: http://www.statcan.gc.ca/pub/89-627-x/89-627-x2007001-eng.htm.

Tarasuk, V., A. Mitchell, and N. Dachner. 2014. *Household Food Insecurity in Canada 2012*. At: http://proof.utoronto.ca/wp-content/uploads/2014/05/Household_Food_Insecurity_in_Canada-2012_ENG.pdf.

Thompson, S. 2005. "Sustainability and vulnerability: Aboriginal Arctic food security in a toxic world." In F. Berkes, R. Huebert, H. Fast, M. Manseau, and A. Diduck, eds., *Breaking Ice: Renewable Resource and Ocean Management in the Canadian North*, 47–70. Calgary: University of Calgary Press.

———, A. Gulrukh, M. Ballard, B. Beardy, D. Islam, V. Lozeznik, and K. Wong. 2011. "Is community economic development putting healthy food on the table? Food sovereignty in northern Manitoba's Aboriginal communities." *Journal of Aboriginal Economic Development* 7, 2: 14–39. At: http://www.edo.ca/edo-tools/jaed.

United Nations. 2008. *United Nations Declaration on the Rights of Indigenous Peoples*. At: http://www.un.org/esa/socdev/unpfii/documents/DRIPS_en.pdf.

United Nations Committee on Economic, Social and Cultural Rights. 1999. "CESCR general comment No. 12: The right to adequate food (Art. 11)." At: http://www.refworld.org/pdfid/4538838c11.pdf.

Voisey's Bay Nickel Company. 1997. "Residual environmental effects." In *Voisey's Bay Mine/Mill Project Environmental Impact Statement*. At: http://www.vbnc.com/eis/chap20/chap203.htm.

World Forum on Food Sovereignty. 2001. *Final Declaration of the World Forum on Food Sovereignty*. Havana, Sept.

Wyatt, S., S. Merrill, and D. Natcher. 2011. "Ecosystem management and forestry planning in Labrador: How does Aboriginal involvement affect management plans?" *Canadian Journal of Forest Research* 41, 11: 2247–58. doi:10.1139/x11-126.

The Retail Food Environment and Household Food Provisioning Strategies in the Rural Region of Bonne Bay on Newfoundland's West Coast

Kristen Lowitt & Barbara Neis

INTRODUCTION

As concerns with healthy eating and sustainable food systems continue to rise, the subject of food environments has gained popularity among researchers and policy-makers (Health Canada, 2015). This field of study seeks to understand the conditions that influence how people access, choose, prepare, and eat food (Pouliot and Hamelin, 2009). The food environment has also become a widely used conceptual framework for understanding food access, availability, and utilization in the context of food security (Health Canada, 2015).

Within the field of food environments, a large body of work has focused on retail food environments. Here, most attention has focused on food access, in terms of the number and types of food outlets available to a community, along with the supply (availability), quality, and cost of healthy foods in these stores (Penney et al., 2015). Emerging work on food environments in Canada indicates that rural and remote regions face a unique set of food access challenges including overall higher food prices, poorer access to food stores, especially for non-motorized households, and poorer availability of fresh fruits and vegetables (Health Canada, 2015; Lawn, Robbins, and Hill, 1998; Nova Scotia Participatory Food Costing Project, 2010; Pouliot and Hamelin, 2009; Travers et al., 1997). Coinciding with these trends, rural residents in Canada have lower consumption of fruits and vegetables and higher obesity rates compared to urban residents (Canadian Institute for Health Information, 2006; Health Canada, 2015).

In Newfoundland and Labrador, investigating rural retail food environments is especially important because of the high proportion of the population (60 per cent) living in rural areas (Canadian Rural Revitalization Foundation, 2015). In line with national trends, studies in the province confirm that food prices are higher in rural regions, the fresh food supply is more limited, and rates of overweight and obesity are rising (Twells et al., 2014; Newfoundland and Labrador Department of Health and Community Services, 2011). Compounding these challenges is a provincial food supply vulnerable to disruption as food is shipped long distances by ferry and truck before reaching most communities.

In this chapter, we focus on the rural region of Bonne Bay on Newfoundland's west coast as a case study for examining how households adapt to these challenges in food access and availability. This study responds to two important gaps in the food environments literature. First, the majority of food environments research to date has focused on urban areas to the neglect of rural regions (Health Canada, 2015); and second, most food environments research has focused on geographic food access in terms of the spatial distribution of food stores, with much less attention to how households interact with retail food environments, including the potential strategies they use to overcome limitations in access (Cummins, 2007). We draw on household interview (n = 37) and survey data (n = 307) collected in the Bonne Bay region in 2011 to identify the key food provisioning strategies households use to adapt to the constraints posed by the local food retail environment. We understand food provisioning as encompassing the acquisition, preparation, cooking, eating, and disposal of food (Marshall, 1995). It extends research about food choice by looking at the socio-cultural and environmental contexts in which food consumption takes place (Delormier, Frohlich, and Potvin, 2009). We focus primarily on strategies that households in Bonne Bay use for acquiring and overcoming limitations in access to retail foods as an important component of food provisioning.

METHODS: CASE STUDY

Bonne Bay is a fjord located in Gros Morne National Park on Newfoundland's west coast. The region consists of five communities, including Rocky Harbour, Norris Point, Woody Point, Glenburnie/Birchy Head/Shoal Brook, and Trout River, with a year-round population of about 3,000 people (see Figure 8.1). With the exception of Trout River, located just beyond the park boundary, these towns have been surrounded by Gros Morne National Park since it was established in 1972. A highway through the park connects the region with small towns on the Northern Peninsula. The larger population centres of Deer Lake and Corner Brook, with populations of approximately 5,000 and 20,000 people respectively, are located to the south.

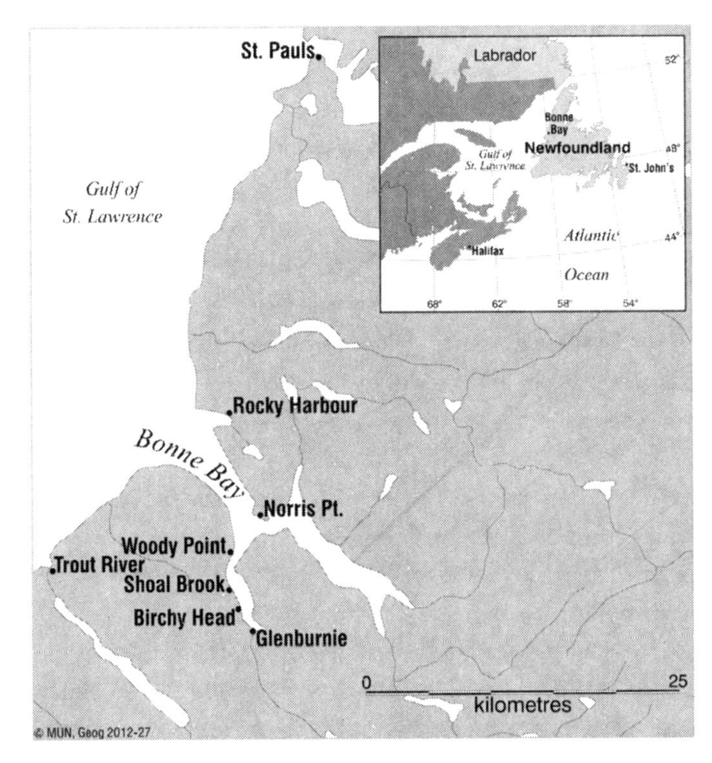

Figure 8.1. Map of study area, the Bonne Bay region, Newfoundland. (Cartography by MUN Geography 2012-27)

Bonne Bay, like some other parts of the west coast, was originally settled for a combination of fishing and forestry opportunities (Mannion, 1977). For generations, these remained the main economic activities. Today, fishing and tourism are key economic sectors, both of which have a high level of seasonal employment. Tourism has assumed an important role because of the region's location within a national park and also in response to the decline of the cod and salmon fisheries and related substantial downsizing that has taken place in fisheries employment in the region over the past 20 years.

Incomes are low in the region. Bonne Bay is located in Economic Zone 7 as defined by the Newfoundland and Labrador Statistics Agency. This zone covers the western coast of the Northern Peninsula. Per capita income in this zone was $27,500 in 2012, below provincial and national averages (Conference Board of Canada, 2015; Newfoundland and Labrador Statistics Agency, 2015a). The incidence of Employment Insurance (EI) in this zone was the second highest in the province in 2013, with 64.4 per cent of the labour force receiving EI (Newfoundland and Labrador Statistics Agency, 2015b).

Over time, social and economic changes in the region have led to changes in patterns of food access. Traditionally, households used a system of occupational pluralism in which they provided for themselves necessities they could produce, including food, based on a seasonal round of activities involving fishing, hunting, and gardening (Omohundro, 1994). Foods they couldn't produce they usually obtained on credit from the merchant store (Ommer, Turner, MacDonald, and Sinclair, 2007). More rapid changes to traditional patterns of food provisioning came following Newfoundland's joining Canada in 1949 and the influx of modern goods, services, and imported foods in the 1950s (Omohundro, 1994). An important shift in the Bonne Bay area happened when the road was put through connecting communities in the area to Deer Lake in the late 1960s. For those with road transportation, this provided access to grocery stores for the first time.

Throughout the 1970s and 1980s, some full-service grocery stores operated in the Bonne Bay region. More recently, here as elsewhere, consolidation in the food retail industry has led to the construction of fewer but larger food stores and loss of some local full-service grocery stores, which has meant many rural populations have had to adjust to new requirements for more extended travel to obtain groceries (Hendrickson et al., 2008, Morton et al., 2005). Currently, the nearest supermarkets for Bonne Bay residents, as well as major retail outlets for clothes, hardware, and other goods, are located in the larger centres of Deer Lake and Corner Brook, respectively about 75 and 125 kilometres to the south. Deer Lake has two supermarkets. Corner Brook is the largest population centre on the west coast with several supermarkets and some specialty food stores.

Within the region there are a number of small grocery and convenience stores that sell a very limited selection of fresh fruits and vegetables, some dairy products, as well as frozen meat (Lowitt, 2009). The stores are normally served by deliveries once a week, but up to twice a week in the summer due to additional business from tourists (Lowitt, 2009). Alongside these conventional food outlets, there are also four fish plants in the region that do some seasonal retail sales, including one with a retail seafood shop attached. One local pharmacy also sells some food items. Alternative food outlets, including a farm stand and farmers' markets, operate on a seasonal basis. Table 8.1 shows the key retail outlets in Bonne Bay (up to date as of December 2015).

Table 8.1. Key retail food outlets in the Bonne Bay region.

Small grocery and convenience stores	
Woody Point	3Ts Store: Fruits and vegetables, dairy, frozen meat, fresh lobster in season, frozen cod and halibut. Pete's One Stop: Fruits and vegetables, dairy, frozen meat.
Glenburnie/Birchy Head/ Shoal Brook	Roy Young Limited: Fruits and vegetables, dairy.
Trout River	Hanns Confectionary Ltd.: Fruits and vegetables, dairy, frozen meat.
Norris Point	C & J Rumbolt Ltd.: Fruits and vegetables, dairy, frozen meat, tinned goods, bakery. Rudy's Pub & Grub: Fruits and vegetables, dairy, tinned goods.
Rocky Harbour	Endicotts Crafts and Convenience: Fruits and vegetables, dairy, frozen meat, tinned goods. C &J Rumbolt Ltd.: Fruits and vegetables, dairy, frozen meat, bakery. Cloverfarm: Fruits and vegetables, dairy, frozen meat, tinned goods. Gros Morne Trading Post: Fruits and vegetables, dairy, frozen meat, tinned goods. Earle's Video & Convenience: Bakery.
Fish plants	
Woody Point	3Ts Ltd.: Limited sales at the plant during the fishing season.
Trout River	Allen's Fisheries: Limited sales at plant during the fishing season.

Rocky Harbour	Harbour Seafoods: Fresh and frozen seafood (seasonal).
Other	
Glenburnie/Birchy Head/ Shoal Brook	Farmers' market (seasonal)
Norris Point	Pharmacy: Fruits and vegetables, dairy, tinned goods. Community garden at Bonne Bay Cottage Hospital. Howell Farm: Fresh vegetables and eggs (seasonal).

DATA COLLECTION

This study draws on household interview and survey data collected in 2011 in the Bonne Bay region. In-depth semi-structured interviews were undertaken with members of local households about their food provisioning practices (n = 37). Interviewees were asked to describe what the household eats in a regular week, where the food they eat comes from, and how important seafood is to their diet. Households were selected to meet a range of characteristics including household size, ages of household members, and socio-economic status (Kuzel, 1992; see Table 8.2). Local gatekeepers helped identify and recruit households to participate in the study.

Upon completion of interviews, these were transcribed. We tried to transcribe these as near verbatim as possible to maintain the original language and flow of the interviews (Poland, 1995). As such, some unique aspects of local dialect appear in the quotations presented in this chapter. Following transcription, these were then thematically analyzed with the assistance of the NVivo software program (Berg, 2004). We used a process of open coding to identify themes in the interview data (Berg, 2004; Dewalt and Dewalt, 2002). We began with broad themes related to different stages of food provisioning, then added other themes, such as food traditions and food knowledge, that did not fit into these categories but seemed deserving of consideration. We continued until the point of theoretical saturation was reached, at which time no new significant themes emerged and no new information was found that added substantially to an understanding of the existing categories (Creswell, 2007; Lacey and Luff, 2001).

Table 8.2. Characteristics of households in study region.

Socio-demographic Characteristics	Study Households (N = 37)	Per cent of Households*
Community		
Rocky Harbour	9	24%
Norris Point	11	30%
Woody Point	3	8%
Glenburnie/Birchy Head/Shoal Brook	4	11%
Trout River	9	24%
Cow Head	1	3%
Number of household members		
One	11	30%
Two	15	41%
Three	2	5%
Four	7	19%
Five or more	2	5%
Ages of household members*		
Children (under 19 years)	12	32%
Young	6	16%
Younger middle age	10	27%
Older middle age	11	30%
Senior	13	41%
Number of active income earners in household		
One	5	14%
Two	12	32%
More than two	1	3%
None Unemployed Retired from work	 1 18	 3% 49%
Sources of household income		
Year-round employment	10	27%
Seasonal employment	10	27%
Private pension	11	30%
Fixed income	9	24%

**Percentages may not total 100 per cent due to rounding. **Quantified data on income and age were not collected. Rather, during the interviews a description of the household*

— including economic situation, number and approximate age of members, and family history — was elicited. Ages of household members are described in the table in descriptive categories rather than numerical ranges. While there are debates about how to define the category "seniors," in this study they were defined according to the Statistics Canada legal definition of 65 years of age and above. The "per cent of households" column for age and income is greater than 100 per cent because some households have members in more than one category (Lowitt, 2013b).

Second, an anonymous survey about seafood consumption was distributed by mail to all residential post office boxes in the Bonne Bay region. The response rate was 27 per cent (307 surveys were returned). The survey collected information about frequency and types of seafood consumed, sources for obtaining seafood, ways of eating seafood, and satisfaction with availability, affordability, and quality of seafood (see Lowitt, 2013a, for details). The survey included multiple choice and Likert-type ranking scale questions.

RESULTS

In this section, we first present the main barriers to retail food access described in the household interviews. We then draw on interview and survey data to identify the key food provisioning strategies that households have developed for overcoming constraints related to retail food access.

Barriers to Retail Food Access

High food prices and a lack of consistent availability of fresh fruits and vegetables in local stores were described by interview informants as key barriers to retail food access. In terms of price, some residents said they would like to support local stores but were constrained because of the higher prices. Sam [pseudonym] explained:

> Well you know those small stores can't compete with those big grocereterias. Probably here you might pay $1.40 for a litre of orange juice, but probably sometimes you go to Deer Lake and there's [a] sale on, 88 cents a litre. I mean that's a big difference. You pretty much got to buy — you like to buy in your community — but when it comes to them prices you pretty well got to buy.

In addition to higher costs, the availability of supplies and different types of fresh fruits and vegetables is limited and purchasing them depends on knowing the delivery schedule. Lynn, who shops for her family of four, explained, "Sometimes, like bananas, where they rot so quick you might go down on a certain day and the ones that are there they're not really fresh because over three or four days they'll get soft. But then when the truck comes in, maybe once a week, you can get fresh ones again."

As a result of these constraints, most households did the bulk of their food purchasing at supermarkets outside the region, with supplemental purchases at local stores. A shopping trip to supermarkets in Deer Lake and Corner Brook every two weeks was a regular part of the food routine for most households. Residents who had moved to the region from larger centres described having to adjust to fewer shopping trips and greater distances to supermarkets. Ellen, who recently moved with her family to Rocky Harbour from urban Ontario, said:

> I've noticed that every two weeks we try to go to Deer Lake or Corner Brook. That's an adjustment for us too. I'm used to going to get what I want and coming home, going after work, picking up a few things, and coming home, right? But, here you actually have to plan a full day to go get groceries and other things.

Long distances to supermarkets were a particular barrier to food access for households without a vehicle, including most seniors interviewed. Sally, a senior who lives alone and no longer drives, explained, "So Jane got a car, and Stephanie got a car, and Walt got a car. I got to stay friendly with everybody." Seniors comprise a growing share of aging rural populations throughout the province (Moazzami, 2014), and accounted for 41 per cent of the households interviewed in this study.

Strategies for Ensuring Food Access

Households have developed food provisioning strategies for adapting to limitations in the retail food environment. Key strategies described in the interviews include bulking up on food; purchasing frozen and canned fruits and vegetables; purchasing food on sale; and combining supermarket shopping trips with other appointments and activities. For those without access to a vehicle, utilizing social networks of friends and family is crucial to ensuring food access. We also found that purchases from non-conventional retail sources, such as fish plants, as well as reliance on self-provisioned foods, were important to overcoming limitations in food access from conventional retail outlets such as grocery stores and supermarkets.

A key food provisioning strategy for many households was "stocking up" or "bulking up" on food from supermarkets in order to have foods available that couldn't be purchased in local stores and to save money by not paying higher prices locally. For example, one interviewee, Lynn, who lives with her husband and two children, explained:

> If the ketchup's going on sale for $2.99, hey I'm buying two. There's only one income. But on one income it's working for me because look at our pantry. There's two boxes of Triscuits, and bananas, and there's two, three jars of peanut butter. I don't have to run to the corner store and say oh the Miracle Whip and pay $5 for it, right, when I got it for $2.99 last week.

Bulking up on food was particularly important for the winter months when poor driving conditions make trips to supermarkets more difficult. For example, Pat described starting to bulk up for the winter as early as June: "The other day I went down there [Walmart] and started stocking up for the winter, next winter. Son says to me, 'This is only June!'"

For many households, bulking up on food also meant buying frozen and canned foods. For example, Joanie said, "I think most of the time I use vegetables for meals — roast dinners and stuff — I use frozen, that way they'll always be there. I do like fresh vegetables. When it comes to salads, I pick up very, very fresh salads and we have to eat them in the first couple of days." Ellen described buying more frozen foods compared to when her family lived in Ontario, closer to supermarkets. She said:

> in Ontario we had a lot more fresh cause you go pick it up whenever we wanted it. Whereas here, you have to freeze a lot more. Where we lived was very close to one of the major supermarkets in Ontario — Fortino's. We could go in anytime and get whatever we wanted. But here it's a lot more frozen things.

Most households bought canned and frozen produce in supermarkets. Seniors who didn't drive were more likely to purchase canned fruits in local stores when they went on sale. Pharmacies also stock some foods, including milk and tinned goods, and were especially important sites of food access for seniors.

Despite the fact that many households purchased frozen and canned fruits and vegetables, most said they preferred the taste of fresh. Others said it was harder to plan meals without fresh produce. Debbie, a mother of three, described herself as a bulk shopper for canned and frozen produce but said, "It's still hard to plan unless you've got staples – mushrooms, onions, peppers, grapes, apples. They should be available but they're not. . . . My deep freeze now is full. But in my mind I have no groceries – no apples, no strawberries, no grapes."

Further, while stocking up on frozen vegetables was common, purchasing frozen meat was not. Some frozen meats are available in local stores. However, most households preferred to buy fresh meats in supermarkets for reasons of taste; many also said fresh meat costs less. For seniors, having access to only frozen meat in local stores was a particular problem because of inappropriate portion sizes. As one senior woman said, "We can't get meat here, only frozen [meat]. In Deer Lake you can buy it [fresh] – you can buy a big chunk and cut it up into meal-size pieces and freeze it."

The practice of bulking up on food changed with seasonal changes in income. Nearly one-third of households interviewed had an income earner who was seasonally employed. These households stocked up on purchased foods during the spring and summer months when the household income was higher. For example, Cathy had two young children and her husband was a crew member on a commercial fishing boat. She said, "Usually spring of the year before he goes fishing, that's when our food stores go down. You're waiting on the fish. This [summer] is time now when he makes the money. I stocks up on stuff I need." Cathy's family waited on the fish not only to have it as part of their diet but also as a source of extra money to buy food. Lynn's husband also worked seasonally. She said, "Bill's working now, I stock up."

In addition to stocking up on food, buying food on sale was important to offsetting the fuel costs to drive long distances to supermarkets. Most households, regardless of their income level, made purchases from multiple stores to pick up items on sale. Tom's description of food purchasing was common for many households.

> We usually have a list. We look at the flyers, pick out the specials. If you see meats on special you make sure you buy enough for a couple of weeks. Cause usually they run in a sort of a cycle. Meats and vegetables and that. You watch for the specials and pick out the ones you need.

Buying food on sale was especially important for households with limited incomes. Jane, a single woman who relied on income assistance, only bought food on sale. However, she pointed to supply challenges that could threaten this strategy: "Problem here is you do find something on sale, then the truck or the ferry doesn't come in and you've gone all that way [to Deer Lake]." Seniors on fixed incomes described "making do" with food purchases and only buying foods on sale. Further, members of many senior households did not drive and relied on family and friends to take them to the supermarket or purchase food for them. For example, Edith, who lives alone without a vehicle, said, "My daughter and them are always going [to the supermarket], so I tell them pick me up this or that. Usually I get what I want."

To offset transportation costs, household members also tried to combine supermarket shopping with work meetings and other appointments in larger centres. Deanna explained, "I would generally incorporate it [grocery shopping] into another trip and over the years my work practices have let me be on the road a lot." Some did grocery shopping in multiple locations as they made trips across the island. Mary said, "When I'm travelling from St. John's, I make stops along the way in Grand Falls, Gander, get what I can. [I] never, ever purposefully go to Deer Lake for groceries. I have enough appointments and reasons to go out of town; I buy it [groceries] as I go."

Lastly, purchases from non-conventional retail outlets, including fish plants, as well as reliance on self-provisioned foods, were important to overcoming limitations in food access from conventional retail outlets such as grocery stores and supermarkets. Compared to fresh produce, households were more satisfied with the availability of seafood in the Bonne Bay region. For example, Ellen described eating more seafood, which she usually purchased from local fish plants, since returning to Rocky Harbour from Ontario:

> now that we're home I find we're eating more fish now than we did in Ontario cause it's a lot fresher, right? So we get it fresh.

> Right now I've got halibut steaks in my fridge to cook for supper for tonight. So that kind of thing — halibut, salmon, trout, cod.

Results from the seafood survey indicate that local seafood is an important part of the diet for most households but it is eaten most frequently in the summer, during the fishing season and when fish plants and a local seafood retail shop are open, with some fish frozen and salted to keep for the winter (see Lowitt, 2013a). Fish plants were ranked by surveyed households as the main source of local seafood, with friends and family as the second most common source. Friends and family were also ranked as the most preferred source for reasons of quality, price, and traceability. As Mary explained, "I know the fishermen, a couple of them are friends, I basically know where to go. I know who's going to give me top-quality, skinned fillets, dried. I can put in an order for fillets fresh or dried."

In addition to local seafood, self-provisioned foods from gardening, hunting, and gathering were important for providing fresh food in season and lessening reliance on purchased foods. Nearly all households interviewed were engaged in some form of food self-provisioning or ate self-provisioned goods given to them by other households (Table 8.3). Moose has been an especially important source of local meat since these were introduced to the island in the first part of the twentieth century (Omohundro, 1994). In the last four years, more moose licences have become available since Parks Canada opened up a moose hunt within the national park to reduce an inordinately high moose population.

Table 8.3. Household participation in food self-provisioning.

Types of Food Self-provisioning Activities	Households Involved in Self-provisioning (N = 37)	
	N	%
Fish (recreational or commercial)	16	43%
Grow vegetables and fruits	15	41%
Hunt	12	32%
Harvest edibles (e.g., berries)	4	11%
Raise animals	3	8%

These self-provisioned foods were an important part of "bulking up" on food for the winter months. For example, when asked about food provisioning in the winter, Elaine said, "Well I mean this is where we shop around town. ... But I mean usually our fridge is full anyway because we get our moose and everything." While self-provisioning provided an important source of fresh, frozen, and preserved local food, households also described other motivations for these activities, including maintaining food traditions, knowing where their food comes from, and as a source of recreation and enjoyment (see Lowitt, 2014). However, some generational differences in self-provisioning activities emerged. In particular, some younger families described a lack of skills for preserving or preparing local foods, or knowing where to go to pick berries or fish. This sometimes constrained them from using these foods. For example, Michael, who lives with his wife and young son, said, "A lot of my generation we don't have cooking skills. Don't know what to do with it [fish]." Others described relying on their parents or grandparents to access traditional foods. For example, Deatra explained: "the older generation is gonna go out; they [younger people] won't know how to do it. If anything happens to you, Father, where am I gonna get my [salt] fish. Could get it in the store but wouldn't be the same as Father's."

DISCUSSION

Bonne Bay residents face limitations in retail food access characteristic of many rural regions, including long distances to supermarkets, higher food prices, and a lack of fresh foods in local stores. In Bonne Bay, these challenges are amplified by the vulnerable nature of the food supply to an island province. Households have developed a particular set of food provisioning strategies for adapting to this retail environment. These include bulking up on food, buying food on sale, substituting frozen for fresh food, combining grocery shopping with other appointments, and, for non-motorized households, utilizing social networks of friends and family to access food stores. We also found that non-conventional food outlets, such as fish plants, and self-provisioned foods were important strategies used by some local people to overcome limitations in the local retail food environment. This study shows that food environment research needs to pay more attention to the range of food sources, including retail stores, non-conventional food outlets, and self-provisioning, all of which are important to ensuring rural food access (Cummins, 2007; Pouliot and Hamelin, 2009).

Drawing on the results from this study, we point to several key areas of policy attention important to building healthy food environments that can support rural food security. Achieving these aims will require cross-sectoral policy approaches that bring together the relevant domains of agriculture, health, transportation, fisheries, and economic development (Canadian Rural Revitalization Foundation, 2015; MacRae, 2011). While we draw on examples from Bonne Bay, the recommendations are relevant to other rural regions in the province and, potentially, elsewhere that face similar challenges. Throughout, we also provide specific suggestions for further research.

First, a more consistent supply of fresh fruits and vegetables is needed in rural areas. In many rural regions, small independent stores are having an increasingly hard time maintaining their market share with implications for food availability (Pouliot and Hamelin, 2009). As Pouliot and Hamelin (2009: 2057) argue, "from a population health perspective it is necessary to ensure that the changing nature of the retail sector leads to an increase in the fresh fruit and vegetable supply and their access."

Second, this study supports previous research indicating that social capital, in terms of trust and social connectivity, has the potential to help vulnerable households meet their food sufficiency needs at the same time that it may also help support collective action towards achieving food security (Dean and Sharkey, 2011; Martin, 2004). Social capital tends to be particularly strong in rural communities (Dean and Sharkey, 2011). Newfoundland and Labrador has a strong history of social norms of reciprocity, including around food sharing, as an important means of surviving in an isolated region (Ommer, 2004; Omohundro, 1994). At a community level, social capital can help support collective action initiatives for food security (Pigott, 2009; Morton et al., 2005; Sloane et al., 2003). There is a need for policy to support the social capital already in place in rural communities by encouraging the formation of new community partnerships based on co-operation and shared norms of civic responsibility (Morton et al., 2005). Community groups can serve as important resources for food access by advocating to keep local grocery stores open or stock more fresh foods, by supporting better transportation to food stores, and by offering food skills workshops about local and traditional foods (Harris and Barter, 2015; Morton et al., 2005). At the same time, they can provide importance spaces for community input into policy decisions (Morton et al., 2005). Here, future research could help to identify the social networks and different types of social capital operating in rural communities with a view to informing how these could be strengthened to support collective action towards food security (see Saint Ville, Hickey, Locher, and Phillip, 2016).

Third, alongside investing in the social structure of rural communities, policy change is needed to address income-related food insecurity. Income is the main determinant of household food security (McIntyre, 2003). National data show that households reliant on government benefits, including Employment Insurance, are more likely to experience food insecurity (Statistics Canada, 2015). Further, populations in rural communities are aging, with research indicating that single-member households on Canada's public pensions, including Old Age Security and Canada Pension Plan, often lack the necessary funds for a nutritious diet (Green, Johnson, and Blum, 2008). Addressing income-related food insecurity will require ensuring adequate living wages, including access to Employment Insurance; indexing the personal allowance portion of income assistance rates to reflect the actual cost of a nutritious diet; and reviewing public pension systems to ensure income adequacy (Nova Scotia Participatory Food Costing Projects, 2010).

Lastly, the broader restructuring of rural regions needs to be considered with a view to supporting food security. For example, the number of fish plants declined in the province from 221 plants in 1990 to only 117 in 2005 (Government of Newfoundland and Labrador, 2006; Schrank, 2005). In 2010, there were 121 fish processor licences, including only 10 in-province fish retail processing licences and 27 active buyer's licences. Downsizing in the fishing industry is continuing in the province, including reductions in the number of registered fish harvesters in the province by almost 50 per cent from 17,118 in 1997 to 8,717 in 2014 (Professional Fish Harvesters Certification Board data, personal communication, Mark Dolomount, Aug. 2015). The rate and consequences of industry downsizing will vary across regions and communities and will have consequences not only for employment and household income to purchase food but also for local seafood purchasing options for harvesters. Future research on coastal communities needs to look at the distribution of small-scale fisheries and at fish plants as parts of local food systems infrastructure, as well as at trends and opportunities for new product development. The government of Newfoundland and Labrador recently changed the regulations to allow fish harvesters to sell a portion of their catch directly to consumers from their vessels. This legalizes a practice that already existed but will also probably encourage it in the future so long as small-scale fisheries continue to operate in rural areas. Future research could examine the impacts of this policy change on local consumption of seafood and the potential for new direct marketing opportunities to support local fish markets (as part of farmers' markets), and community-supported fisheries (see Lowitt, Mount, Khan, and Clement, forthcoming; Lowitt, 2011).

ACKNOWLEDGEMENTS

The authors acknowledge funding support from the Social Sciences and Humanities Research Council (SSHRC) and the Community-University Research for Recovery Alliance project at Memorial University.

REFERENCES

Berg, B. L. 2004. *Qualitative Research Methods for the Social Sciences*, 5th ed. Boston: Pearson Education.

Canadian Institute for Health Information. 2006. *How Healthy Are Rural Canadians? An Assessment of Their Health Status and Health Determinants*. At: http://www.phac-spc.gc.ca/publicat/rural06/pdf/rural_canadians_2006_report_e.pdf.

Canadian Rural Revitalization Foundation. 2015. *The State of Rural Canada Report*. At: http://sorc.crrf.ca/.

Conference Board of Canada. 2015. "Per capita income." At: http://www.conferenceboard.ca/hcp/details/economy/income-per-capita.aspx.

Cummins, S. 2007. "Neighbourhood food environment and diet — Time for improved conceptual models?" *Preventive Medicine* 44, 3: 196–97. doi:10.1016/j.ypmed.2006.11.018.

Dean, W. R., and J. R. Sharkey. 2011. "Rural and urban differences in the associations between characteristics of the community food environment and fruit and vegetable intake." *Journal of Nutrition Education and Behaviour* 43, 6: 426–33. doi:10.1016/jneb.2010.07.001.

Dewalt, K. M., and B. R. Dewalt. 2002. *Participant Observation: A Guide for Fieldworkers*. Walnut Creek, Calif.: Altamira Press.

Government of Newfoundland and Labrador. 2006. *Fishing Industry Renewal — A Discussion Paper*. At: http://www.fishaq.gov.nl.ca/industry_renewal/pdf/fo114_5_2006e.pdf.

Green, R., P. Williams, C. Johnson, and I. Blum. 2008. "Can Canadian seniors on public pensions afford a nutritious diet?" *Canadian Journal on Aging* 27, 1: 69–79. doi:10.3138/cja.27.1.69.

Harris, C., and G. Barter. 2015. "Pedagogies that explore food practices: Resetting the table for improved eco-justice." *Australian Journal of Environmental Education* 31, 1: 12–33. doi:10.1017/aee.2015.12.

Health Canada. 2015. "Measuring the food environment in Canada." Ottawa. At: http://www.hc-sc.gc.ca/fn-an/nutrition/pol/som-ex-sum-environ-eng.php.

Hendrickson, M., J. Wilkinson, W. Herrernan, and R. Gronski. 2008. "The global food system and nodes of power." Social Science Research Network. At: http://papers.ssrn.com/sol3/papers.cfm?abstract_id=1337273.

Kuzel, A. J. 1992. "Sampling in qualitative inquiry." In B. Crabtree and W. Miller, eds., *Doing Qualitative Research*, 31–44. London: Sage.

Lawn, J., H. Robbins, and F. Hill. 1998. "Food affordability in air stage communities." *International Journal of Circumpolar Health* 57 (Supplement 1): 182–88.

Lowitt, K. 2009. *A Community Food Security Assessment of the Bonne Bay Region*. CURRA, Memorial University of Newfoundland. At: http://www.curra.ca/documents/CFS%20Assessment%20Report_%20Final_Oct%2009.pdf.

———. 2011. "Examining the foundations for fisheries–tourism synergies and increased local seafood consumption in the Bonne Bay region of Newfoundland." CURRA, Memorial University of Newfoundland. At: http://www.curra.ca/documents/CURRA_Fisheries-Tourism_Full_Report_Nov_2011_Final%20revised.pdf.

———. 2013a. "Examining fisheries contributions to community food security: Findings from a household seafood consumption survey on the west coast of Newfoundland." *Journal of Hunger and Environmental Nutrition* 8, 2: 221–41. doi:/10.1080/19320248.2013.786668.

———. 2013b. "An examination of rural and coastal foodscapes: Insights for the study of community food security and sustainable food systems." Doctoral dissertation, Memorial University of Newfoundland. At: http://www.curra.ca/documents/PhD_Thesis_Kristen%20Lowitt_Final%20version_Sept%202013.pdf.

———. 2014. "A coastal foodscape: Examining the relationships between changing fisheries and community food security on the west coast of Newfoundland." *Ecology and Society* 19, 3: 48. doi:10.5751/ES-06498-190348.

———, P. Mount, A. Khan, and C. Clement. Forthcoming. "Governing challenges for local food systems: Emerging lessons from agriculture and fisheries." In *Conversations in Food Studies.* Winnipeg: University of Manitoba Press.

Mannion, J. 1977. "Settlers and traders in western Newfoundland." In J. Mannion, *The Peopling of Newfoundland: Essays in Historical Geography,* 234–78. St. John's: Memorial University of Newfoundland.

Martin, K., B. Rogers, J. Cook, and H. Joseph. 2004. "Social capital is associated with decreased risk of hunger." *Social Science & Medicine* 58, 12: 2645–54. doi:10.1016/j.socscimed.2003.09.026.

McIntyre, L. 2003. "Food security: More than a determinant of health." *Policy Options* 24, 3: 46–51.

Moazzami, B. 2014. "Strengthening rural Canada: Fewer and older – the coming population and demographic challenges in Newfoundland and Labrador." Essential Skills Ontario and Literacy Newfoundland and Labrador. At: http://strengtheningruralcanada.ca/file/Strengthening-Rural-Canada-Fewer-and-Older-The-Coming-Demographic-Crisis-in-Rural-Newfoundland-and-Labrador1.pdf.

Morton, L., E. Bitto, M. Oakland, and M. Sand. 2005. "Solving the problems of Iowa food deserts: Food insecurity and civic structure." *Rural Sociology* 70, 1: 94–112. doi:10.1526/0036011053294628.

Newfoundland and Labrador Department of Health and Community Services. 2011. "Newfoundland and Labrador nutritious food basket." At: http://www.foodsecuritynews.com/nutritious-food-basket.html.

Newfoundland and Labrador Statistics Agency. 2015a. "Community accounts." At: http://nl.communityaccounts.ca/acct_wellbeing_rank.asp?_=vcTKnZOWgYmgzLOXS6rCvKGej7KclrR9wI.vvZWli73Haai-xcaIpZjMj3.uxcqiq7xiy6mcu2d3wdGniq9vcINaW72mvbeXjMijsXddqrWypm8_.

———. 2015b. "Community accounts." At: http://nl.communityaccounts.ca/acct_wellbeing_rank.asp?_=vcTKnZOWgYmgzLOXS6rCvKGej7hlTMi-soquu5CnnJfKkZ3Gu7.PrpaEi8i4zL6Yh8qUv59lfJHD0sCucnpjXWyQmsWourWhjL57aLSpnqt7.

Nova Scotia Participatory Food Costing Project. 2010. *Can Nova Scotians Afford to Eat Healthy? Report on 2010 Participatory Food Costing.* Participation Action Research and Training Centre on Food Security. Halifax: Mount Saint Vincent University. At: http://www.foodsecurityresearchcentre.ca/storage/docs/food-costing/Food%20Final%202010_Food%20Final.pdf.

Ommer, R. 2004. "Informal rural economies in history." *Labour* 53L 127–57.

———, N. J. Turner, M. MacDonald, and P. Sinclair. 2007. "Food security and the informal economy." In Christopher C. Parish, Nancy J. Turner, and Shirley M. Solberg, eds., *Resetting the Kitchen Table: Food Security, Culture, Health and Resilience in Coastal Communities*, 115–28. New York: Nova Science.

Omohundro, J. 1994. *Rough Food: The Seasons of Subsistence in Northern Newfoundland.* St. John's: ISER Books.

Penney, T., H. Brown, E. Maguire, I. Kuhn, and P. Monsivais. 2015. "Local food environment interventions to improve healthy food choice in adults: A systematic review and realist synthesis protocol." *BMJ Open* 5, 4. doi:10.1136/bmjopen-2014-007161 .

Pigott, K. 2009. *Cultivating Community Partnerships to Improve Access to Healthy Food.* CHNET-Works Fireside Chat, June. Waterloo, Ont.

Pouliot, N., and A. Hamelin. 2009. "Disparities in fruit and vegetable supply: A potential health concern in the greater QuébecCity area." *Public Health Nutrition* 12, 11: 2051–59. doi:10.1017/S1368980009005369.

Saint Ville, A., G. Hickey, U. Locher, and L. Phillip. 2016. "Exploring the role of social capital in influencing knowledge flows and innovation in smallholder farming communities in the Caribbean." *Food Security* 8: 535. doi:10.1007/s12571-016-0581-y.

Schrank, W. 2005. "The Newfoundland fishery: Ten years after the moratorium." *Marine Policy* 29, 5: 407–20.

Sloane, D. C.., A. L. Diamant, L. B. Lewis, A. K. Yancey, G. Flynn, L. M. Nascimento, . . . M. R. Coursineau. 2003. "Improving the nutritional resource environment for healthy living through community-based participatory research." *Journal of General Internal Medicine* 18, 7: 568–75.

Smith, D. M., S. Cummins, M. Taylor, J. Dawson, D. Marshall, L. Sparks, and A. S. Anderson. 2010. "Neighbourhood food environment and area deprivation: Spatial accessibility to grocery stores selling fruits and vegetables in rural and urban settings." *International Journal of Epidemiology* 39: 277–84. doi:10.1093/ije/dyp221.

Statistics Canada. 2013. "Living arrangements of seniors." At: http://www12.statcan.ca/census-recensement/2011/as-sa/98-312-x/98-312-x2011003_4-eng.cfm.

———. 2015. "Food insecurity in Canada." At: http://www.statcan.gc.ca/pub/82-624-x/2015001/article/14138-eng.pdf.

Travers, K., A. Cogdon, W. McDonald, C. Wright, B. Anderson, and D. MacLean. 1997. "Availability and cost of heart healthy dietary changes in Nova Scotia." *Journal of Canadian Dietetic Association* 58: 176–83.

Twells, L., D. Gregory, J. Reddigan, and W. Midodzi. 2014. "Current and predicted prevalence of obesity in Canada: A trend analysis." *Canadian Medical Association Journal* 2, 1:, E18–E26.

Walker, R. E., C. R. Keane, and J. G. Burke. 2010. "Disparities and access to healthy food in the United States: A review of food deserts literature." *Health and Place* 16, 5: 876–84. doi:10.106/j.healthplace.2010.04.013.

Part III

Sustainable Fisheries, Aquaculture, Apiculture, and Agriculture

Capelin! © Dave Howells

Bringing Seafood into Food Regime Analysis: The Global Political Economy of Newfoundland and Labrador Fisheries

Paul Foley & Charles Mather

INTRODUCTION

For centuries, the harvesting, processing, and international trading of fish by people living in coastal communities have fundamentally shaped Newfoundland and Labrador society. The fishing industry was dominated by cod production and exportation for nearly 400 years, and by the production and export of shellfish in 1990s and early 2000s in the wake of the historic collapse of cod and other groundfish. Current fundamental ecological changes in the Northwest Atlantic Ocean suggest that cod and other groundfish are making a comeback (Mather, 2013; Rose and Rowe, 2015). This has prompted industry players, government officials, and academics to begin work preparing for another challenging transformation in the fishing industry and the effects on people and communities that rely on it. Despite the tragedy of the cod collapse and the legacy of ecological and social problems that came with it, the fishing industry clearly will continue to remain the economic foundation of most rural coastal areas of the province for the foreseeable future (Dean, Wareham, and Walters, 2001: 5; Neis, Ommer, and Hall, 2014). Moreover, recent studies suggest fisheries offer potential for creating space for grassroots and community-based responses to food security and sovereignty challenges in the province. This chapter suggests that a critical food regimes perspective can be used to identify deep, structural, political-economic challenges and contradictions that may guide the transformation of Newfoundland and Labrador fisheries into more sustainable and socially just directions.

Although the Newfoundland and Labrador fishery is one of the most researched fishing areas in the world, few studies have examined it explicitly from a food regime (or even a food studies) perspective. Neis's (1991) work on the restructuring of the cod sector in the Northwest Atlantic in the 1970s and 1980s provides insight into understanding fisheries as food systems. She explores changes in groundfish production and consumption through the lens of the French regulation school, and traces a shift in the regime of accumulation from Fordist production to flexible specialization. Although Neis's work was not framed explicitly by food regime theory, her emphasis on changes in the consumption of fish and her attention to household–factory relations provides important insights that are relevant to food regime theory. In addition, the emphasis Neis places on ecological processes, which play an important role in shaping the shift in the regime of accumulation, aligns well with recent debates on food regime theory (e.g., Campbell, 2009). Reade Davis (2014) provides a more recent analysis of changes in the province's fishery sector associated with debates on the potential for the return of cod after more than two decades of moratoria. His work is also sensitive to cod as food, which he situates within a broader global whitefish market. This market is shaped by new developments in aquaculture and by the seamless substitutability of one whitefish species for another. As Davis (2014: 717) explains:

> Growing openness of trade has resulted in a situation in which commercial fishing and fish processing operations in Newfoundland must now struggle to compete for market share with producers of similar products around the world, many of which are able to process cod or substitute species at a lower cost.

These challenging market conditions for seafood explain why many in Newfoundland and Labrador are worried about a regime shift from shellfish back to cod. Our paper builds on and contributes to this research base that positions fish as food in the politics of global production and consumption relations.

The limited number of studies on Newfoundland and Labrador fisheries that situate fish in global food regimes is matched by the almost total absence of fish within food regime theory. Indeed, despite being recognized as "among the most influential framing tools in agrofood studies" (Magnan, 2012: 372), we were unable to find a single study that analyzes fish using food regime theory. Part of the explanation for the absence of studies on fish is the strong agrarian base of food regime analysis. As Bernstein (2015) explains, food regime theory emerged out of dissatisfaction with 1970s and 1980s agrarian political economy that failed to engage with the increasingly globalized nature of food production and consumption (also see Campbell and Dixon, 2009). As a result, food regime theory has been applied to *agricultural* production rather than food production more broadly. An important goal of this paper is to begin exploring the circumstances under which fish can be understood through food regime theory.

Food regime analysis was developed in the late 1980s by social scientists as an approach to analyzing agricultural change in a context of national and international power relations that shape production and consumption. After sparking some debate in the 1990s, as well a notable critique (e.g., Goodman and Watts, 1994), the analytical approach has experienced resurgence since the late 2000s (Campbell and Dixon, 2009; McMichael, 2009; Magnan, 2012; Bernstein, 2015). Proponents of the approach argue that key processes of global food production and consumption can be explained by understanding historical and political actors, relations, institutions, and structures. Three food regimes span from colonialism to the contemporary period of "green capitalism." These reflect the colonial-diasporic food regime from about 1870 to 1914, the mercantile-industrial food regime from about 1945 to 1970, and the corporate or corporate-environmental food regime since about the 1980s, although there is no agreement on a strict periodization. In the chapter, approximations are made in the dates to accommodate scholarly differences in opinion (see, for example, differences between Friedmann, 2005: 227; Friedmann, 2009: 335; and McMichael 2009: 141).

The food regime approach is instructive not only because it enables a social scientific framework of fisheries as food systems, but also because it enables an analysis of global relations, in which Newfoundland and Labrador seafood production is embedded. A global analysis is appropriate given that, historically, the vast majority of the seafood produced in the province has been exported and the export-dependent nature of the industry continues. In 2014, approximately 90 per cent of the province's seafood was exported to more than 40 countries, with export values reaching $950 million the following year (DFA, 2015: 23; DFA, 2016).

Figure 9.1 shows the top market destinations of Newfoundland and Labrador seafood products based on export value; these markets received nearly 80 per cent of total exports by value.

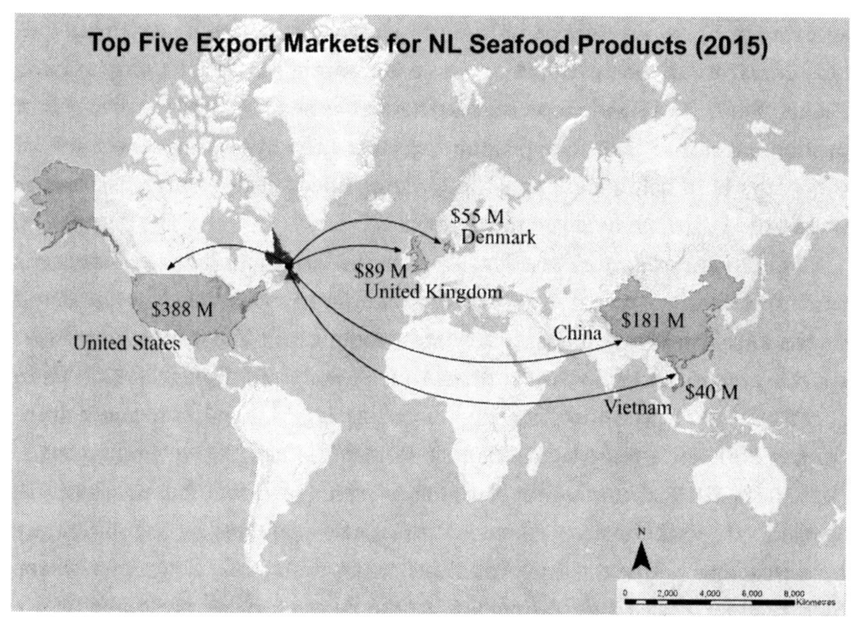

Figure 9.1. Top five export markets by value in 2015 for Newfoundland and Labrador seafood products. (Cartography by Myron King) *Source: Data from Newfoundland and Labrador Statistics Agency.*

The need for a global scale analytical approach is made clear by the deep integration of Newfoundland and Labrador seafood production into the global economy. The food regime approach is similar to staples theory, in that both have a world-historical perspective that emphasizes the international political economy of commodity production and consumption. The food regime approach also echoes studies that link Newfoundland and Labrador fisheries to the pressures of the wider commercial world, as well as studies that more explicitly analyze the province's fisheries in the context of globalization (Newell and Ommer, 1999; Neis et al., 2005; Bavington and Kay, 2007). However, food regime analysis is explicitly food-centric; it explores ways in which food production and consumption are shaped by, and in turn shape, the global political economy. In this context, the two central questions asked in this chapter are: How can food regime analysis help us understand the trajectory of seafood production and trade in the province? And, what lessons does the Newfoundland and Labrador fish sector provide for advancing scholarly debates about food regime analysis?

In answering these questions, two contributions to the literature emerge. First, this chapter adds to the Newfoundland and Labrador food studies literature by incorporating food regime analysis to explain transformations in the province's fisheries. Second, the chapter contributes to the food regime analysis literature by expanding fisheries and seafood into the food regime approach, which is overwhelmingly informed by research on agriculture. Although the food regime literature tends to adopt an explicitly normative stance (against corporate agribusiness), the purpose of this chapter is not to selectively gather and deploy empirical evidence to verify food regime analysis. Instead, a case study of Newfoundland and Labrador fisheries is used to investigate the questions posed by the analytical framework of food regime analysis, and those specifically relevant to the current corporate-environmental food regime. The remainder of the chapter is structured as follows: first, we provide an overview of food regime theory and identify key issues in the contemporary debate about food regimes. Second, we analyze Newfoundland and Labrador fisheries within the lens of food regime theory, specifically organized around the three food regimes. Third, we conclude the chapter by returning to our two core questions about the value of food regime analysis for understanding Newfoundland fisheries and the value of fisheries for food regime analysis.

FOOD REGIME ANALYSIS

Harriett Friedmann first formulated the "food regime" concept nearly three decades ago (Friedmann, 1987), with the first systematic formulation by Friedmann and McMichael (1989). The approach explores the role of agriculture and food in the development of the capitalist world economy and in the international state system since the 1870s. The approach links forms of food production and consumption with geopolitical historical trends and transformations. While significant international trade in food commodities predated the 1870s, the distinguishing feature that established world food regimes was the establishment of "a world price for staple foods" (McMichael, 2013: 24). Friedmann and McMichael (1989) identified two food regimes, the first from 1870 to 1914 during British hegemony in the world economy, and a second from 1945 to 1973 during US hegemony in the world economy. A third food regime since the 1980s under corporate neo-liberal globalization appears to be "in the making," although there are debates about whether this represents a stable third regime (Friedmann, 2009). Food regime analysis has included a strong normative dimension that challenges and critiques the dominant food regimes at the centre of analysis.

A key contribution of food regime analysis is that it identifies how food production and consumption are embedded in, and shaped by, global power relations that include colonialism and imperialism (McMichael, 2009). Although early literature contained an emphasis on the structural dynamics of periodization, with state and capital as the dominant analytical categories, recent engagement of the food regime concept is far more attentive to transformation and uncertainty. New categories such as social movements receive more analytical attention (McMichael, 2013: 7; Bernstein, 2015). Bernstein (2015: 2) suggests eight analytical dimensions of food regime analysis that lead to three key research questions:

1. Where and how is food produced in the international economy of capitalism?
2. Where and how is what type of food consumed, and by whom?
3. What are the social and ecological effects of the international relations of food production?

A key aspect of food regime theory involves identifying distinct periods in the global food system from the late nineteenth century to the present. The current food regime theory literature overwhelmingly focuses on identifying the dynamics of the third food regime, while remaining open to the possibility that this regime has not yet stabilized. Drawing from a range of different commodities and geographical contexts, researchers have pointed to the contours of what might be an emerging third food regime. Significantly, Harriett Friedmann and Philip McMichael — originally responsible for food regime theory — differ on what they see as the key elements of the third food regime. Friedmann (2005: 229) has proposed the idea of a corporate-environmental regime shaped by a "standoff" between agrifood corporations and environmental and other movements:

> a new round of accumulation appears to be emerging in the agrifood sector, based on selective appropriation of demands by environmental movements and including issues pressed by fair trade, consumer health, and animal welfare activists.

In other words, the third food regime is characterized and shaped by large corporations responding to the pressures of environmental and social activists through corporate control of global food chains. In these privately regulated global food chains, large corporate food companies are able to impose new conditions on food producers considered more desirable by environmental and social activists. An outstanding example of this compromise in the fish sector is Marine Stewardship Council (MSC) certification. The MSC emerged as a sustainability certification mechanism through a joint effort by Unilever, one of the world's largest food producers, and the World Wide Fund for Nature (WWF). It provides an excellent illustration of what Friedmann calls a corporate-environmental regime, or green capitalism. While the MSC represents an important example, Friedmann claims that this form of compromise (which may be viewed as a standoff between powerful corporations and environmental and other movements) represents a driver, or potential pivot of the third food regime. Consumers may benefit from this standoff, but the outcomes for small-scale producers, particularly in the Global South, may not be as positive. The imposition of new and stringent standards for northern consumers is likely to "deepen longstanding processes that dispossess and marginalize peasants and agrarian communities" (Friedmann, 2005: 257).

McMichael's (2009) analysis of the third food regime stresses the power that corporations have to shape the global food system in a world increasingly influenced by neo-liberal policies. In contrast with the Friedmann formulation, there is no meaningful standoff between environmental movements and agrifood corporations. Instead, agrifood corporations have become increasingly powerful in a global political economy that has become liberalized and where nation-state functions have been largely privatized. In this way, McMichael's formulation situates the third food regime squarely within a process of neo-liberal globalization.

Bernstein (2015) has broken down McMichael's formulation into four key processes operating at the global scale. First, agriculture has become deregulated, which has led to the corporatization of agro-exports and the casualization of farm labour. In this liberalized economic environment, world food prices do not reflect costs. This, in turn, has led to the increasing vulnerability of farmers in many parts of the world. Second, the third food regime is characterized according to "accumulation by dispossession," a term David Harvey uses to describe the contemporary nature of capitalism, which situates key processes in agrifood and land-based restructuring. Dispossession is exemplified by the rapid progression of land grabs in the Global South that shift production from food into staples destined for animal feeds and biofuels. It may have potentially devastating implications for food security.

Bernstein highlights the "ecological impacts" of the contemporary food regime, in that food production is destructive to soils and contributes in significant ways to climate change through greenhouse gas emissions. Other ecological impacts of the third food regime include those associated with genetically modified organisms (GMOs) and bio-piracy, both of which are arguably compromising the biological diversity of farming regions across the world, but especially in the developing world. Finally, according to Bernstein, McMichael points to key questions around how food is produced and consumed, and for whom? In answering this question, he identifies two contrasting approaches to food production and consumption: the generic production of "food from nowhere" for poor consumers in the Global South, and a new progressive "place-based" alternative that he describes as "food from somewhere." This alternative vision of food production and consumption is where McMichael sees the potential for strong and ecologically sound alternatives to the corporate food regime. These alternatives are currently being pursued by agrarian and land-based social movements, such as La Vía Campesina food sovereignty movement (discussed in the Introduction to this volume), that provide the foundation for a radical alternative to the current globalized food system. As Bernstein has argued, McMichael's formulation "demonstrates how definitive and, at the same time, how encompassing the arguments are" (Bernstein, 2015: 16).

While many authors have explored the contours and drivers of an emerging third regime, most contributors to the debates are reluctant to claim that a third regime is stable or can be definitively specified. This hesitancy, in part, is a response to the difficulty of analyzing and describing a global food regime that is in considerable dynamic flux. As Le Heron and Lewis argue, the concern with identifying a stable food regime is that it will obscure the "diversity and fluidity of the relations, actors, metrics, translations and contexts" (Le Heron and Lewis, 2009: 346). Friedmann (2009: 335) has been most vocal on this issue, suggesting that we need to ask a series of critical questions of an emerging regime before it can be considered stable:

> Are the relations stable and durable, is there a central pivot (as there was in the first and second food regime), and are there converging interests between states, corporations, producers and consumers? And what are the institutional foundations for the regime?

The effect of this thinking on food regime theory has been profound. It has led researchers to shift attention away from imposing strict periodic time frames for food regimes. Instead, food regime theory is used as a lens or framework to understand the global food system. As McMichael (2009: 148) writes:

> The "food regime" can be considered to be simply an analytical device to pose specific questions about the structuring processes in the global political-economy, and/or global food relations, at any particular moment. Here the "food regime" is not so much an episodic structure, or set of rules, but becomes a method of analysis.

NEWFOUNDLAND AND LABRADOR FISHERIES IN THE DEVELOPMENT OF A GLOBAL FOOD REGIME

In the first section of the chapter, we traced the origins of food regime theory and examined debates and changes in focus within the literature. Perhaps the most significant development has been a shift away from attempts to detail the precise contours of a food regime, particularly within the third food regime. Instead, researchers are now using food regime theory as an analytical framework and methodological orientation to ask questions about the relations of food production, food consumption, and global capitalist accumulation. In other words, food regime theory has become a framework/ lens through which to analyze the global food system (McMichael, 2009). This section examines the historical development of fisheries in Newfoundland and Labrador set against the backdrop of the food regime periodization as a heuristic device.

The First Food Regime

The first food regime, which in McMichael's formulation lasted from 1870 to about 1940, is also called "the settler-colonial food regime" or "the colonial-diasporic food regime." During this time fully commercial farming was created in settler territories, such as Canada, the US, and Australia, based on availability of family labour (Friedmann, 2004, 2005: 235). The evolution of European fishing in the province was embedded in similar patterns of the settler-colonial food regime, with British settlement and commerce emerging as the dominant pattern in the nineteenth century when the first food regime became consolidated.

The emergence of household commercial fishing was a defining feature in settler-colonial Newfoundland and Labrador fisheries. During the sixteenth century, the predominantly seasonal migration to North America's oldest industry, the cod fishery, began to see a new pattern of unregulated English settlements. These settlements were established and sustained by the exchange of cod for wine, in the broader context of the early development of a modern consumer economy mediated by metropolitan merchants and international competition (Pope, 2004). By the early eighteenth century, the migratory fishery declined as the white settler population of Newfoundland and Labrador grew. Permanent settlement resulted in the establishment of household production, which dominated rural development well into the late twentieth century (Ommer, 2002: 25). The cod fishery remained arguably the most important dimension of European commercial activity in North America for centuries (Pope, 2004: 13–14) and cod was clearly the historic staple of the North Atlantic political economy (Innis, 1940).

The role of the state was limited but not absent during this period. Acts of the British Parliament were passed for the control or regulation of the Newfoundland fishery as early as 1788, but formal resource management ideas and practices did not emerge until the late nineteenth century. In 1888, during the "Responsible Government" period, a Fisheries Royal Commission proposed establishing a centralized bureau devoted to fisheries research and assistance in order to address the perceived problem of integrated resource planning and development. An independent Fisheries Commission was subsequently established in 1889, which conducted research, prepared reports, proposed rules and regulations, and made suggestions for the proper curing of fish. The first department devoted fully to fisheries was established in 1898 as the Department of Marine and Fisheries. From about 1888, during the Responsible Government period, until 1949 when Newfoundland and Labrador joined Canada, the primary concern of the province's fisheries administration involved the control and development of production and marketing in the salt fish industry. The objective of government was to maximize export earnings from the fishery so that surplus labour could be accommodated. This included taking steps to develop new processing activities (fish freezing), improve quality of products, and modernize fishing vessels and gears. No consideration was given to limiting the catches of any species or the numbers of people participating in the overall fishery (Vardy and Dunne, 2003: 106–07).

In addition to its integration in the international state and capitalist system, Newfoundland and Labrador fisheries were characterized by regimes of accumulation containing non-capitalist domains during the early development (Neis, 1991: 152). The most notable non-capitalist social relations included household production, as well as household reliance on subsistence agriculture in many communities. By the middle of the twentieth century, however, state and capital facilitated a process of industrialization that began to replace the household production-based salt cod fishery with increasingly industrialized and mechanized deep-sea trawlers and frozen fish plants that employed paid labour (Ommer and Sinclair, 1999).

The Second Food Regime

The second food regime, lasting from about 1945 to 1970, is characterized by two patterns, mercantilism and industrialism. Mercantilism resulted in "national agricultures," which were systems characterized by price supports and export subsidies that, combined with new comprehensive foreign aid regimes, led to competitive dumping and trade competition (Friedmann, 1993: 32). The industrialization of agriculture was characterized by the increasing role of global agribusiness in food production and distribution, as well as greater mechanization and increasingly common use of chemicals.

The industrialization of Newfoundland and Labrador fisheries coincided with the emergence of the second food regime and a shift in markets from Great Britain, which dominated the first food regime, to the US, which dominated the second food regime. During World War II, the majority of exports of the province's frozen fish products shifted from Great Britain to the US. Advances in "quick-freezing" technology, the rise of public cold storage plants, and the expansion of home refrigerators transformed the US food industry and helped alter social relations of production in food-producing regions. Demand for frozen food products, especially in the US, fuelled the expansion of the fishery (Wright, 1997: 728). According to Neis, this transformation was rooted in the Fordist regime's relationship to nature and was influenced by external, largely US mass markets, including the fish stick revolution (Neis, 1991). The state, under the Commission of Government in the 1930s and the 1940s and subsequently with the federal and provincial governments following Confederation, played a major role in facilitating these transformations (Wright, 1998).[2] While Confederation changed the constitutional status of responsibility for how fisheries were managed, the direction of fisheries development continued on a path of industrialization and North American orientation that had begun with the advent of World War I (Wright, 1998).

The development of infrastructure for frozen food marketing that characterized the rise of Fordism in North America and Europe also altered the structure of the fishery (Neis, 1991: 147). According to Miriam Wright, "massive industrialization" was "the most striking" aspect of the province's fishery in the decades following World War II (Wright, 1997: 727). Industrialization began in earnest in the 1940s, transforming the fishery from a salt fish trade based on merchant credit to a vertically integrated frozen fish industry. The fishery transformed from a household-based production of salt fish to the mass production of semi-processed blocks of fish fillets. The production of dried salt fish by individual families that was exported to Southern Europe, South America, and the Caribbean during the first food regime declined, while the production of frozen cod fillets and blocks that were exported to the US rapidly expanded. Newfoundland and Labrador firms did not control block markets, which were instead controlled by US food conglomerates (Neis, 1991: 161). The inshore, household fishing society based on merchant credit was transformed into a cash-based economy with fishers selling their catch to vertically integrated harvesting/processing firms, while other members of fishing communities went to work at frozen fish plants or on offshore trawlers owned by the vertically integrated fishing companies (Wright, 1997: 728). Fish harvested both by corporate-owned fresh fish trawlers and by small, artisanal inshore fishers was processed in onshore plants using freezing and filleting technologies developed by factory-freezer trawlers (FFTs) to produce fish blocks for the fish stick market (Neis, 1991: 161).

The Northwest Atlantic fishery rapidly expanded after World War II and was largely unregulated until 1977. Both international and domestic industrial organization in the fishery facilitated intensive competition among nations for the resource. The inshore, household-based fishery that relied on communities to regulate resource access within a three-mile limit coexisted with the largely unregulated corporate trawlers from Spain, Portugal, France, the Soviet Union and other Eastern bloc countries, Japan, Cuba, West Germany, and Britain (Neis, 1991: 147–48). By 1974, 1,076 Western European and Communist bloc fishing vessels fished off the North American coast (Neis, 1991: 156). The profitability of Canadian and provincially based factory-freezer trawlers was organized around serving mass markets for standardized products, thereby producing more fragmented jobs embedded in standardized technologies and separating workers from their communities for prolonged periods (Neis, 1991: 157). As Neis explains:

> The profitable operation of the FFTs also depended upon the existence of mass markets and on corporate control over those markets. Such markets for fish, created in the early postwar period, were primarily located in the retail sector where mass produced, semi-processed, frozen fish products were sold to housewives, small restaurant owners and firms offering food service to institutions. Control over these markets was based primarily on the relative cheapness of the mass produced commodities and on the use of advertising to create demand for certain brand name products. However, in all three markets consumers could, relatively easily, substitute alternative protein products or cheaper fish products if prices were to increase substantially. (Neis, 1991: 158)

The state played a key, but contradictory, role in this transformation, supporting the frozen fish industry that expanded through the 1980s (Wright, 1997). This reflects the relationship between capital and the state in fisheries development in Newfoundland and Labrador. Prior to 1974, federal and provincial governments responded to accumulation crises (caused by stock depletion) by financing local companies, constructing trawlers, and encouraging multinational food conglomerates to set up operations in the province. The multinationals were unsuccessful and soon were supplanted by regional firms (Neis, 1991: 148).

The state also helped protect the inshore harvesting and processing sectors. In an effort to prevent the corporate consolidation of fishing rights in the inshore, small-scale fishery, the federal Minister of Fisheries and Oceans, Romeo LeBlanc, introduced the Owner-Operator Policy and a Fleet Separation Policy in the 1970s. The Fleet Separation Policy applies to fishing vessels less than 65 feet in length and is designed to prevent the issuance of inshore licences to corporations, including processing companies. The Owner-Operator Policy applies to licence holders using vessels that are less than 65 feet in length and requires licence holders to be present on their vessels and personally fish their licences. The goal of the two policies is to promote distribution of fishing access and to restrict vertically integrated fishing companies from owning smaller vessels and consolidating their control of the commons. These state policies provided some important protections against the growing power of corporations in the Atlantic fishing industry.

In terms of processing, the Canadian Constitution grants provincial legislatures the authority to legislate on property and civil rights. This includes the authority to license processing facilities that operate on provincial lands, as well as the implied power to specify the types of species, the types of products, and the location to which the licence applies (Dunne, 2003: 131). The provincial government first required fish processing plants to obtain operating licences in 1975, but with the crisis of the offshore sector in the early 1980s the government imposed a freeze on the issuance of additional licences (Dunne, 2003). During the period of the second food regime, the provincial government passed the Fish Inspection Act, which has been amended to include Minimum Processing Requirements (MPR) legislation, and introduced legislation mandating Collective-Bargaining Price Setting. The MPR legislation, through the

Fish Inspection Act, required over 30 species caught by provincial fishers to be processed within the province before being exported. Until the mid-1970s, and arguably beyond, the policy approach of the Newfoundland and Labrador government towards the fish processing sector was based on agendas of industrial development and modernization. Public subsidies were distributed to encourage companies based in Newfoundland and Labrador to expand capacity in order to ward off outside competitors, with a loans guarantee program lasting into the late 1980s. These national and provincial development agendas can be understood as the "national-provincial fisheries" counterpart to "national agricultures" identified by Friedmann and McMichael in the second food regime.

In the 1970s and 1980s, the crisis in the Atlantic fishing industry facilitated a process of restructuring that shifted the fishery from a Fordist regime towards a more flexible or post-Fordist regime of accumulation. The process was facilitated by ecological factors (Neis, 1991). In 1977, the Canadian state introduced an exclusive economic management zone (EEZ) extending jurisdiction to 200 miles from shore. New regulation, stock recoveries, government subsidies, and predictions of global food shortages combined to entice investment in the fishery (Neis, 1991: 148). The increased market segmentation for fish products caused by the expansion of the fast-food industry in the US deepened the contradiction between capital accumulation and ecological limits (Neis, 1991: 155). Fish stock depletion, combined with the emergence of a conservation-oriented regulatory regime, and the rapid expansion of the fast-food and food service industries undermined the profitability of the FFT technology. This process occurred in Canada and globally, thereby initiating a general process of global restructuring that arguably continues today. In addition, competition from other fishing countries intensified and the rapid expansion of US chicken production encouraged consumers to switch from fish to chicken, both of which reduced fish prices in the early 1980s. In Newfoundland and Labrador, the changes resulted in the near collapse of the large, vertically integrated fish companies like Fishery Products and Nickerson-National Sea, which were most wedded to Fordism.

The failure to take natural barriers into account "contributed to the crisis in Fordism in the fishery and these have continued to hamper efforts to establish a new effective regime of accumulation, not only in the North Atlantic, but globally as well" (Neis, 1991: 168). Ommer (2002: 21) writes that almost 400 years after the expansion of large-scale transatlantic commercial fisheries in the Northwest Atlantic:

> The resource-based communities of Canada are in crisis. Their ability to survive is now in question, not only because their resource bases have been foundering (for a variety of reasons), but also because the exigencies of deficit reduction and long-term recession have combined to cut away much of the state support that had kept them relatively secure in the past. This is as true in other parts of the world as it is in Canada, and it raises serious questions about the viability of rural communities in the current post-industrial era of globalization.

The Third Food Regime?

The second food regime underwent transformation with the demise of "national agricultures" and the rise of globalization, or neo-liberal globalization. As discussed above, discourse about a possible third food regime is characterized by differences in opinion between McMichael and Friedmann, the former suggesting a "corporate food regime" and the latter suggesting a "corporate-environmental food regime" (Bernstein, 2015). In this section we consider the evidence for both formulations in Newfoundland and Labrador's dynamically changing fish sector.

Corporate food regime

The rapid spread of neo-liberal orthodoxy is the foundation and driver of McMichael's corporate food regime. In practice, this is facilitated by the liberalization of markets and the privatization of state regulations that allow corporate capital to shape food production and consumption. There is strong but uneven evidence that this shift transpired within Newfoundland and Labrador's fishing sector, particularly in light of the global financial crisis in the late 2000s. In this section we explore the efforts to liberalize markets and reduce state protection. We also note the strong resistance to these changes within the fish sector in the province and within Atlantic Canada more broadly. In other words, while the corporate food regime's effects are being felt in the local fish sector, there is considerable resistance to the spread of neo-liberal orthodoxy.

Ecological and social transformations associated with the cod and groundfish closures to some degree militated against corporate consolidation. The industry restructuring that occurred during the post-1992 cod and groundfish moratoria period lessened the role of the offshore fishery. The changing profiles of the inshore harvesting and processing sectors are changing corporate dynamics (Dean, Wareham, and Walters, 2001), where the federal Owner-Operator and Fleet Separation policies prevent formal vertical integration and corporate consolidation of fishing rights. Specifically, the collapse of groundfish and the expansion of the shellfish sector in the post-1992 period saw the role of certain firms with a groundfish dependency decline and the expanded involvement of firms with a major stake in the shellfish sector (Dean, Wareham, and Walters, 2001: 100).

However, during the 1990s and early 2000s, three forms of corporate consolidation linked to globalization had impacts on the province's seafood production. These broader trends of corporate consolidation continue today. The first form consists of transformations in the international division of labour. The growth of China as a manufacturing area has been principally driven by the availability of capital to open manufacturing plants and to provide labour cost advantages (Dunne, 2003: 12). The second example consists of the consolidation of global seafood buyers, particularly large food retailers and food service corporations (Dunne, 2003). According to the province's Fish Processing Policy Review, a single large buyer can now purchase what once had been acquired by as many as five or six independent retail chains. As a result, large buyers now have significant pricing power and influence over production (Dunne, 2003: 13). Consolidation in the food service industry has also resulted in increasing the buyer power over producers. Referencing the US, as Dunne explains:

> Generally, a company will develop relationships with one or two primary suppliers for a given commodity, and then encourage that supplier to pack under their specifications. When this happens, a large portion of a processor's business can come to depend on a single customer, again with ramifications for pricing. (Dunne, 2003: 13)

The consolidation of buyer power results in increased vulnerability for Newfoundland and Labrador producers.

Third, the consolidation of buyers has facilitated mergers among seafood producers. Large global seafood companies have aggressively expanded by either buying other production companies or entering into agreements that expand market coverage. These shifts are prompting calls for Newfoundland and Labrador processors to amalgamate in order to compete with global companies that are able to offer buyers a more stable resource supply (Dunne, 2003; Clift, 2011).

These three forms of corporate consolidation have shaped an ongoing debate in the province about restructuring the province's fishery in ways that challenge the highly distributed and community-based character of the harvesting and processing sectors (which include thousands of small-scale, independently owned harvesting enterprises with dozens of relatively small fish processing plants spread across the province). Various economic crises in the province's fish sector in the 2000s led to new and apparently urgent calls for a restructuring of the province's fish sector. Some analysts pointed to deeper structural problems in the industry's ability to compete globally, which, for them, required a new and concerted effort to restructure the fishing sector. One of the outcomes of a crisis originating in the shrimp fishery starting in 2009 was a process of research and consultation called the "Memorandum of Understanding" (MOU) (Clift, 2011). The provincial government, the major processors, and the Fish, Food and Allied Workers Union (FFAW), an organization representing both processing workers and harvesters, endorsed the MOU process. This process lasted several years, involved extensive research, and consisted of many meetings with relevant stakeholders around the province. The MOU found the industry to be largely "unviable" and recommended drastic downsizing or "rationalization" of both harvesting and processing sectors, which would particularly impact remote fish-dependent coastal communities (Clift, 2011). When the much-anticipated final report was released in 2011, however, it was quickly dismissed by the provincial government, which was unwilling to commit to the large projected costs associated with restructuring — as much as $750 million — and was clearly uncomfortable with the social ramifications likely to follow the MOU's recommendations (McLeod, 2011).

The MOU's recommendation to restructure Newfoundland and Labrador's fishery to support vertically integrated corporate structures clearly failed. Yet, the discourse of rationalization and restructuring to better align with vertically integrated market logic continues to be articulated locally and nationally. Within the province, some academics and policy analysts continue to call for a deep rationalization of the fishing sector in order to become internationally competitive in global markets. A recent analysis co-authored by the chair of the MOU process argues for a new "three R" approach: rationalization, restructuring, and reorientation. The authors note that Newfoundland and Labrador's fish sector has recovered remarkably well from the financial crisis of the late 2000s, and that there has been ongoing rationalization in harvesting, processing, and employment. Despite these developments and the apparent financial resilience of the local fish sector, the authors argue that the level of restructuring "at this pace is insufficient to allow the industry to achieve the type of production efficiencies and financial performance ... required to allow it to remain competitive in the international marketplace" (Clift and Cooper, 2014: 38). Clift and Cooper (2014: 40) add that the pace of restructuring is "insufficient to allow local processing firms to compete more effectively against the well-financed, vertically integrated seafood companies that operate in Western Europe." For those promoting the neo-liberal globalization agenda, financial success in global markets does not seem to dampen the call for deeper restructuring and rationalization, with the attendant social impacts on harvesters, processing workers, and coastal communities.

At the national scale there have been similar efforts to transform the fishery through the language of rationalization and restructuring. In 2012, the Department of Fisheries and Oceans (DFO) released a discussion document called "The Future of Canada's Commercial Fisheries." Similar to Newfoundland and Labrador's MOU, this document called for a strong focus on profitability and global competitiveness. Absent from this discussion, as progressive policy analysts noted, was any commitment to the policies that would have protected coastal communities and small-scale harvesters from restructuring and dispossession of fish resources. Notably absent were the Fleet Separation and Owner-Operator policies, which prevent vertical integration and ensure the viability of an independent fish harvesting sector. There was also no mention of long-standing policies of adjacency and coastal community sustainability, which have shaped resource allocation policies especially in Atlantic Canada (Foley, Mather, and Neis, 2015; Foley and Mather, 2015).

The response by fisheries organizations representing independent harvesters was vehement. The various harvester organizations mobilized into a new national advocacy organization called the Canadian Independent Fish Harvesters' Federation (*The Telegram*, 2013). The new group successfully convinced the federal government to reassert its commitment to social protection policies such as the federal Owner-Operator Policy and the Fleet Separation Policy, despite the strongly neo-liberal orientation of the then-governing Conservative Party of Canada.

Although local and national efforts to push a corporate food regime for the province's fisheries may have stalled in some policy arenas, other policy developments are set to enhance corporate interests and to increase power within the sector. The erosion of policies aimed at protecting food production employment and infrastructure in coastal communities may be exacerbated by a bilateral trade agreement designed to bring about more dramatic market liberalization and deregulation of state functions. The Canada-European Union Comprehensive Economic and Trade Agreement (CETA), ratified in 2017, has been lauded by both the provincial government and the FFAW as an important new opportunity for the province's fish (export) sector. CETA effectively eliminates all EU tariffs for Newfoundland and Labrador fish exports, which will make it easier to export higher-value processed fish, as opposed to raw product. As Song and Chuenpagdee (2015: 448) write: "CETA is expected to create new opportunities for provincial seafood producers with respect to secondary processing, brand building and marketing strategies to deliver high-quality, premium products to EU markets."

Yet, according to an important and detailed analysis by Scott Sinclair of the Canadian Centre for Policy Alternatives (2013), the new trade deal may challenge existing national and provincial regulations aimed at protecting small-scale harvesters and onshore processing jobs that are important to the sustainability of coastal communities. Indeed, Sinclair has argued that long-standing policy commitments of adjacency and historical dependence may be challenged under a bilateral trade agreement that sees Canadian and provincial regulations as contrary to the principle of "national treatment," a provision that ensures that Canadian and EU commercial interests are treated equally. Regulations that protect domestic fishery interests — including minimum processing and fleet separation provisions — may be seen as discriminatory by EU corporations and under international law.

Sinclair (2013: 5) is concerned that CETA may lead to the erosion of policies that "help spread the benefits of the fishery more widely among smaller, independent fishers and coastal communities." Indeed, the government of Newfoundland and Labrador agreed to phase out Minimum Processing Requirements (MPRs), the provincial regulation that requires fishing companies to process fish in the province rather than export it for processing overseas, in response to federal pressure to conform to CETA principles. While the province has been reluctant to entertain exceptions to the MPR regulations in recent decades, large companies have also found ways to leverage promises of secure processing jobs in return for temporary or even permanent exemptions from MPR rules. Not surprisingly, these actions have been extremely divisive, pitting coastal communities against each other in the struggle to maintain local processing jobs.

The evidence we have provided suggests that the provincial government is facing strong pressures from corporate interests and the policy analysts that support neo-liberal globalization. These pressures have been particularly evident since the financial crisis of the late 2000s, and are being articulated at a range of scales: at the provincial level through the MOU process and ongoing debates about restructuring and rationalization; at the national scale through new attempts to privatize, commodify, and expand markets for fish harvesting rights especially relevant to Atlantic Canada; and now, internationally, through a free trade agreement that promises much in terms of tariff liberalization, but comes with many new obligations that run counter to the policies that have protected small-scale harvesters and coastal communities from neo-liberal globalization.

Corporate-environmental food regime

Some proponents of the third food regime concept suggest that a key element of the regime consists of the transformation of agrifood supply chains, particularly with respect to the leading role and power of supermarkets in the management of supply chains. The food system involves a shift in control over management of the chains, from the manufacturing sector to the retail sector dominated by large supermarket chains such as Walmart, Tesco, and Carrefour. In this food system, retailers require more flexible production organized around a wide array of product criteria, based on convenience, choice, health, wellness, freshness, and innovation manifested in ready-made meals and other convenience foods (Burch and Lawrence, 2005). Other increasingly powerful segments of the market include the food service sector, such as companies that supply food to schools, universities, hospitals, prisons, and restaurants (Burch and Lawrence, 2007), and the financial sector and private capital markets (Burch and Lawrence, 2009).

Friedmann anticipates the possible emergence of a corporate-environmental food regime based on the convergence of environmental politics and retail-led reorganization of food supply chains. The third food regime, according to Friedmann, is based on "selective appropriation of demands by environmental movements, and including issues pressed by fair trade, consumer health, and animal welfare activists" (Friedmann, 2005: 229). It is shaped by the convergence of environmental politics and corporate repositioning, particularly supermarket revolution and retail-led reorganization of supply chains (Bernstein, 2015: 13). Whereas the second food regime was characterized by the consolidation of state regulation, the third food regime is characterized by the growing role of environmental social movements in the private transformations of agrifood supply chains.

The rapid expansion of the food service industries and the development of competing fast-food chains in the 1970s increased the demand for specialized products. In fisheries located within the province, this had the effect of facilitating a shift away from Fordist mass production towards more "flexible," specialized batch production in the 1970s and 1980s (Neis, 1991). This required the reorganization of labour in response to changing market demand, including increasing the labour time required for production of more specialized packs and for grading (Neis, 1991: 166).

The potential emergence of the corporate-environmental food regime in NL fisheries is perhaps clearest in the area of NGO-led third-party environmental certification systems. As noted above, Newfoundland and Labrador fisheries over the last decade have become deeply integrated into the most influential international environmental certification program for fisheries, the Marine Stewardship Council (MSC) certification and eco-labelling program. Reflecting the high point of neo-liberal optimism in the market and a crisis of legitimacy in state-based regulation, the MSC was created by the corporate giant Unilever and the World Wide Fund for Nature[3] (WWF) in 1997 in the United Kingdom. Early leaders cited the cod collapse as a key motivation for creating a market-oriented approach to addressing a fisheries management crisis (Sutton, 1998). Like the development of private third-party certification standards in the agrifood sector, the structural power of retailers in European and US markets is the most important factor driving producer uptake of MSC certification (Ponte, 2012). These structural pressures became directly manifested in the province's fisheries by the mid-2000s when the Newfoundland and Labrador-based processors in the Association of Seafood Producers, with support from federal and provincial agencies, entered the northern shrimp fishery into third-party assessment against the MSC environmental standard for sustainable fishing. The processing agency was responding directly to expressions of interest from European buyers connected to large retailers to get MSC-certified (Foley, 2012, 2013). With northern shrimp certified in 2007, the processing association continued to engage the MSC by successfully acquiring certification for snow crab by 2013. This marked the 200th fishery certified to the MSC and, with shrimp already certified, brought the two most lucrative fisheries in the province into the MSC's global initiative (MSC, 2013). Other fisheries are in various stages of assessment, including parts of the historic and rebounding cod fishery. A Fishery Improvement Project (FIP) for the southern Newfoundland cod fishery began in 2011 under the leadership of the WWF, and the fishery subsequently entered a third-party assessment for Marine Stewardship Council (MSC) certification.

There are other social relations in Newfoundland and Labrador fisheries that exhibit more grassroots social movement characteristics that appear as potential alternatives to the corporate-environmental food regime. The fisheries are not only engaged in the corporate-environmental food regime of the MSC, but also increasingly engage in other certification, traceability, and alternative marketing initiatives, strategies, and alliances. The MSC is one powerful player in a broader, complex movement commonly referred to as the "sustainable seafood movement," which is driven by a complex and dynamic mix of corporate actors and various types of international environmental non-governmental organizations (NGOs). However, the limited influence of environmentalism as a social movement is demonstrated by the lack of international NGO presence in Newfoundland and Labrador historically, with the WWF hiring a provincially based employee only in the last decade. As in other jurisdictions where industry and state actors have developed alternative certifications, in part due to dissatisfaction with the MSC (Foley and Hebert, 2013; Foley and Havice, 2016), Newfoundland and Labrador producer organizations have sought out alternative means to assess and communicate seafood attributes in order to engage in broader ideas and organizations affiliated with international environmentalism. A core motivation in producers' engagement in alternatives strategies and alliances is a basic interest in economic benefits, which are ambiguous in the MSC.

Emerging strategies and alliances share characteristics with the "food from somewhere" movement. An important example includes new traceability technologies that allow fishers to bridge geographic distances between producers and consumers (Parlee and Wiber, 2011). One new initiative was recently developed along Newfoundland's southwest coast with Gulf of St. Lawrence halibut and lobster fisheries. The traceability project started in 2013 through a partnership between the provincial FFAW union and the British Columbia-based NGO EcoTrust. The initiative currently uses a consumer-facing seafood tagging and tracing system called "ThisFish," a system developed by EcoTrust that allows consumers to trace individual fish products back to fish harvesters. With more than 250 harvesters engaged in the project, the initiative is designed "to help Newfoundland tell its unique story to the world" (ThisFish and Ecotrust Canada, 2013). Other strategies include integrating alternative fishing methods with unconventional marketing and trade networks. Two examples from Fogo Island illustrate alternatives to corporate-controlled supply chains. The first pilot project uses an experimental method for catching cod with pots, which is designed to minimize damage to the ecosystem and the food product. A second pilot project, called "Fogo Island Fish," focuses on handline-caught cod sold directly to high-end chefs and restaurants, with 20 Toronto restaurants participating in the early stages of the program. The niche market method allows direct feedback between chefs and harvesters on fish quality and is designed as part of a broader set of initiatives to revitalize the rural and remote island's economy (CBC News, 2016). Examples of other alternative trade networks include calls for the development of community-supported fisheries, similar to community-supported agriculture initiatives, in Newfoundland and Labrador (Lowitt, 2009); the recent provincial stakeholder engagement in the launch in 2013 of Slow Fish Canada, a movement spawning from the international Slow Food movement (Ebel and Adler, 2013); and "Great Fish for a Change," developed through the MUN-based Too Big To Ignore (TBTI, 2016) research project.

These alternative methods of catching, processing, and marketing fish contrast with the industrial fishing methods and mass export strategy that characterized the pre-cod moratorium fishing in the early 1990s. They signal attempts to integrate social and ecological dimensions of food production, trade, and consumption, with the underlying goal to capitalize on the artisanal and community-embedded nature of social relations in small-scale Newfoundland and Labrador fisheries. The examples above are consistent with research showing how producers in other food sectors have exerted power to mobilize "bottom-up," alternative institutions of certification, eco-labels, and alternative trade networks that are locally embedded and globally connected (Friedmann and McNair, 2008). The emerging patterns of trade in Newfoundland and Labrador also appear consistent with research on the potential third food regime in other parts of the world. Examining how Australian and New Zealand food production systems have contributed to, or have been reshaped by, an emerging third food regime, Campbell (2009) points to the significance of flows and feedback of information from producers of agricultural commodities to consumers of food products. These "information flows and feedbacks between consumers and distant ecologies" have led to new emphases on food quality and food safety, a concern to reduce the environmental impact of food production, and a strong commitment to taste and locality (Campbell, 2009: 316). These new quality conventions are regulated through a pervasive audit culture and a strong commitment to certification and traceability. While the goal is to provide a more profitable "food from somewhere," these systems sit alongside poor-quality food production chains that continue to deliver "food from nowhere" to the vast majority of consumers. In this sense, the third food regime is characterized by two contrasting food systems, one that provides high-quality food through certified and audited food chains to wealthy consumers in the Global North, and a second that is less regulated and provides food to a larger number of poorer consumers.

CONCLUSIONS

This chapter began with two questions: How can food regime analysis help us understand the trajectory of seafood production and trade in Newfoundland and Labrador, and what lessons does the province's fish sector provide for advancing scholarly debates about food regime analysis?

Food regime analysis can help us understand the trajectory of seafood production and trade as deeply embedded in a global political economy of food. First, NL's commercial fishing and seafood industry continues to remain predominantly export-oriented. In other words, it stays deeply interconnected with global market forces and the political institutions that affect market regulation and transformation. Second, the social and political relations of Newfoundland and Labrador fisheries have transformed in ways that more or less correspond with the three "ideal-type" historical food regimes:

1. The consolidation of the Newfoundland and Labrador settler society embedded in European markets through the colonial-diaspora period.
2. The post-World War II industrialization and mechanization of "national-provincial fisheries" driven by demands from the US food sector.
3. The corporate-led neo-liberal globalization transformations towards flexibility and specialized products through the integration of quality, health, and environmental considerations.

Our argument is not that food regime theory perfectly explains Newfoundland and Labrador fisheries or that the periodization of food regime analysis corresponds neatly with provincial fisheries. However, our analysis does suggest that food regime theory provides an insightful lens through which to explain and understand transformations in the province's fisheries over time, and that the use of the three-regime typology in this context is relatively defensible. We invite researchers to engage critically with this approach in studies of seafood in Newfoundland and Labrador and beyond.

Our second objective was to explore how the Newfoundland and Labrador fisheries in particular, and fisheries more generally, might advance scholarly debates about a food-centric analytical approach that has surprisingly ignored seafood, one of the most important food commodities globally. Newfoundland and Labrador fisheries provide an ideal case study through which to "test" food regime theory by considering one of the most important and early commodities embedded in European, primarily British, capitalist expansion. The development of Newfoundland and Labrador fisheries over time, including the historic cod collapse of the 1980s and 1990s, provides evidence that fisheries share important similarities with agricultural food commodities. Two potential differences include the role of the state and ecological limits in fisheries. First, the substantial role of the state in fisheries regulation and development has been distinguished from other food sectors (Wilkinson, 2006). One factor that distinguishes the state's role from other agrifood sectors is that it allocates fish resources, especially since the international extension of state sovereignty from 12 to 200 miles. The declaration of the 200-mile EEZ fundamentally consolidated fisheries resources under state control in Canada and elsewhere. This change resulted in extensive state interventions in both domestic industrial expansion and conservation. The state's role in shifting the industry towards industrialized and highly mechanized production systems has contributed to the decline in traditional fisher knowledge and in turn undermined the capacity for people to support transitions to local food production and provisioning that depends on such knowledge (Chapter 10, this volume). The role of the state is substantial, though contested, in development and conservation efforts. In addition, ecological limits played a significant role in the transition from Fordism to post-Fordism (Neis, 1991), which is closely related to the transition from the second to third food regimes.

We were particularly interested in the question of what the case of the province's fisheries might tell us about the potential emergence and characteristics of a third food regime. Anticipating the demise of the second food regime, Friedmann and McMichael (1989: 113) propose two alternatives, one of which included the promotion of regional, local, and municipal politics of decentralization to reconnect and redirect local food production and consumption. The advocacy of smaller farming based on agro-ecological principles emerged as a central component of the resistance to the current global food system under the umbrella of food sovereignty (Bernstein, 2015). Internationally popular policy instruments are being appropriated by producers in Newfoundland and Labrador as a way to define those instruments on their own terms in ways that serve their specific, place-based interests and identities. These initiatives, from the traceability program organized by labour–NGO collaboration to alternative marketing networks, provide examples of the place-based "food from somewhere" alternatives to the corporate food regime that McMichael sees as promising. McMichael sees alternatives to the corporate food regime in the form of land-based NGOs like La Via Campesina. In the fishing sector the alternatives to the corporate food regime exist through networks of progressive scholars, and through organizations that support independent harvesters and vibrant fish-dependent coastal communities. The production of seafood from coastal seas adjacent to rural and remote communities offers important opportunities for creating space for grassroots responses and solutions to food security and sovereignty challenges, as discussed in other chapters in this book. Some (but not all) of the alternatives emerging in Newfoundland and Labrador are indeed challenging the corporate food regime. These initiatives, arguably, are not yet radically transformative. Newfoundland and Labrador fisheries appear to be shaped by, and subsequently shape, the processes consistent with an emergent, yet highly dynamic and contradictory third food regime. The regime includes a complex ensemble of social forces, including organized labour in forms of "accumulation from below." The initiatives are not directed towards food sovereignty per se, though there is potential for transformation into forms of food sovereignty.

In summary, in Newfoundland and Labrador fisheries, we see the existence of corporate food regimes and corporate-environmental food regimes, as well as alternative networks and trade relations that sit uneasily within and alongside the corporate food system. As in other food sectors such as the dairy industry (Pritchard, 1996), the persistence of organized small-scale producers supported by social protection policies and a strengthening of transnational corporate influence are mutually compatible. How, then, do Newfoundland and Labrador fisheries pertain to the question of the third food regime? In short, we see corporate, neo-liberal globalization and the corporate-environmental food regime, meaning that both Friedmann and McMichael are correct. But we also see continued and powerful struggles and still other social relations and networks that are more ambiguous in nature. The case of Newfoundland and Labrador points to the continued ability of the state to work against the vertical integration of capitalist development, and recent circumstances in the province suggest that it is in a situation of transformation more consistent with the third food regime.

NOTES

1. Some products exported to these destinations undergo further processing and re-export to other destinations, including Canada and Newfoundland and Labrador.

2. After Confederation with Canada in 1949, Newfoundland and Labrador became a different kind of state, one that still involved a fierce nationalism. This nationalistic impulse subsequently influenced decision-making around the fisheries. While management responsibility shifted to the Canadian government, conservation activities remained limited to the nearshore area until the late 1970s. Much of the focus for the 1950s and 1960s remained on marketing, quality control, and fisheries development (Vardy and Dunne, 2003: 92).
3. The World Wildlife Fund (WWF) was renamed the World Wide Fund for Nature (WWF) in 1986, but has retained its earlier name in Canada and the United States.

REFERENCES

Bavington, D., and J. Kay. 2007. "Ecosystem-based insights on Northwest Atlantic fisheries in an age of globalization." In M. Schechter, W. Taylor, and L. Wolfson, eds., *Globalization: Effects on Fisheries Resources*, 331–63. Cambridge: Cambridge University Press.

Bernstein, H. 2015. "Food regimes and food regime analysis: A selective survey." *BRICS Initiative for Critical Agrarian Studies (BICAS)*, Apr. At: http://www.iss.nl/fileadmin/ASSETS/iss/Research_and_projects/Research_networks/BICAS/BICAS_WP_2-Bernstein.pdf.

Burch, D., and G. Lawrence. 2005. "Supermarket own brands, supply chains and the transformation of the agri-food system." *International Journal of Sociology of Agriculture and Food* 13, 1: 1–18. At: http://espace.library.uq.edu.au/view/UQ:74986/UQ74986_OA.pdf.

——— and ———. 2007. "Supermarket own brands, new foods and the reconfiguration of agri-food supply chains." In D. Burch and G. Lawrence, eds., *Supermarkets and Agri-food Supply Chains: Transformations in the Production and Consumption of Foods*, 100–28. London: Edward Elgar.

——— and ———. 2009. "Towards a third food regime: Behind the transformation." *Agriculture and Human Values* 26: 267–79. doi:10.1007/s10460-009-9219-4.

Campbell, H. 2009. "Breaking new ground in food regime theory: Corporate environmentalism, ecological feedbacks and the 'food from somewhere' regime?" *Agriculture and Human Values* 26, 4: 309–19. doi:10.1007/s10460-009-9215-8.

——— and J. Dixon. 2009. “Introduction to the special symposium: Reflecting on twenty years of the food regimes approach in agri-food studies.” *Agriculture and Human Values* 26, 4: 261–65. doi:10.1007/s10460-009-9224-7.

CBC News. 2016. “How Toronto chefs got hooked on fish from Fogo Island.” At: http://www.cbc.ca/news/canada/newfoundland-labrador/hooked-on-fogo-fish-1.3371146.

Clift, T. B. 2011. *Report of the Independent Chair: MOU Steering Committee: Newfoundland and Labrador Fishing Industry Rationalization and Restructuring*. St. John’s: Government of Newfoundland and Labrador, Department of Fisheries and Aquaculture. At: http://www.fishaq.gov.nl.ca/publications/mou.pdf.

——— and T. Cooper. 2014. “The three R’s: Rationalization, retrenchment, reorientation: The Newfoundland and Labrador fishery after the MOU.” *Newfoundland Quarterly* 107, 2: 37–40. At: https://www.mun.ca/harriscentre/reports/nlquarterly/MemPre-NQ-Fall2014.pdf.

Davis, R. 2014. “A cod forsaken place?: Fishing in an altered state in Newfoundland.” *Anthropological Quarterly* 87, 3: 695–726.

Dean, L. J., H. Wareham, and D. S. Walters. 2001. *Report of the Special Panel on Corporate Concentration in the Newfoundland and Labrador Fishing Industry*. St. John’s: Government of Newfoundland and Labrador, Department of Fisheries and Aquaculture Newfoundland. At: http://www.fishaq.gov.nl.ca/publications/archives/report_the_special_panel_on_corporate_concentration.pdf.

Department of Fisheries and Aquaculture (DFA). 2015. *Seafood Industry Year in Review 2014*. St. John’s: Government of Newfoundland and Labrador, Planning Services Division, Department of Fisheries and Aquaculture. At: http://www.fishaq.gov.nl.ca/publications/pdf/SYIR_2014.pdf.

———. 2016. *Seafood Industry Year in Review 2015*. St. John’s: Government of Newfoundland and Labrador, Planning Services Division, Department of Fisheries and Aquaculture. At: http://www.fishaq.gov.nl.ca/publications/pdf/SYIR_2015.pdf.

Dunne, E. B. 2003. *Final Report: Fish Processing Policy Review*. Fish Processing Policy Review Commission. St. John’s: Government of Newfoundland and Labrador.

Ebel, S., and D. Adler. 2013. “Launching ‘Slow Fish Canada’.” *Small Scales*, 9 Dec. At: https://smallscales.ca/2013/12/09/sfc/.

Foley, P. 2012. "The political economy of Marine Stewardship Council Certification: Processors and access in Newfoundland and Labrador's inshore shrimp industry." *Journal of Agrarian Change* 12, 2/3: 436–57. doi:10.1111/j.1471-0366.2011.00344.x.

———. 2013. "National government responses to Marine Stewardship Council (MSC) fisheries certification: Insights from Atlantic Canada." *New Political Economy* 18, 2: 284–307. doi:10.1080/13562367.2012.684212.

——— and E. Havice. 2016. "The rise of territorial eco-certification: New politics of transnational sustainability governance in the fisheries sector." *Geoforum* 69: 24–33. doi:10.1016/j.geoforum.2015.11.015.

——— and K. Hébert. 2013. "Alternative regimes of transnational environmental certification: Governance, marketization, and place in Alaska's salmon fisheries." *Environment and Planning A* 45, 11: 2734–51. doi:10.1068/a45202.

——— and C. Mather. 2016. "Making space for community use rights: Insights from 'community economies' in Newfoundland and Labrador." *Society & Natural Resources* 29, 8: 965–80. doi:10.1080/08941920.2015.1089611.

———, ———, and B. Neis. 2015. "Governing enclosure for coastal communities: Social embeddedness in a Canadian shrimp fishery." *Marine Policy* 61: 390–400. http://dx.doi.org/10.1016/j.marpol.2014.11.009.

Friedmann, H. 1987. "The family farm and the international food regimes." In T. Shanin, ed., *Peasants and Peasant Societies*, 2nd ed, 247–58. Oxford: John Wiley and Sons.

———. 1993. "The political economy of food: A global crisis." *New Left Review* 197 (Jan./Feb.): 29–57. At: https://newleftreview.org/I/197/harriet-friedmann-the-political-economy-of-food-a-global-crisis.

———. 2005. "Feeding the empire: The pathologies of globalized agriculture." *Socialist Register 2005: Reloaded* 41: 124–43. At: http://socialistregister.com/index.php/srv/article/view/5828#.V7TbOo-cF3x.

———. 2005. "From colonialism to green capitalism: Social movements and emergence of food regimes." In F. H. Buttel and P. McMichael, eds., *Research in Rural Sociology and Development*, vol. 11, 227–64. Amsterdam: Elsevier.

———. 2009. "Discussion: Moving food regimes forward: Reflections on symposium essays." *Agriculture and Human Values* 26, 4: 335–44. doi:10.1007/s10460-009-9225-6.

——— and P. McMichael. 1989. "Agriculture and the state system: The rise and decline of national agricultures, 1870 to the present." *Sociologica Ruralis* 29, 2: 93–117. doi:10.1111/j.1467-9523.1989.tb00360.x.

——— and A. McNair. 2008. "Whose rules rule? Contested projects to certify 'local production for distant consumers'." *Journal of Agrarian Change* 8, 2/3: 408–34. doi:10.1111/j.1471-0366.2008.00175.x.

Goodman, D., and M. Watts. 1994. "Reconfiguring the rural or fording the divide? Capitalist restructuring and the global agro-food system." *Journal of Peasant Studies* 22. 1: 1–49. doi:10.1080/03066159408438565.

Innis, H. A. 1940. *The Cod Fisheries. The History of an International Economy.* Toronto: University of Toronto Press.

Le Heron, R., and N. Lewis. 2009. "Discussion: Theorising food regimes: Intervention as politics." *Agriculture and Human Values* 26, 4: 345–49. doi:10.1007/s10460-009-0226-5.

Lowitt, K. 2009. *A Community Food Security Assessment of the Bonne Bay Region.* CURRA, Memorial University of Newfoundland. At: http://www.curra.ca/documents/CFS%20Assessment%20Report_%20Final_Oct%2009.pdf.

Magnan, A. 2012. "Food regimes." In J. M. Pitcher, ed., *Oxford Handbook of Food History*, 370–88. Oxford: Oxford University Press.

Marine Stewardship Council (MSC). 2013. "Newfoundland and Labrador snow crab becomes 200th fishery to achieve MSC certification." 18 Apr. At: https://www.msc.org/newsroom/news/newfoundland-and-labrador-snow-crab-fishery-becomes-200th-fishery-to-achieve-msc-certification.

Mather, C. 2013. "From cod to shellfish and back again? The new resource geography and Newfoundland's fish economy." *Applied Geography* 45: 402–09. doi:10.1016/j.apgeog.2013.06.009.

McLeod, J. 2011. "MOU report released, then rejected by government." *The Telegram*. At: http://www.curra.ca/fishery_MOU.htm.

McMichael, P. 2009. "A food regime genealogy." *Journal of Peasant Studies* 36, 1: 139–69. doi:10.1080/03066150902820354.

———. 2013. *Food Regimes and Agrarian Questions*. Halifax: Fernwood.

Neis, B. 1991. "Flexible specialization: What's that got to do with the price of fish?" *Studies in Political Economy* 36, 1: 145–75. doi:10.1080/19187033.1991.11675446.

———, R. Ommer, and P. Hall. 2014. "Moving forward: Building economically, socially and ecologically resilient fisheries and coastal communities." *A Policy Booklet*, Apr. At: http://www.curra.ca/documents/CURRA-Booklet-FINAL-WebRes.pdf.

Newell, D., and R. E. Ommer, eds. 1999. *Fishing Places, Fishing People: Traditions and Issues in Canadian Small-scale Fisheries*. Toronto: University of Toronto Press.

Ommer, R. 2002. "The interdisciplinary eco-research project: An overview." In R. E. Ommer, ed., *The Resilient Outport: Ecology, Economy and Society in Rural Newfoundland*. St. John's: ISER Books.

——— and P. R. Sinclair. 1999. "Outports under threat: Social roots of systemic crisis in rural Newfoundland." In R. Byron and J. Hutson, eds., *Local Enterprise on the North Atlantic Margin*, 253–75. Aldershot: Ashgate.

Parlee, C., and M. G. Wiber. 2011. "Who is governing food systems? Power and legal pluralism in lobster traceability." *Journal of Legal Pluralism and Unofficial Law* 43, 64: 121–48. doi:10.1080/07329113.2011.10756672.

Ponte, S. 2012. "The Marine Stewardship Council (MSC) and the making of a market for 'sustainable fish'." *Journal of Agrarian Change* 12, 2/3: 300–15. doi:10.1111/j.1471-0366.2011.00345.x.

Pope, P. E. 2004. *Fish into Wine: The Newfoundland Plantation in the Seventeenth Century*. Chapel Hill: University of North Carolina Press.

Pritchard, W. N. 1996. "The emerging contours of the third food regime: Evidence from Australian dairy and wheat sectors." *Economic Geography* 74, 1: 64–74. doi:10.1111/j.1944-8287.1998.tb00105x.

Rose, G. A., and S. Rowe. 2015. "Northern cod comeback." *Canadian Journal of Fisheries and Aquatic Sciences* 72, 12: 1789–98. doi:10.1139/cjfas-2015-0346.

Sinclair, S. 2013. *Globalization, Trade Treaties and the Future of the Atlantic Canadian Fisheries*. Ottawa: Canadian Centre for Policy Alternatives.

Song, A. M., and R. Chuenpagdee. 2015. "A principle-based analysis of multilevel policy areas on inshore fisheries in Newfoundland and Labrador, Canada." In S. Jentoft and R. Chuenpagdee, eds., *Interactive Governance for Small-scale Fisheries: Global Reflections*, vol. 13, 435–56. Cham, Switzerland: Springer International. doi:10.1007/978-3-319-17034-3_23.

Sutton, M. 1998. "New hope for marine fisheries." In *Fish Stakes: The Pros and Cons of the Marine Stewardship Council Initiative: A Debate from the Pages of SAMUDRA Report*. Chennai, India: International Collective in Support of Fishworkers. At: http://aquaticcommons.org/263/.

The Telegram. 2013. "Fishermen organizations create new national federation." 11 Dec. At: http://www.thetelegram.com/News/Local/2013-12-11/article-3539720/Fishermen-organizations-create-new-national-federation/1.

ThisFish and Ecotrust Canada. 2013. "Traceability to help Newfoundland tell its unique story to the world." At: http://thisfish.info/generic/article/newfoundland-traceable-seafood/.

Too Big to Ignore (TBTI). 2016. At: http://toobigtoignore.net/.

Vardy, D., and E. Dunne. 2003. "New arrangements for fisheries management in Newfoundland and Labrador." *Report to the Royal Commission on Renewing and Strengthening Our Place in Canada*, Mar. At: http://www.exec.gov.nl.ca/royalcomm/research/pdf/Vardy.pdf.

Wilkinson, J. 2006. "Fish: A global value chain driven onto the rocks." *Sociologia Ruralis* 46, 2: 139–53. doi:10.1111/j.1467-9523.2006.00408.x.

Wright, M. 1997. "Frozen fish companies, the state, and fisheries development in Newfoundland, 1940–1966." *Business and Economic History* 26, 2: 727–37.

———. 1998. "The background to change in the Newfoundland cod fishery at the time of Confederation." *Newfoundland Studies* 14, 2: 253–65.

10

Experts in the Field: Using Fishers' Ecological Knowledge (FEK) in Primary Food Production

Myron King

INTRODUCTION

It is well understood that fishing practices and experiences are part of traditional, place-based knowledge and that they provide a valuable role in food acquisition. As long as people have known there were fish in the sea, they have harvested food from it and they have built their lives around it. Fishing for food, along with all the related food system characteristics and functions, forms a bigger picture over time and space. From a local perspective, the Newfoundland and Labrador fishery can be traced back hundreds of years, with generations of residents originating from the same communities over the centuries. For settlers in coastal communities, or "outports" as they became known, fishing remains an important traditional and economically necessary way of life. Fish both as a food source and as the economic engine of rural communities permeates the province's history and the social and economic fabric of the people. Fishers typically gained knowledge about their profession at an early age. They advanced their expertise and the profession itself through their experiences and sharing it with younger generations. The expert knowledge developed and honed through the daily fishing profession, along with the subtle nuances accompanying this traditionally steeped food-gathering activity, has been referred to as "fishers' ecological knowledge" or FEK. This terminology recognizes the ecological knowledge of experienced people involved with fishery practices worldwide, and was often dismissed by early scientists. García-Quijano (2007) notes that the knowledge of those who are directly engaged in food production is being increasingly recognized as valuable for understanding change, ecosystem management, and conservation. With the use of resources so important to the evolution of knowledge and security of food for humankind, it follows that generational FEK often comprises a subset of TEK or LEK.

LINKAGES BETWEEN TEK, LEK, AND FEK

Traditional ecological knowledge (TEK) has been defined as "a cumulative body of knowledge, practice, and belief, evolving by adaptive processes and handed down through generations by cultural transmission, about the relationship of living beings (including humans) with one another and with their environment" (Berkes, Colding, and Folke, 2000: 7). TEK is often used to describe the knowledge of Indigenous peoples with regard to their expertise or "knowledge learned through experience" in and around natural resources. It also reflects knowledge as a result of insight and intellectual activity in traditional contexts, including know-how, practices, and learning (Leidwein, 2006). Meanwhile, local environmental or ecological knowledge (LEK) is a more generalized term that is not directly aligned with Indigenous knowledge or oriented towards a particular culture or professional practice. Researchers have also used TEK and LEK interchangeably, using both terms to refer to the detailed knowledge about traditional resources and environments evident in resource harvesting families (Berkes, 1993, 1999; Freeman and Carbyn, 1988; Johannes, 1981; Murray, Neis, and Petter Johnsen, 2006; Neis and Felt, 2000). In concentrating on the fishery and the knowledge surrounding the activity of fishing, it is therefore desirable to form this study directly around FEK, helping to increase the depth of literature for FEK that now exists.

The body of knowledge called FEK and specialized to the fisher has been growing rapidly since the 1960s and 1970s. In her study of marine ecowebs involving fishers from Newfoundland and Labrador, Neis (1992) found that FEK could provide invaluable insights. She also found FEK to be informed by a fisheries success — that is, the understanding by local fishers of the relationship between social and ecological factors, which must be accounted for by management to help fill the gaps on ocean ecosystem knowledge. This knowledge is recognizable. You can observe it when you spend time aboard a vessel steaming out to sea to fish. The knowledge is present in the fishery market where captains share stories of storms encountered. It is on the maps that fishers paint in our minds and on our televisions in the evening. As we look at the knowledge of the fisher, we endeavour to understand its roots and structure, examining how it might be categorized within a fishing and food production framework.

The study at hand examines the mental mapping capabilities that many fishers are known to carry. It is difficult to grasp how deeply these maps are intuitively formed through the fishers' experiences. Geographical information systems (GIS) mapping interviews allow the fishers' mental mapping cognition to be recorded digitally, and some analyses applied. Doing so provides a glimpse of the fishers' "knowledge-in-action," a crucial component of fishers' knowledge that they use every day. This study looks at how the practice of fishing has evolved, to answer the questions: What happened to the one-time small, local community fisher? How has he or she changed in relation to a global food production market shift? The study concludes with a discussion about how fishing for small-volume community food production has evolved to large export-targeted catches, and how the Newfoundland and Labrador fishing industry has changed along the way.

These knowledge systems can be powerful, and they have the potential to be used in a variety of ways, including in community-based initiatives and natural resource management (Crate, 2006). Generally viewed as unrestricted by geographical or cultural boundaries, FEK can be found in abundance throughout the world among peoples where there is a history of fishing activity. Such knowledge of fish behaviour is often based on personal observation while a fisher is on fishing grounds, and has been compiled generationally (Ruddle, 1994). FEK is a social, technical, and cultural product (Neis, 1992). A person's FEK grows directly from the regular exploitation of resources for food and profit, even though it may be executed differently and in various places amid a myriad of social and environmental factors. An expert fisher at one geographical location will likely have common core characteristics similar to an expert fisher at another location, although the detailed knowledge might be quite different. FEK also contributes to idiosyncratic community food practices. These nuances, along with the perception that fisher knowledge is mostly "anecdotal," can make the study of FEK difficult. Nonetheless, scholars have often called for greater use of FEK in management planning and resource policy (Felt, 2010; Griffin, 2009; Hartley and Robertson, 2008; Hutchings and Ferguson, 2000; Neis and Felt, 2000;

Macnab, 2000). It is important to understand FEK thoroughly if its use is recommended for fisheries management and primary food production.

Over the past few decades, researchers, governments, and community organizations within Canada and across the world have paid increasing attention to TEK, LEK, and FEK. Prominent examples include Barbara Neis's (1992) study of FEK and the cod stock assessment in Newfoundland, Gisli Pâlsson's work on fishers' knowledge (2000), and Barbara Neis and Lawrence Felt's edited volume specifically dedicated to linking the knowledge of "fisher folks" with science and fisheries management (Neis and Felt, 2000). The research of Fikret Berkes, Johan Colding, and Carl Folke on the rediscovery of traditional knowledge has had a profound impact on adaptive management (see Berkes, Colding, and Folke, 2000), while Grant Murray (Murray et al., 2008) and Robert Johannes (1981) helped lay foundational TEK research, each carrying a subset of material dedicated to the fisher. Internationally, Alpina Begossi (2008), Gilden and Conway (2002), and Hartley and Robertson (2006, 2008), among others, have provided significant and lasting contributions to the TEK, LEK, and FEK literature.

This eastern Canadian study builds on the FEK literature by increasing the awareness, acceptance, and understanding of FEK and by emphasizing the benefits it provides to resource management and food production. In the study at hand, the fishers' ecological knowledge has been recorded during mapping exercises using geographical information systems (GIS) technology and compared against verifiable scientific data. The comparison is used to validate the FEK technique and to show, in part, its continued value for fisheries management. Through literature review and case study, this chapter demonstrates a direct linkage between FEK and the present-day understanding of fishing resources. These linkages are conceptualized from the viewpoint of the outport community, Conception Bay North, NL.

FEK IN NEWFOUNDLAND AND LABRADOR

Across Newfoundland and Labrador, some communities have been involved in the fishery for hundreds of years. These were once tightly knit, usually isolated, communities where people relied heavily on the land and sea, particularly for food and trade goods. This has also been the case for the collection of communities that comprise Conception Bay North, situated in eastern Newfoundland, as shown in Figure 10.1.

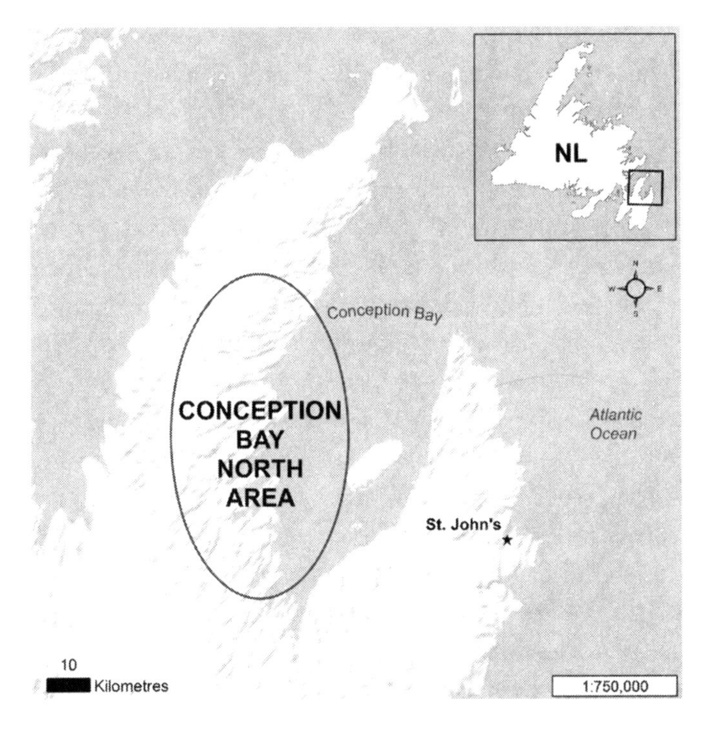

Figure 10.1. Map of study area, Conception Bay North, NL. (Cartography by Myron King)

Conception Bay North boasts a long tradition of involvement in the fisheries. The Portuguese Captain Gaspar Corte-Real first sailed into these unknown waters in the year 1500, naming it "Baie de Cos eicam" (Conception Bay), and envisioning what Andrews (1997: 2) calls "a panorama of undisturbed wilderness inspiring visions of bounty." The place names of Conception Bay communities often bear witness to early European fishing origins – Spaniard's Bay, Port de Grave, Carbonear, and Portugal Cove to name a few.

During the early settlement years, multi-generational families were usually involved in the fishery as the primary source for their livelihoods. For many families, fish was the main dietary staple, and it was consumed several times per week. If a father fished for food and pay, often his sons would become fishermen as well. Female family members were also often heavily involved by mending nets, processing the catch, painting the boats, and many other needed tasks. Some women fished with their husbands. For fishing families, it was "all hands on deck" for sharing the workload. A fisher's knowledge accompanied the intergenerational transition, and knowledge was passed from one family member to the next. Fishers elsewhere have also noted that fishing knowledge is passed on within family groups (McKenna, Quinn, Donnelly, and Cooper, 2008; Ruddle, 1994).

In the geographically remote fishing outports, access to technology, formal learning, external foods, and a wider sense of society was limited. This makes communities like Conception Bay North an ideal location to study generationally transitioned knowledge and FEK. However, modern changes stemming from technology and science can also transform a fisher's knowledge beyond the traditional fisheries background usually offered in this type of historic setting. Improved transport networks also lead to greater food availability from external resources. Murray, Neis, and Petter Johnsen (2006) note that the knowledge of fish harvesters co-evolves with fishing practices, and that it is embedded into a dynamic socio-ecological network extending beyond an individual fisher.

For generations, families relied almost entirely on fish to fulfill daily food requirements. FEK was borne out of necessity, yet it still evolved as part of a technologically advancing world. Pálsson (2000) advises that a fisher's decision-making patterns are more the result of practical knowledge than of calculations or mental reflection. In other words, these abilities are the product of experience.

FEK and Science

As is often debated, FEK is not always aligned or in agreement with scientific knowledge and practices. In an industry where FEK has been deeply affected by technological advances, the gap between practices based on FEK and those based on scientific understanding will likely continue to narrow. Another contributing factor is the evolution of fishing from a community food production resource with smaller quantities to trade, into larger quantities of export-targeted catches, as discussed in Chapter 9 by Foley and Mather. Hence, understanding the convergence of FEK and science is very important for addressing escalating global problems like overfishing. It may also serve as an opportunity for fishers and scientists to work together and to draw the community more inclusively into fisheries management and policy-making. The feedback from fishers, combined with the community's desired level of co-operation and integration, can help shed light on how to successfully integrate FEK with science. These outport community fisher evaluations, along with any follow-up selection, inference, analysis, or transformation, as Murray, Neis, Palmer, and Schneider (2008) noted, can help improve over the long-term the realization of best-possible FEK-science integration. The combination of the two systems can also result in the capture of finer scales of ecological information (Nenadovic, Johnson, and Wilson, 2012).

METHODS

A semi-structured interview approach was used to explore the FEK of career fishers in the area of Conception Bay North. The interview process comprised two parts.

Part one involved a multi-categorical interview instrument. The instrument consisted of 12 categories consisting of 50 different questions, each formulated to draw out the fishers' knowledge in relation to their fisheries expertise. More information about the survey instrument can be found in King (2012).

Table 10.1. Categories used for interview instrument.

Category	Number of Questions
Group I	
Introduction	2
Fishing procedure	5
Time	3
Food and bait	2
Bycatch	1
Fishing gear	6
Group II	
Fish health and habitat	7
Fishing environment (e.g., ice)	6
Group III	
Fisher learning	6
Community support	5
Management awareness	5
Closure	2

The 12 interview categories ranged from fishing procedures and species familiarity to regulatory awareness, food production value, and fisheries management. Table 10.1 shows the three groups of question categories, along with the number of questions per category. The questions were designed and categorized initially in consultation with two career fishers, who helped outline the various duties and fishing aspects related to the profession they had mastered. Each question was therefore created with the experiential component of fishers' knowledge in mind. This allowed the fishers who were subsequently interviewed to elaborate beyond basic responses with their individual experiences. The fishers were encouraged to share their responses with others. This question style allowed the fishers to provide as much detail as possible and to facilitate a comfortable, open discussion. Each fisher was free to skip any questions from the interview instrument.

To begin the interviewing process, a group of four known, active fishers were contacted and asked to participate in the initial round of the interviews. These four participants then shared names and contact details of other fishers in the area. In some cases, they directly contacted the other fishers to encourage participation. This snowball interviewing methodology helped to enlist additional fishers for interviewing. With consent from each fisher, the interview was also recorded via digital video for transcription and follow-up analysis. The interview analysis was conducted using a thematic approach, and the occurring themes were repeated by the fishers in answering each question.

Part two involved a follow-up mapping session with each fisher. The purpose of the mapping session was to draw upon the fishers' knowledge of geo-spatial information such as species location, fishing routes, and observed ocean weather patterns. As with the interview instrument session, there was freedom for fishers to respond individually, with some control over the components being mapped, as well as the scale. To encourage information-sharing, coloured pencils and a total of 24 different black-and-white paper maps of varying scale were provided. Each map promoted a certain set of marine and sea floor characteristics. The areas ranged from inshore areas around the eastern side of Newfoundland (primarily Conception Bay and Trinity Bay) up to and including offshore areas out to the 200-mile exclusive economic zone (EEZ) and just beyond. Some maps showed the Grand Banks area and Northwest Atlantic Fisheries Organization (NAFO) division areas. During preliminary interview preparations, the target group usually remarked on these features with emphasis. The large number of maps provided a high freedom of choice for the fishers to feel more comfortable during the mapping session, which in turn allowed them to maximize their own individual level of mapping effort.

Noting the constraints in GIS mapping methodology experienced by Macnab (2000), the easy assortment of paper maps and use of coloured pencils were presented as a seemingly "low-tech" approach in order to maximize each fisher's comfort with the interview. The mapping sessions took 30–60 minutes in addition to the interview time. *ArcGISArcMap* software from ESRI was used for the display and spatial analysis of fishers' mapping information. In order to digitize the fishers' data, each scanned paper map was geo-referenced against a Newfoundland and Labrador base map in *ArcMap*, to ensure an exact match between the paper and digital references. For each point of interest the fishers entered on their paper maps, a point feature was digitized on the computer map at the same location. The same method was applied for lines or enclosed areas the fishers created, by making use of line features or polygon features in *ArcMap*. In this way, all of the fishers' mapping data were entered into a digitized GIS format.

RESULTS

As shown in Table 10.1, results were tabulated using the knowledge, experiences, and influences of each fisher within three categories:

1. Fishing procedures, time, and gear;
2. Fish health and habitat, fishing environment;
3. Learning, community and food production, and fishery management.

Each grouping was further divided into relevant fisher experiential criteria, which depended on the subject matter under the specific category. A checkmark was inserted if a fisher discussed or displayed knowledge and experience pertaining to that topic. Further results from the interviews are now discussed.

Sampling

A total of 11 fishers were interviewed from the Conception Bay North towns of Spaniard's Bay, Port-de-Grave, Bareneed, and Hibb's Cove. The average interview time was 80 minutes, with the shortest interview being 55 minutes and the longest at 107 minutes. All interviews were done at fishers' homes in their respective communities. No fisher skipped questions during the interview and all participated in the follow-up mapping session. Of those interviewed, five were still active fishers and six were retired from fishing. The fishers varied in age from 21 years to 77 years. All fishers reported having roots in the fishery (meaning all or most were multi-generational fishers), and all fishers reported having inshore fishing experience.

All but one fisher had offshore fishing experience, generally defined as greater than or equal to 20 miles from land. All fishers had multiple years of experience fishing various species in Newfoundland and Labrador coastal waters. Their levels of responsibility varied, with some fishers having the additional roles as fishing captain and/or fishing enterprise owner/operator.

Fishing Procedure, Time, and Fishing Gear

A variety of questions were used with the intention of drawing out fishers' knowledge with respect to their regular operation and fishing procedures. The questions allowed fishers to expound on procedural fishing and boating knowledge and to elaborate further, if warranted. For example, when asked about "the main fishing-related activities you and others do before, during, and after a fishing season, which make up the bigger picture," the fishers answered directly, explaining what they believed was suitable in quality and length of response. For the fishers who tended to answer succinctly, it was easy to spur the discussion a little further with casual prodding, like this exchange related to the question above:

> *Fisher:* You got to make, do your gear. If you are crabbin' then you make your crab pots. Repair crab pots. You have to make sure the boat is painted and everything, any repairs to the boat and that she's scrubbed up.
> *Interviewer:* What about when you get in from a fishing trip? What kind of activities going on then?

> *Fisher:* Well after fishing season is over, it's same thing. You have to make sure the gear is ready for next year and do any repairs to the boat. You have to make sure the boat is painted up.

Such exchanges regarding fishing procedure, fishing time, and other fishing-related activities were repeated for each fisher. Since each fisher had had multiple years on the ocean, the answers to these questions were generally straightforward and provided without hesitation. Fishers explained how fishing is a year-round job, not just a summer event. Some of the major off-season duties include boat and gear maintenance, as well as fisheries-related training. Fishing times on the ocean also depended on species sought and distance travelled, ranging from a few days to a couple weeks per trip. Fishers with experience on more than one vessel and with more than one type of species or fishing gear spent additional time answering the questions. For example, one fisher progressed through the years from fishing inshore on an open, 17-foot boat using hand-hauled nets for cod, to finishing his career on a 65-foot longliner outfitted with hydraulics for hauling pots and trawls to capture cod, shrimp, crab, herring, capelin, and swordfish. Through this period of evolving experience, the fisher also noted how the initial catches were just enough to feed family and to trade/sell for other necessities. At the time of the final fishing trips in the same fisher's career, all catches were targeted for processing and subsequent export. This evolution in experience for the fisher also helps to show how changes occurred in the food production and acquisition for fisher families over the same time period. This grouping of questions was popular with the fishers, since many of them enjoy talking about the variety of boats and equipment involved. More than one fisher would add a small story or anecdote to an answer, smiling or laughing as they spoke.

Fish Health, Habitat, and Fishing Environment

The second grouping of questions pertained to knowledge generally viewed as secondary to the act of fishing. These included questions about fish health or biology, fish habitat or location, and the fishing environment. For fish health, questions centred on snow crab (*Chionoecetes opilio*), which is currently the most heavily exploited species. The fishers' understanding of a healthy crab, including size and condition optimal for harvesting, was evaluated through a series of detailed questions about snow crab maturity, population density, and signs of disease. The fishers displayed direct, detailed ecological knowledge about the health and well-being of snow crabs, including the ability to recognize species characteristics related to optimal harvest times.

Questions related to the fishing environment involved navigation amid storms, ice flows, and harsh North Atlantic weather. The fishers' answers for environmental-type questions were similar to those for questions on fishing procedure. This was not unexpected because the act of fishing and the fishing environment are closely interwoven. The value of using that experience at the peaks and troughs of extreme weather situations was evident, as well. Consider the following description of events from one fisher. As the storm worsens, his actions are critical to ensure a safe outcome:

> The weather started to get worse so we slowed her in about 11 a.m. Then, at one o'clock, we were clocking the wind at 70 knots. We were already heading in to the loft. You alter your course five degrees, so that turned her a little bit and the storm is coming in more or less on her bell, instead of right straight on her stem. I said haul her back to 1200 RPM, as we were still doing around 1500 RPM. There were times we were doing two knots ahead and there were times we were a knot backwards. The slower you could go. It was better for you to do that than it is to give it to her because a boat is only going to stand so much. We never did and had no damage.

Similar experiences were shared by other fishers, pointing to a common mindset in their knowledge of the environment and how to operate within it. This was particularly evident among those who were also fishing captains, who bear additional responsibilities commensurate with that position. For the fishers interviewed, the scale of their respective fisheries did not appear to factor into their knowledge about the environment, navigation, or quality of catch. These professional characteristics, as they relate to experiential depth of knowledge, were recognizable for every fisher.

Fisher Learning and Community Food Production

The third grouping of questions examined tertiary fishing expertise and knowledge. These questions were oriented towards understanding the fishers' evolutionary learning process, along with their opinions and awareness of the wider fishery support structure. Elements of the fishers' awareness and experience in relation to food production and fishery management were also explored. Fishers shared their formal and informal educational experiences, as well as other sources of fishing knowledge. Older fishers tended to leave school at an early age, entering the fishery to help sustain the family, while younger fishers usually had a minimum of a high school education. All fishers sampled were men with families, living in historically tightly knit fishing communities, and all had roots in the fishery. Understandably, these common backgrounds influenced the respondents' answers to some of the questions in this group. All the fishers believed their community highly supported their fishing activities. Specific examples included the well-known yearly boat-lighting celebration in Port-de-Grave and special services for fishers' at the local church before the beginning of each fishing season. Fishers also commented on how fishing was once done to put food on the table and to sell in order to buy other needed items. The interviewees' opinions of management effectiveness and awareness of management's role in the fishery were explored through a final set of questions, including discussion about the 1992 cod moratorium.

GIS Mapping

A total of 101 *ArcMap* GIS features were created and digitized from 32 scanned paper maps generated by the interviewees. From the large amount of fisher mapping information collected, three significant knowledge themes were chosen for further GIS amalgamation. Figure 10.2 reflects a consolidated geographical information map, showing a variety of hallmark features within the Conception Bay vicinity, as noted by the fishers. This map concentrates on the inshore herring fishing locations, cod trawling locations, and the fishers' display of other knowledge such as turtle sightings and known shipwreck locations.

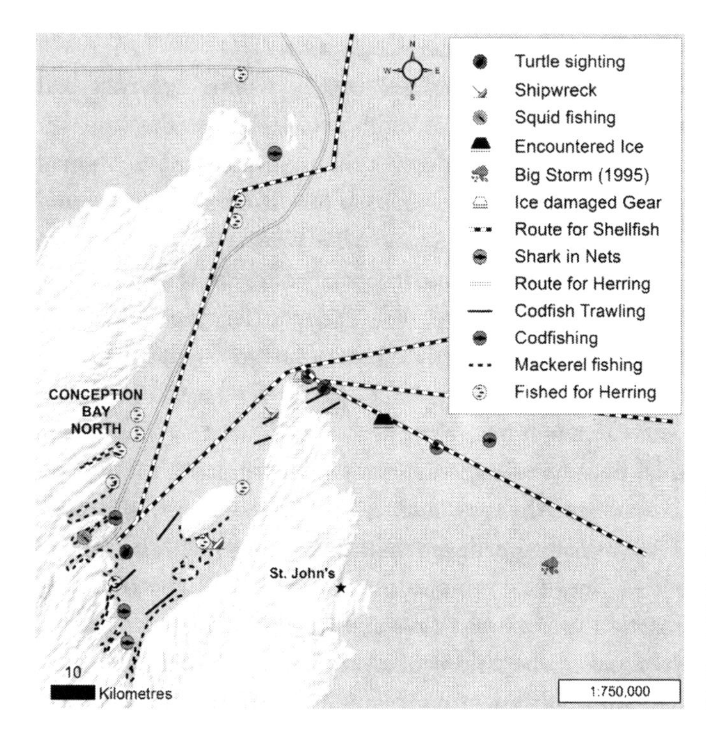

Figure 10.2. Consolidated inshore GIS map showing features identified through FEK. (Cartography by Myron King)

Currently, the snow crab fishery, perhaps the largest fishery on Newfoundland and Labrador's east coast, has grown to exceed expectations and, consequently, has changed the face of fishing within many Newfoundland and Labrador coastal communities. All interviewees participated in the crab fishery, so their combined results are presented in Figure 10.3. As part of this mapping display, Canada's 200-mile EEZ and the relevant NAFO divisions are included. These divisions were also present on the paper maps created by the fishers. The offshore backdrop is a bathymetry raster created using XYZ coordinate data sourced from the Scripps Institute of Oceanography, allowing for the discernment of the continental shelf and historically significant fishing banks around the province. The positions for crab fishing, as indicated by the fishers, were used for Figure 10.3 with a kernel density spatial analysis applied to aid in visualization.

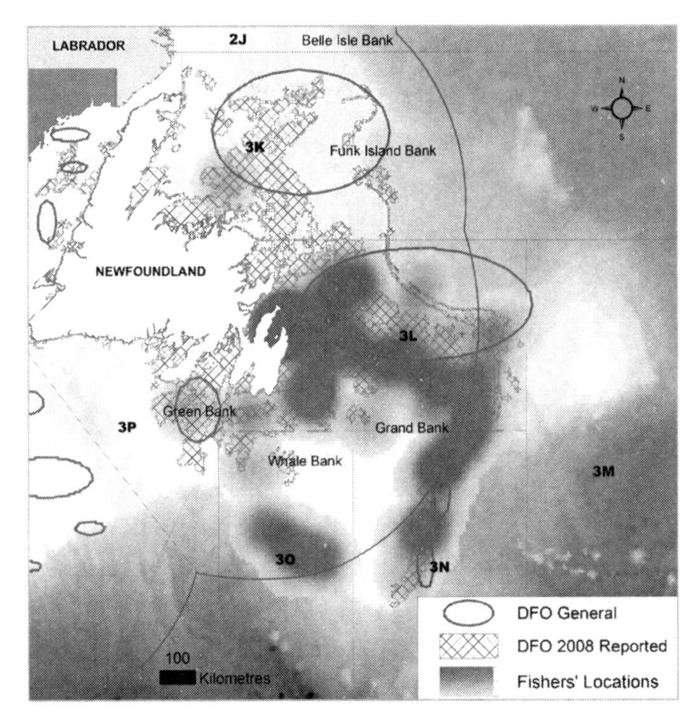

Figure 10.3. Offshore map of Newfoundland and Labrador combining crab-fishing locations identified by the fishers and governmental data. (Cartography by Myron King)

The final fishers' map (Figure 10.4) highlights another lucrative Northwest Atlantic shellfish species. Shrimp (*Pandalus borealis*) is often exploited on the same trips as snow crab and it is therefore popular among the fishers. To construct the map, a similar base template was employed to highlight the shrimp-fishing positions indicated by the 11 fishers through the mapping sessions. A kernel density spatial analysis was applied for visualization and further comparison.

Crab and shrimp fishing topped the list of current Newfoundland and Labrador species sought by the fishers. Fishing trends and changes in market demand for these products influenced fishing practices over the decades. Elder fishers commented that these species were usually discarded many years ago when they were found in nets or traps during fishing. This speaks directly to the adaptive nature of fishers, since a different species of fish or shellfish means that the entire regime of food production shifted, further affecting how the fisher approaches the fishery.

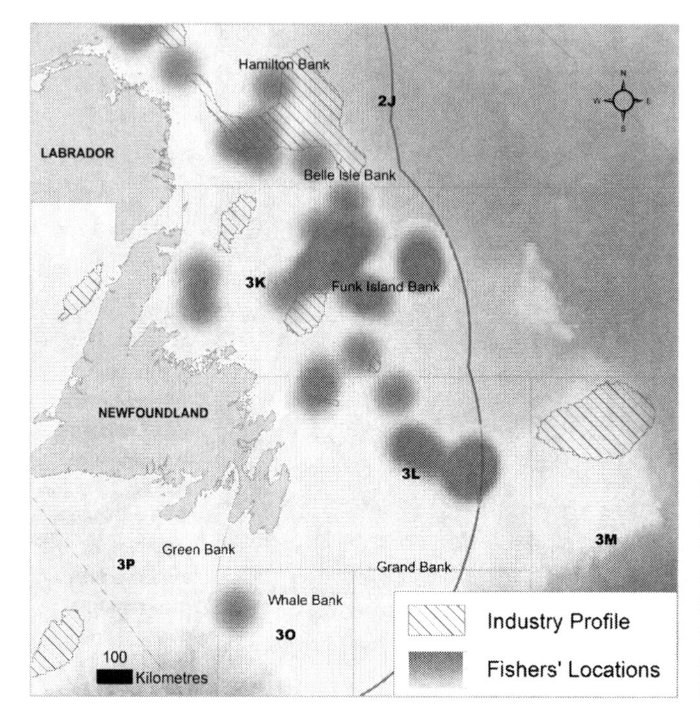

Figure 10.4. Fishers' shrimp-fishing locations compared with industry-defined shrimp areas. (Cartography by Myron King)

Science-based source data features that were added to the Figure 10.3 crab map reflected the general crab-fishing zones, as indicated by government sources (Hartwig, 2009). This included detailed and instrument-based spatial distribution of commercial crab-fishing efforts. The Figure 10.4 shrimp map also includes the location of Newfoundland and Labrador shrimp-fishing grounds. The resulting science-based dataset features allow for comparison between the fishers' mapping knowledge of their crab- and shrimp-fishing locations with the general scientific-based governmental information sources using a geo-spatial approach.

DISCUSSION

With a detailed qualitative analysis completed for the interviewed fishers, a number of important points can be made regarding the fishers' knowledge. First, it is important to recognize the validity of the approach. Scientists attempting to study FEK have often noted the difficulty in applying analysis or drawing conclusions from it, given that some consider FEK and science-based production two separate, parallel knowledge systems. Murray et al. (2008) also recognize this, noting that while fishers and scientists rely on observations, they are not always the same observations. Other scholars believe that a knowledge system should be studied on its own merits and not based on preconceptions originating elsewhere.

FEK has often been called "anecdotal" and therefore perceived to be less useful to scientists. However, FEK should not be dismissed, even at a "pure" science level where fishers' knowledge may conflict with biological and physical data (Hallwass, Lopes, Jurano, and Silvano, 2013; Ruddle, 1994; Neis, 1992). Clearly, through the many studies now available on TEK and FEK, there are avenues where the two knowledge systems (FEK and scientific) can provide greater value when used together. Such a necessary combination was alluded to by Neis (1992), especially in dealing with complex, intricate, socio-economic driven management systems like a specialized fishery.

For example, the two approaches are complementary when the focus of research is to gather as much relevant information as possible before making a resource management decision. The researchers' inclusion of the fisher and FEK also provides the scientist an excellent opportunity for investment in local fishing communities (Gilden and Conway, 2002; Hartley and Robertson, 2006; Yochum, Starr, and Wendt, 2011; Johannes, Freeman, and Hamilton, 2000). Using scientific knowledge alone may even inhibit decision-makers, rendering them ineffective at the local level (Bethel et al., 2011). An inclusive approach that integrates both FEK and scientific knowledge provides additional validation that may ensure that the best possible information set is available.

Wider stakeholder participation in the management process can also have longer-lasting, more effective outcomes (Nenadovic, Johnson, and Wilson, 2012). Providing fishers equal opportunity to submit expertise alongside science for resource management has profound implications for food production and marketing. As an extension within the context of food sovereignty, providing equality to fishers to use their knowledge in resource management would help increase Newfoundland and Labrador's, and thus Canada's, food sovereignty portfolio. This is consistent with La Vía Campesina's concept of food sovereignty as "the right of nations and peoples to control their own food systems, including their own markets, production modes, food cultures and environments" (Wiebe and Wipf, 2011: 4), and as further discussed in the Introduction to this volume.

To further elaborate, food sovereignty hinges fundamentally on the goal of "redress[ing] the abuse of the powerless by the powerful, wherever in the food system that abuse may happen" (Patel, 2007: 302). Neis and Felt (2000) recognize that more powerful interests have historically not only decided what production modes, food cultures, and environment decisions were made, but also have controlled the very policy-making apparatus (i.e., the constructed context where the decision was considered). This control often left the typical fisher outside of management's focus. In contrast, by directly empowering local fishers and recognizing their knowledge and social networks, forward-thinking policy-makers can subsequently help increase Canadian food sovereignty and fisheries governance.

In Chapter 9, Foley and Mather explore two possible pathways for the future of Newfoundland and Labrador fisheries in the coming years. One path predominately highlights corporate interests, with little regard for social and community-based food production value. This "not-from-here" mentality further diminishes the role of fishers and their knowledge within local food contexts. Alternatively, the pathway where corporate interests are placed second to the social networks of food management at the coastal community level could strengthen fisher–manager relationships and reinforce the recognition and use of FEK. This empowerment is the key to reversing the trend of centralized, corporate-driven, top-down authority seen in the past with fisheries management, and can help to revitalize small communities by placing emphasis on the value of local food production.

Fishers' Knowledge Is Valuable for Community Food Production

There is high value in the knowledge of seasoned fishers. The fishers interviewed in this study have travelled near and far across many miles of ocean over the years, continuously re-evaluating their surroundings. In the realm of food production, fishers have been involved in primary food production for their entire lives. The need for food in early fishery days drove local outport fishers to do their jobs constantly and effectively, at the risk of starvation. Thus, a bad fishing season had a high impact, affecting not just fish availability as food, but its availability as a trade good. Fishers in the community were also procuring fish for the community's needs, since not all families were a part of the fishery. Local knowledge systems, including FEK, can also help provide a framework of reference for solving food problems (Rajasekaran and Whiteford, 1993). In the example at hand, imported food and goods were scarce; having fish on the family table carried many families through long winters. Fast-forward 200 years and fishers remain an important component of local food production in many outports. Having shrunk in number — due to advanced technologies, expanding transport networks, and increased global markets — fishers still often can be the best local source of fresh fish for small, "mom-and-pop" style restaurants, small markets, and niche markets like those described in other chapters throughout this book.

Studies of traditional food systems elsewhere by L. Filippo D'Antuono (2013) have found that local raw ingredients comprised the majority of local food systems. Local markets, food production chains, and the local people they serve benefit from locally available fish. FEK can be an important asset, particularly if we hope to achieve food security at the local household level (Rajasekaran and Whiteford, 1993). The increasing local use of fishers' knowledge in multiple jurisdictions, much like a political commodity (Dubois et al., 2016), is a strong asset for locally focused food production development and planning. A similar concept, food sovereignty planning, is further discussed in the Introduction to this volume.

FEK as an Indicator of Ecological Change

Fisher observations can also be used to evaluate larger ecological changes. In studying FEK over time, both Eddy et al. (2010) and Hallwass (2013) found that FEK can provide insight into changes in stock abundance on intergenerational time scales and prior environmental states. Fishers' knowledge can be useful as an early warning sign for such changes (Rochet et al., 2008). In this study, the fishers were asked several questions on the behaviour of crab and regarding long-term observations in several areas. One fisher observed: "We would go for deeper water right up until the mid-90s, you wouldn't think about putting your pots in less than 120 fathom of water, now after that it's 60 fathom, even 40 fathom now to get the crab, this is how it has changed."

As previously discussed, in merely the past 20 years, crab has become the province's top fishery product. Some would say that it came along just in time to save many fishermen from losing everything following the cod-fishing closures. The year-to-year crab fishery has its up and downs, though, which has not gone unnoticed, as described by another fisher: "You know, we might be down for one year but then the next year is way up there again, because you have managed the quota to a sustainable level and you are not going to destroy it."

This ability of fishers to capture important aspects of a specific species, in this case crab, is similar to the findings of Murray et al. (2008) on FEK and cod migrations in the northern Gulf of St. Lawrence. Those authors showed that for topics that are poorly understood, such as the continued migration of cod northward along the Labrador coast, FEK can shed some light on the broader ecological picture. Often, FEK may be the only "data" available, depending on the time scales involved (Johannes, Freeman, and Hamilton, 2000). Consider the following statements from two fishers, along with the earlier fisher observation regarding crab populating shallower waters:

> Years ago when we went out, in 1985 say, if you could find a hole and put your pot down in the hole, you would get lots of crab. You would never get any up on the shoal'er ground.

> Some people always used to fish in deeper water, 80–90 fathom, sometimes deeper. This year we fished up the shoals, 25–27 fathoms and the crab catch rates up there was sometimes better than what they were out in 80–100 fathom.

The statements from these three fishers, in aggregate, suggest a fisheries trend — the crab is being caught in shallower waters. Whether or not this is already known or being studied from a fisheries science perspective, such an assessment would now have the supporting observations of the people on the ocean actually harvesting the species. Indeed, fishers' knowledge plays an important role because fishers can possess detailed knowledge about fish behaviour and ecology (Begossi, 2008; Silvano and Valbo-Jørgensen, 2008).

FEK as a Dynamic Learning Process with Global Orientation

The interviews demonstrated that fishers were engaged in a lifelong learning process, and their FEK was both dynamic and adaptive. While empirical knowledge is a mainstay for fishers, the interviewees also identified other resources of lifelong influence. These included learning by doing, generational transfer, learning via fisheries management, and learning through formal coursework. All fishers identified their father or another elder fisher as mentors from whom they received expertise. As was discovered by Carbonell (2012) in his study of Spanish fishers, the Conception Bay fishers also learned a great deal from media sources. In the case of fishers who started their careers at a young age, their development of fishing expertise may also involve a knowledge-based learning framework consistent with the traditional human development of knowledge. This "clustering," as referred to by Danovitch and Keil (2004), means these expert fishers began specific, detailed, non-random organization of their fishery knowledge early in life when they were likely still forming the organizational frameworks that eventually became the basis or core for their lifelong learning process.

The elder fishers showed progressions over 30–50 years, a period of time that required adaptive change and knowledge about several fishing activities such as technology, fishing distance, expanding public awareness, increasing boat size, increasing regulation (both in the restrictive sense and in their knowledge of it), increasing market reach, increasing fishing efficiency, increasing catch volumes, and increasing local market participation. These findings parallel the conclusions of Murray, Neis, and Petter Johnsen (2006). In their study, the authors called this continuous change "a shift along a continuum from local ecological knowledge (LEK) towards globalized harvesting knowledge (GHK)" (Murray, Neis, and Petter Johnsen, 2006: 1). The authors also note that during this time there was a reduction in community reliance on local fishers for food production. However, the global awareness also has potential to revitalize local seafood marketing and sales. Recently, the sale of fresh seafood in small quantities at the local level by fish harvesters, as recommended by Smith et al. (2014), is becoming a reality.

For the Conception Bay North study, every interviewee discussed extensively the expanding contexts of the fishery towards a globally based knowledge set. Note this comment from a fisher about Canada as a player in the global crab fishery market:

> It's all the global market, with better prices. I know Alaska crab, Alaska caught a lot so their crab is on the market and the US, when they went to the recession they stopped buying the crab so that opened — like say for Canada and the US had the free trade, when they stopped buying it opened, we had all this excess product and all of a sudden China wants more — like all

> these countries that had hardly bought any because US had a monopoly on it but now it's diversifying everywhere, you are just diversifying where your product is going, not like it affects us, like it affects us at the end product, but like the plants just buy it and they diversify everything among globally we'll say.

This discussion shows how the present-day fisher has "GHK." In the old days, a fisher might have only fished for cod and sold it to a merchant just after docking his boat. Now a fisher's expertise can span several fisheries, destined for several different world markets, and integrated into a variety of food products. Fishers now need to know about food resource mapping and food certification standards and practices. Furthermore, if the fisher has developed into a fishing captain or fishing business owner/operator, then such globally situated knowledge is necessary for career success.

Mapping with the Fishers

Exploring the mapping process with the fishers proved to be an interesting experience, full of recollections spawned by spatially oriented fishing conversations. The fishers refrained from using logbooks, professional fishing charts, or computers to help them during the data collection phase. They relied on their memories and the knowledge gained from years of experience on the water. The fishers' ability to recall fishing location information with spatial and temporal referencing (sometimes from many years past) is an excellent display of their deeply developed mental mapping proficiency. Each fisher recalled the species, catch size, year, location fished, and (in many cases) the route taken to the fishing location. Other geographical points of interest were shared, with references to on-the-water experiences like turtle or whale sightings. These tended to be sightings that stuck out from typical fishing trips.

As shown in Figure 10.3, there was good agreement between the fishers' mental recollection of their crab-fishing locations around Newfoundland and Labrador and the governmental general assessment of crab-fishing locations, based on 2008 spatial distribution results. Since the government's online information is available to anyone via the Internet, it is possible that the fishers used the Internet as a source for the location information rather than experience. However, this is unlikely since the fishers were asked to share their personal crab-fishing experiences, rather than simply where to find snow crab. The fishers also provided very detailed information, which further supports personal experience. Similarly, for the shrimp map presented in Figure 10.4, the fishers' knowledge showed some congruence with the published industry data. The agreement between the fishers' maps and the government's scientifically based published maps reinforces findings by McKenna, Quinn, Donnelly, and Cooper (2008) that it is unwise to disregard LEK just because it is not acquired by high-technology, scientific methods.

FEK Complements Science and Fisheries Management

As has been found by other researchers (Smith et al., 2014; Begossi, 2008; Berkes, Colding, and Folke, 2000; Hartwig, 2009; Mackinson, 2001; Murray et al., 2008), this study shows that FEK can complement science. FEK can supplement long-term scientific observations, and the historical knowledge might also fill information voids. This combined approach allows for better-informed fisheries management decisions. Sometimes it may be appropriate to use both knowledge systems together, in order to effectively bridge the gaps in perspective and understanding (Huntington, 2002). One real benefit to utilizing FEK is having a strengthened arsenal for consultation. Indeed, if Newfoundland and Labrador cod fisheries managers had listened more to inshore fishers regarding the collapsing cod stocks in the 1980s, perhaps the collapse may have been a little less painful and expansive (Johannes, Freeman, and Hamilton, 2000). One of the active fishers interviewed participates in the snow crab co-management board, where he offers insight both as a fisher and as an experienced businessman on board decisions about everything from quota recommendations to local food production awareness ideas. This is an example of how the complementary nature of FEK and science may be used successfully within the province. Another fisher agreed, stating:

> These last few years they have come around, more or less sitting down with the fisherman and, especially with crab, sitting down and drawing up the management plan and it is really called a management plan and I think that is the reason why, in our area, it has worked pretty good.

This is not unlike other directives by the government of Canada. Papik, Marschke, and Ayles (2003) describe a similar management committee striving for a complete management plan for regional rivers in British Columbia. While these are examples of empowering fishers to share their knowledge, there is room for improvement, as pointed out by a third experienced fisher who was seemingly frustrated by recent restrictions: "We live in a different day and I'm only allowed to do what they like and seems like some department may not listen to what fishermen say, even though he's a hands-on man who spends his lifetime there at it."

Using FEK with science has been described by the scientific community as "fishers' knowledge integration." It is viewed as an important road map to the new way managers and policy-makers must operate in order to manage the ever-evolving fishing industry sustainably and successfully. Fisheries management should strive to reach the goal of equal knowledge status, where the expertise of the fisher is recognized and employed alongside the popular scientific theoretical thinking of the day. Only then can we say we have achieved co-operative research, according to Hartley and Robertson (2008). In his Amsterdam presentation at the 2001 Mare Conference, "People and the Sea," Diegues (2005) concluded that a wider audience of fishery, conservation, environmental, and public policy specialists is becoming increasingly aware of the importance of traditional knowledge and the potential it has to improve marine management. In Newfoundland and Labrador, fishers are already working with scientists to some extent. With continuing efforts to use FEK, government and industry decision-makers can move from simply saying integration is needed to actually measuring the results of successful integration on a more consistent basis.

CONCLUSIONS

FEK is dynamic. It evolves throughout the progression of a fisher's career, and it is influenced by technological developments and socio-economic change. In Newfoundland and Labrador, this evolution changed fishing from a traditional family-oriented, localized process focused on fishing for food and trade into a revolutionized and globalized business undertaking. As noted by Foley and Mather in Chapter 9, there is increasing potential for consumers who seek "food from somewhere" to show direct consumer support for localized knowledge when they make food purchases. Regardless of how future food regimes evolve, FEK integration is recommended as the best path forward to deliver the fishers' knowledge and contributions into sustainable fisheries management.

This study has shown that FEK integration is already in progress throughout Newfoundland and Labrador. Fishers have been increasingly included in consultations for fisheries management and decision-making. More integration and empowerment are necessary, however, if we are to reach a point where the full value of FEK is to be absorbed into decision-making and fisheries policy regimes. At the local level, fishers' expertise with primary food production is a necessary component in the planning and development of revitalized, healthy coastal communities.

This study also documents the level of advanced mapping expertise associated with being a career fisher. By closely depicting the geo-spatial position for different species over a multi-year period, these fishers demonstrate the mental mapping capabilities required to successfully perform diverse skills associated with fishing activities. These capabilities are honed over time. The resultant mapping information can be highly accurate and dependable when validated against governmental sources.

Finally, this study shows that fishing is much more than going to sea and throwing out a net. The modern fisher now participates in formal training efforts and relies on experience as well as strategic use of modern technologies. A fisher remains influenced by the past and the traditional attributes often associated with being a fisher, while also — through fisheries and community network changes — taking the profession to a new level. This is the case in Conception Bay North, where the fishers have a deep history with the fishery yet continue to evolve year after year parallel to the societal, technological, and environmental changes that are continuously reshaping the industry. In other words, today's fisher is informed with both historical and dynamically evolving FEK to successfully navigate an ever-expanding food production network.

ACKNOWLEDGEMENTS

Portions of this chapter are drawn from a Master's research study through the University of Ulster (King, 2012). The author would like to extend a special thank-you to Professor J. A. G. Cooper (University of Ulster) for his support, and especially to the fishers and their families of Conception Bay North, NL, for their generosity and thoughtfulness in sharing their experiences at length.

REFERENCES

Andrews, G. W. 2006 [1997]. *Heritage of a Newfoundland Outport: The Story of Port-de-Grave.* Carbonear, NL: Jesperson Publishers.

Begossi, A. 2008. "Local knowledge and training towards management." *Environment, Development and Sustainability* 10, 5: 591–603. doi:10.1007/s10668-008-9150-7.

Berkes, F. 1993. "Traditional ecological knowledge in perspective." In J. Inglis, ed., *Traditional Ecological Knowledge: Concepts and Cases.* Ottawa: Canadian Museum of Nature/International Development Research Centre (IDRC), International Program on Traditional Ecological Knowledge Research Centre.

———. 1999. *Sacred Ecology: Traditional Ecological Knowledge and Resource Management.* Philadelphia: Taylor and Francis.

———, J. Colding, and C. Folke. 2000. "Rediscovery of traditional ecological knowledge as adaptive management." *Ecological Applications* 10, 5: 1251–62. doi:10.1890/1051-0761(2000)010[1251:ROTEKA]2.0.CO;2.

Bethel, M., L. Brien, E. J. Danielson, S. B. Laska, J. P. Troutman, W. M. Boshart, . . . M. A. Phillips. 2011. "Blending geospatial technology and traditional ecological knowledge to enhance restoration decision-support processes in coastal Louisiana." *Journal of Coastal Research* 27, 3: 555–71. doi:10.2112/JCOASTRES-D-10-00138.1.

Carbonell, E. 2012. "The Catalan fishermen's traditional knowledge of climate and the weather: A distinctive way of relating to nature." *International Journal of Intangible Heritage* 7: 61–75.

Crate, S. A. 2006. "Elder knowledge and sustainable livelihoods in Post-Soviet Russia: Finding dialogue across the generations." *Arctic Anthropology* 43, 1:40–51. doi:10.1353/arc2011.0030.

Danovitch, J., and F. C. Keil. 2004. "Should you ask a fisherman or a biologist? Developmental shifts in ways of clustering knowledge." *Child Development* 75, 3: 918–31.

D'Antuono, L. F. 2013. "Traditional foods and food systems: A revision of concepts emerging from qualitative surveys on-site in the Black Sea and Italy." *Journal of the Science of Food and Agriculture* 93, 14: 3443–54. doi:10.1002/jsfa.6354.

Diegues, A. C. 2005. "Traditional fisheries knowledge and social appropriation of marine resources in Brazil." In A. C. Diegues, ed., *Maritime Anthropology in Brazil*, 80–95. São Paulo: NUPAUB-USP/Center for Research on Human Population and Wetlands in Brazil. Originally presented at Mare Conference, "People and the Sea," Amsterdam, Aug.–Sept. 2001.

Dubois, M., M. Hadjmichael, and J. Raakjaer. 2016. "The rise of the scientific fisherman: Mobilising knowledge and negotiating user rights in the Devon inshore brown crab fishery, UK." *Marine Policy* 65: 48–55. doi:10.1016/j.marpol.2015.12.013.

Eddy, T. D., J. Gardner, and A. Perez-Matus. 2010. "Applying fishers' ecological knowledge to construct past and future lobster stocks in Juan Fernández Archipelago, Chile." *PLoS ONE* 5, 11: 1–12. doi:10.1371/journal.pone.0013670.

ESRI. 2016. *ArcGIS, ArcMap* software. At: http://www.esri.com/software/arcgis.

Felt, L. 2010. "It all depends on the lens, b'y: Local ecological knowledge and institutional science in an expanding finfish aquaculture sector." In K. Culver and D. Castle, eds., *Aquaculture, Innovation and Social Transformation*, 167–90. The International Library of Environmental, Agricultural and Food Ethics. Dordrecht, Netherlands: Springer.

Freeman, M. R., and L. N. Carbyn, eds. 1988. *Traditional Knowledge and Renewable Resources Management in Northern Regions*. Occasional Publications No. 23. Edmonton: Canadian Circumpolar Institute.

García-Quijano, C. G. 2007. "Fishers' knowledge of marine species assemblages: Bridging between scientific and local ecological knowledge in southeastern Puerto Rico." *American Anthropologist* 109, 3: 529–36. doi:10.1525/aa.2007.109.3.529.

Gilden, J., and F. Conway. 2002. "An investment in trust: Communication in the commercial fishing and fisheries management communities." *Oregon Sea Grant Publication* (ORESU-G-01-004). Corvallis: Oregon State University.

Griffin, L. 2009. "Scales of knowledge: North Sea fisheries governance, the local fisherman and the European scientist." *Environmental Politics* 18, 4: 557–75. doi:10.1080/09644010903007419.

Hallwass, G., P. Lopes, A. Juras, and R. Silvano. 2013. "Fishers' knowledge identifies environmental changes and fish abundance trends in impounded tropical rivers." *Ecological Applications* 23, 2: 392–407. doi:10.1890/12-04291.

Hartley, T. W., and R. A. Robertson. 2006. "Emergence of multi-stakeholder-driven cooperative research in the northwest Atlantic: The case of the Northeast Consortium." *Marine Policy* 30, 5: 580–92. doi:10.1016/j.marpol.2005.09.006.

——— and ———. 2008. "Stakeholder collaboration in fisheries research: Integrating knowledge among fishing leaders and science partners in northern New England." *Society and Natural Resources* , 1: 42–55. doi:10.1080/08941920802001010.

Hartwig, L. 2009. *Mapping Traditional Knowledge Related to Identification of Ecologically and Biologically Significant Areas in the Beaufort Sea*. Canadian Manuscript Report of Fisheries and Aquatic Sciences 2895. Winnipeg: Fisheries and Oceans Canada, Oceans Program Division. At: http://www.dfo-mpo.gc.ca/Library/339428.pdf.

Huntington, H. P., P. K. Brown-Schwalenburg, K. J. Frost, M. E. Fernandez-Gimenez, D. W. Norton, and D. H. Rosenberg. 2002. "Observations on the workshop as a means of improving communication between holders of traditional and scientific knowledge." *Environmental Management* 30, 6: 778–92. doi:10.1007/s00267-002-2749-9.

Hutchings, J., and M. Ferguson. 2000. "Links between fishers' knowledge, fisheries science, and resource management: Newfoundland's inshore fishery for Northern Atlantic cod, *Gadus morhua*." In B. Neis and L. Felt, eds., *Finding Our Sea Legs: Linking Fishery People and Their Knowledge with Science and Management*, 82–110. St. John's: ISER Books.

Johannes, R. E. 1981. *Words of the Lagoon: Fishing and Marine Lore in the Palau District of Micronesia*. Berkeley: University of California Press.

———, M. R. Freeman, and R. Hamilton. 2000. "Ignore fishers' knowledge and miss the boat." *Fish and Fisheries* 1, 3: 257–71. 10.1111/j.1467-2979.2000.000019.x.

King, M. 2012. "Tracing the roots of Fishers' Ecological Knowledge: A study into the traditional influence, experiential evolution and geospatial observations of local fishers in a small Newfoundland and Labrador outport community." Master's thesis, Ulster University, Coleraine, Ireland.

King, T. D. 1997. "Folk management and local knowledge: Lobster fishing and tourism at Caye Caulker, Belize." *Coastal Management* 25, 4: 455–69. doi:10.1080/08920759709362337.

Leidwein, A. 2006. "Protection for traditional knowledge associated with biological and genetic resources: General legal issues and measures already taken by the European Union and its member states in the field of agriculture and food production." *Journal of World Intellectual Property* 9, 3: 251–75. doi:10.1111/j.1422-2213.2006.00277.x.

MacKinson, S. 2001. "Integrating local and scientific knowledge: An example in fisheries science." *Environmental Management* 27, 4: 533–45. doi:10.1007/s0026702366.

Macnab, P. 2000. "Drawing from experience: Harvester mapping of fishing grounds in Bonavista Bay, Newfoundland." In B. Neis and L. Felt, eds., *Finding Our Sea Legs: Linking Fishery People and Their Knowledge with Science and Management*, 224–35). St. John's: ISER Books.

McKenna, J., R. Quinn, D. Donnelly, and J. A. G. Cooper. 2008. "Accurate mental maps as an aspect of local ecological knowledge (LEK): A case study from Lough Neagh, Northern Ireland." *Ecology and Society* 13, 1. At: http://digitalcommons.usu.edu/unf_research/34/.

Murray, G., B. Neis, C. T. Palmer, and D. C. Schneider. 2008. "Mapping cod: Fisheries science, fish harvesters' ecological knowledge and cod migrations in the northern Gulf of St. Lawerence." *Human Ecology* 36, 4: 581–98. doi:10.1007/s10745-008-9178-1.

———, ———, and J. Petter Johnsen. 2006. "Lessons learned from reconstructing interactions between local ecological knowledge, fisheries science, and fisheries management in the commercial fisheries of Newfoundland and Labrador, Canada." *Human Ecology* 34, 4: 549–71. doi:10.1007/s10745-006-9010-8.

Neis, B. 1992. "Fishers' ecological knowledge and stock assessment in Newfoundland." *Newfoundland Studies* 8, 2: 155–78.

——— and L. Felt, eds. 2000. *Finding Our Sea Legs: Linking Fishery People and Their Knowledge with Science and Management*. St. John's: ISER Books.

Nenadovic, M., T. Johnson, and J. Wilson. 2012. "Implementing the western Gulf of Maine area closure: The role and perception of fishers' ecological knowledge." *Ecology and Society* 17, 1: 20. doi:10.5751/ES-04431-170120.

Pálsson, G. 2000. "'Finding one's sea legs': Learning, the process of enskilment, and integrating fishers and their knowledge into fisheries science and management." In B. Neis and L. Felt, eds., *Finding Our Sea Legs: Linking Fishery People and Their Knowledge with Science and Management*, 26–40. St. John's: ISER Books.

Papik, R., M. Marschke, and G. B. Ayles. 2003. *Inuvialuit Traditional Ecological Knowledge of Fisheries in Rivers West of the Mackenzie River in the Canadian Arctic*. Canada/Inuvialuit Fisheries Joint Management Committee Technical Report Series, Report 2003-4. Inuvik, NWT. At: http://fishfp.sasktelwebhosting.com/publications/FJMC%20Report%20 Series/WSWG_TEK.pdf.

Patel, R. 2012 [2007]. *Stuffed and Starved: Markets, Power and the Hidden Battle for the World's Food System*. Toronto: HarperCollins.

QSR International computer software suite NVivo 9. 2016. At: http://www. qsrinternational.com/nvivo-product.

Rajasekaran, B., and M. Whiteford. 1993. "Rice-crab production in south India: The role of indigenous knowledge in designing food security policies." *Food Policy* 18, 3: 237–47. doi:10.1016/0306-9192(93)90080-U.

Rochet, M.-J., M. Prigent, J. A. Bertrand, A. Carpentier, F. Coppin, J.-P. Delpech, . . . V. M. Trekel. 2008. "Ecosystem trends: Evidence for agreement between fishers' perceptions and scientific information." *ICES Journal of Marine Science* 65, 6: 1057–68. doi:10.1093/icesjsm/fsn062.

Ruddle, K. 1994. "Local knowledge in the folk management of fisheries and coastal marine environments." In C. L. Dyer and J. R. McGoodwin, eds., *Folk Management in the World's Fisheries: Lessons for Modern Fisheries Management*, 161–206. Boulder: University Press of Colorado.

Silvano, R., and J. Valbo-Jørgensen. 2008. "Beyond fishermen's tales: Contributions of fishers' local ecological knowledge to fish ecology and fisheries management." *Environment Development and Sustainability* 10, 5: 657–75. doi:10.1007/s10668-008-9149-0.

Smith, D., K. Vodden, M. A. Woodrow, A. Khan, and B. Fürst. 2014. "The last generation? Perspectives of inshore fish harvesters from Change Islands, Newfoundland." *Canadian Geographer* 58, 1: 95–109. doi:10.1111/j.1541-0064.2013.12053.x.

Wiebe, N., and K. Wipf. 2011. "Nurturing food sovereignty in Canada." In H. Wittman, A. A. Desmarais, and N. Wiebe, eds., *Food Sovereignty in Canada: Creating Just and Sustainable Food Systems*, 1–19. Halifax: Fernwood.

Yochum, N., R. M. Starr, and D. E. Wendt. 2011. "Utilizing fishermen knowledge and expertise: Keys to success for collaborative fisheries research." *Fisheries* 36, 12: 592–605. doi:10.1080/03632415.2011.633467.

11

Sustainable Aquaculture Production

Cyr Couturier & Keith Rideout

INTRODUCTION: CONTEXT FOR SUSTAINABLE AQUACULTURE AND RESEARCH PARADIGMS

Aquaculture is the farming of aquatic plants and animals for food (Rana, 1997). It is the natural progression from hunting to farming for food security and access, begun on terrestrial fields over 10,000 years ago and in our aquatic environs well over 4,000 years ago in Asia (Hickling, 1962). The development of commercial aquaculture, however, has a fairly recent history, with this "blue revolution" commencing in the 1960s, about 20 years after the terrestrial "green revolution" for food production following the rapid population increases after World War II.

Today, aquaculture accounts for more than 50 per cent of the aquatic protein consumed by humans (World Bank, 2013; FAO, 2014a). In 2014, the total global volume of aquatic farmed food was estimated at 73.8 million tonnes (98.1 million tonnes including marine plants) with a value of US$160 billion (FAO, 2014b, 2016). While global supplies from capture fisheries have stagnated since the 1980s, it's expected that with increasing population growth and the demand for healthy, sustainable protein there will be a shortfall of seafood of over 30 million tonnes by 2030 and 60 million tonnes by 2050, when global population is expected to exceed 9 billion (World Bank, 2013; FAO, 2014a). In 2012, farmed seafood surpassed beef as a major protein source for humans (Larsen and Roney, 2013).

On a global scale, aquaculture is responsible to a significant extent for food security, for improved livelihoods in developing countries, and for supplying "all-natural," locally produced healthy proteins (FAO, 2016). Fish from wild and farmed sources is now the major source of animal protein for over 20 per cent of the world's population, with an estimated 3 billion portions consumed daily and over 1 trillion portions annually on the planet (based on 20 kg per person per year of seafood consumed, 100 g portion size; FAO, 2014a, 2016). The vast majority of aquaculture produce is sold locally and fresh year-round. However, there is also a significant trade component for aquaculture produce, particularly for high-value species such as shrimp and salmon. More than 75 per cent of global aquaculture produce is farmed "organically" without chemical fertilizers, pesticides, therapeutants, or genetically modified organism (GMO) inputs. The balance is produced with very little in the way of synthetic fertilizers, therapeutants, or pesticides, while using all natural inputs for both farmed aquatic plants and animals.

Relative to terrestrial agriculture, the environmental impacts from aquaculture are arguably much less. In today's world, agrifoods have significant carbon footprints and are major emitters of greenhouse gases (GHGs), comprising up to an estimated 30 per cent of total emissions (IPCC, 2007; Smith et al., 2008). In addition, well over 70 per cent of freshwater resource withdrawal on the planet is employed for agrifoods production, with the balance used for industry and drinking (UNEP, 2008; FAO AQUASTAT, 2015). There are questions, given current water withdrawal and usage rates, as to whether there will be sufficient water to support the global population by the year 2050.

In this context, it is noteworthy that aquatic farming or "aquaculture" is among the most sustainable of food-producing activities on the planet, typically creating less GHG emissions than the production of most, if not all, animal proteins, and often much lower GHG emissions than the production of many terrestrial protein producers (Costa-Pierce et al., 2012). In fact, many of the organically farmed aquaculture products are net carbon sinks for GHGs (Hall et al., 2011; FAO, 2015). Through the recent Summit on Climate Change (UNFCCC, 2015) in Paris, a call to action was placed on all nations to reduce food-derived GHG emissions by adopting best practices and more organic forms of food production. Increased aquaculture has been highlighted as an important means globally of enhancing the production of protein with a low environmental impact. Aquaculture and fisheries were ranked the highest in most metrics in Canada and among leading OECD nations (e.g., low

GHG emissions, low carbon and freshwater footprints, negligible pesticide, fertilizer, and chemical usage) in terms of sustainable food production (Le Vallée and Grant, 2015). With climate change affecting almost all food production activities globally, including fisheries and aquaculture (Cochrane et al., 2009; Shelton, 2014), it is incumbent upon nations to develop more climate-friendly practices. Aquaculture is suited to meet many of these goals.

In terms of spatial utilization for food production, current world food production occupies nearly 38 per cent of available land, or 11 per cent of the globe's surface land (Bruinsma, 2003). Agricultural production is constrained in many areas due to water availability, as well as by climate change impacts on soil erosion and salt intrusions from rising coastal waters. A major push is ongoing to intensify crop production on existing agricultural lands and to develop crops that are tolerant of less arable land surfaces (e.g., cold or dry tolerant varieties, genetically modified or pest-resistant varieties, etc.). In contrast, aquatic food production occupies much less than 1 per cent of available space on Earth, has very low GHGs in terms of protein output, and employs very limited amounts of freshwater resources (Hall et al., 2011).

Farming practices on our aquatic environs have progressed significantly in the past five decades. Staff, faculty, and students at Memorial University have played a key role in advancing the sector, with research and training in support of production strategies and technologies, both in the province and abroad.

The next few sections of this chapter address the following research questions:

1. Can we produce sustainable aquaculture products in Newfoundland and Labrador?
2. What are the best practices in aquaculture?
3. What are the future prospects for sustainable aquaculture in the province?

The chapter finishes with a brief examination of sustainability concerns facing aquaculture production, and concludes with future prospects.

HISTORY OF SUSTAINABLE AQUACULTURE RESEARCH AND DEVELOPMENT IN NEWFOUNDLAND AND LABRADOR

"We must plant the sea and herd its animals using the sea as farmers instead of hunters. That is what civilization is all about — farming replacing hunting." This quotation is from an interview with the famous oceans explorer and conservationist, Jacques Yves Cousteau, in 1971 (cited in Neill, 2008: 180). Perhaps at that time Cousteau had already begun to see the devastating effects that industrial, large-scale fishing vessels and pollution were having on our ocean ecosystems. Coincidentally, this was about the same time that commercial aquatic farming efforts commenced around the planet, including aquaculture research and development (R&D) in the province of Newfoundland and Labrador.

Prior to the 1970s, either private individuals or government agencies led aquaculture development, and it primarily focused on finfish enhancement activities. The chronology of Memorial University's aquaculture R&D is summarized in Table 11.1. The table provides examples of aquaculture research projects, the investigators involved, and their published work.

The earliest aquaculture efforts were devoted to wild seed collection for mussels and scallops in the late 1960s. By the late 1970s researchers were engaged in developing production technologies for shellfish farming in Newfoundland and Labrador waters (scallops, mussels), as well as examining the potential for finfish culture, mainly salmon and trout.

During the 1980s, concerns about wild stock status in the capture fisheries around the globe saw increased emphasis by researchers on developing seed supply technologies for scallops, halibut, salmon, and cod. Much of this work culminated in the establishment of the world's first commercial sea scallop and cod hatcheries in Newfoundland and Labrador in the mid-1990s. These efforts contributed to commercial hatchery production of Atlantic halibut in the 1990s in Nova Scotia, New Brunswick, and Norway. The 1980s also saw the first commercial sea scallop farm in the world developed in Little Mortier Bay, the first commercial mussel farms in insular Newfoundland, and the establishment of land-based and ocean farmed salmon and trout facilities.

Table 11.1. Chronology of aquaculture R&D activities by Memorial University staff and faculty.

Years	Synopsis of Research and Development Activities in Newfoundland and Labrador	Select Publications by Memorial Faculty, Staff, and Students
1960s	• Shellfish seed supply • Seaweed culture	Morgan 1974 South 1970
1970s	• Shellfish and finfish seed supply • Broodstock development and culture	Felt 2010 Sutterlin et al. 1981
1980s	• Shellfish hatchery research (scallops) • Marine finfish broodstock development (cod, halibut) • Production of GMO fish • Salmonid culture and strain evaluation (char, salmon, trout, triploid fish) • Reproductive control of finfish	O'Neill et al. 1984 Crim et al. 1983 Brown and Colgan 1984 Couturier 1983 Dabinett 1989 Fletcher and Davies 1991 Benfey and Sutterlin 1984
1990s	• Shellfish genetics • Production technology improvements • Shellfish hatchery establishment • Mussel seed supply dynamics • Shellfish carrying capacity • Shellfish physiology • Shellfish health • Marine finfish nutrition	Innes et al. 1999 Couturier et al. 1995 Moret et al. 1999 Daniel et al. 2008 Purchase et al. 2000 Goddard 1995 Halfyard et al. 2000 MacNeill et al. 1999
2000s	• Shellfish production enhancement • Salmonid performance evaluation • Environmental stress and indicators in shellfish and finfish • Fish health mitigation and prevention • Wild and farmed interactions • Salmonid physiology	Gallardi and Couturier 2007 Harding et al. 2004a Parsons and Robinson 2005 Rideout 2006 Fleming et al. 2003 Gamperl et al. 2004 Barker et al. 2002 McLaughlin and Couturier 2005 Vickerson et al. 2007
2010+	• Alternate shellfish aquaculture • Alternate diet formulations for finfish • Environmental performance of finfish and shellfish • Integrated pest management solutions • Adaptations to climate change • Mitigation of invasive species	Hamoutene et al. 2015 Rise et al. 2014 Hixson et al. 2014 Mercier et al.2012 Gallardi et al. 2014 Best et al. 2014 Reid et al. 2015

Note: There are well over 1,000 relevant publications over the 50-year period from Memorial researchers.

Coinciding with the cod moratorium of 1992, there was renewed interest by granting agencies, governments, and coastal communities to diversify seafood production to include several more species. In the 1990s, R&D at MUN focused on oysters, scallops, clams, seaweeds, sea urchins, eels, salmon, trout, cod, wolffish, Arctic char, yellowtail flounder, and Atlantic halibut. Most of the research during this period involved broodstock development, reliable seed procurement for shellfish, and commercially relevant production strategies for the Newfoundland and Labrador environment for scallops, mussels, salmon, trout, and cod. By the late 1990s there were very few commercial and environmentally sustainable aquaculture operations in the province. At the request of governments and the burgeoning industry, R&D emphasis was placed on four key groups deemed to have the best chance of success: native Atlantic salmon, Atlantic cod, blue mussels, and introduced trout. A more detailed overview is provided in the subsection on sustainable finfish aquaculture.

Since 2000, MUN's various research units (School of Fisheries of the Marine Institute; Ocean Sciences, Biology, and Biochemistry departments; Faculty of Engineering and Applied Sciences; Faculty of Medicine) have focused on advancing the commercial production of the four species mentioned above. A broad range of research domains have been addressed, including fish health and nutrition, broodstock development, and production. Commencing about 2005, research has been expanded to include impacts of climate change and coastal production as it relates to shellfish farming. More recent research has focused on the adaptation of marine finfish as a predictor of climate change impacts.

Over the past 15 years, R&D has been underway to develop commercial oyster and clam culture in the province, as well as to examine lobster, snow crab, and whelk culture for a variety of commercial interests. In recent years, the entire shellfish and finfish aquaculture industries have adopted internationally recognized, science-based certification standards for safe food, environmental sustainability, social acceptance, and animal welfare practices. Several MUN faculty and staff participated in the development of these standards, and assisted the industry with implementation.

SUSTAINABLE AQUACULTURE PRODUCTION OF FINFISH IN NEWFOUNDLAND AND LABRADOR

Definitions: Aquaculture, Enhancement, and Sea Ranching

According to the Food and Agriculture Organization of the United Nations (FAO), "aquaculture" can be defined as the farming of aquatic organisms: fish, molluscs, crustaceans, aquatic plants, crocodiles, alligators, turtles, and amphibians (FAO, n.d.). This implies some form of intervention in the rearing process to enhance production, such as regular stocking, feeding, protection from predators, etc., and individual or corporate ownership of the stock being cultivated. For statistical purposes, aquatic organisms harvested by an individual or corporate body that has owned them throughout their rearing period contribute to defining aquaculture, while aquatic organisms that are exploitable by the public as a common property resource, with or without appropriate licences, are the harvest of capture fisheries.

While not technically aquaculture from an FAO statistical perspective, similar techniques (including containment, feeding, and protection from predators) are used in aquatic animal enhancement and ranching activities.

"Enhancement" is defined by the FAO as:

> any activity aimed at supplementing or sustaining the recruitment, or improving the survival and growth of one or more aquatic organisms, or at raising the total production or the production of selected elements of the fishery beyond a level that is sustainable by natural processes. It may involve stocking, habitat modification, elimination of unwanted species, fertilization, or combinations of any of these practices. (FAO Term Portal, n.d.)

Aquaculture practices are often used in enhancement efforts that involve the stocking of eggs or juvenile fish in natural waterways.

"Sea ranching" is a specific type of fisheries enhancement where aquaculture techniques are used to enhance natural productivity. Specifically, it is:

> the raising of aquatic animals, mainly for human consumption, under extensive production systems, in open space (oceans, lakes) where they grow using natural food supplies. These ani-

> mals may be released by national authorities and re-captured by fishermen as wild animals, either when they return to the release site (salmon), or elsewhere (seabreams, flatfishes). (FAO, n.d.)

In the province, many of the commercial aquaculture techniques for the production of trout and salmon came directly out of early restocking efforts in an attempt to replenish the natural fish runs that had been depleted through overfishing or industrialization.

Early Aquaculture in Newfoundland: Fish Introduction and "Fisheries Enhancement"

Initial finfish aquaculture efforts in the nineteenth century were conducted through the Newfoundland Game Fish Protection Society (NGFPS) to increase the numbers of freshwater salmonids available for angling. Many of these early efforts involved the introduced species, brown trout (*Salmo trutta*) and rainbow trout (*Oncorhynchus mykiss*).

Even before the NGFPS received a Crown lease in April 1887 to operate a hatchery at Long Pond (adjacent to the present-day Fluvarium), its founding president, John (Jock) Martin, introduced brown trout (*Salmo trutta*) eggs to Windsor Lake in 1883 (Hustins, 2007). Martin may even have introduced brown trout eggs in the Heart's Content and Carbonear areas in the 1870s (Hustins, 2007). The lease at Long Pond was surrendered in 1894. Interestingly, the annual rental cost to the Society was 25 cents plus the delivery of 10,000 fry to the Crown to stock the lakes and ponds of the Crown's choosing (Hustins, 2007). Hustins (2007) reports that Jock Martin was also responsible for the introduction of rainbow trout to the province in 1887.

In addition to the more commonly known brown and rainbow trout, other salmonid species have been introduced to the island of Newfoundland. In 1886, lake whitefish (*Coregonus clupeaformis*) eggs (200,000) from Lake Erie were introduced by the NGFPS to three ponds (Hogan's, Murray's, and South Side Hills) on the Avalon Peninsula (Scott and Crossman, 1964). Also in 1886, 500,000 salmon trout (i.e., lake trout: *Salvelinus namaycush*) eggs were hatched by the NGFPS and fry were placed in Bay Bulls Big Pond, Salmonier ponds, Quidi Vidi Lake, and Long Pond (Hustins, 2007). There is no evidence that any lake trout populations have survived to the present day.

The NFGPS, in 1889, also placed 40,000 Atlantic salmon (*Salmo salar*) and brown trout (*Salmo trutta*) hybrids in the Quidi Vidi River (Hustins, 2007).

More recently, from 1958 to 1966, pink salmon (*Oncorhynchus gorbuscha*) eggs from British Columbia were introduced to the North Harbour River on the Avalon Peninsula (Scott and Crossman, 1964; Van Zyll de Jong et al., 2004). There is no evidence that this species has established populations in Newfoundland rivers (Van Zyll de Jong et al., 2004) despite the fact that spawning runs were seen in 1969 and 1970, and 800 pink salmon were taken by commercial fishers in 1969 (Scott and Crossman, 1973). In 1979, experimental aquaculture cage sites with pink salmon were set up in the discharge of the Conne River and at Hermitage (Apold et al., 1996).

Enhancement Activities

Since the 1950s, various enhancement techniques have been used to improve Atlantic salmon fisheries on the island of Newfoundland. According to Van Zyll de Jong et al. (2004), enhancement techniques may include fishways; adult transfer; artificial spawning channels; ongrowing of hatchery-reared juveniles in lake cages;[1] river stocking with hatchery-reared fry; cage-rearing of wild smolt from grilse stage through release; and habitat enhancement. Of these, three techniques (ongrowing, stocking, cage-rearing) use aquaculture practices oriented towards improving the health of wild Atlantic salmon populations.

Modern Salmonid Aquaculture Efforts

Figure 11.1 indicates total finfish aquaculture production in Newfoundland and Labrador for the 1986–2015 period. The significant majority of finfish production in all years is due to Atlantic salmon and steelhead trout, with smaller contributions by other species (e.g., Atlantic cod, Arctic char, etc.) in certain years.

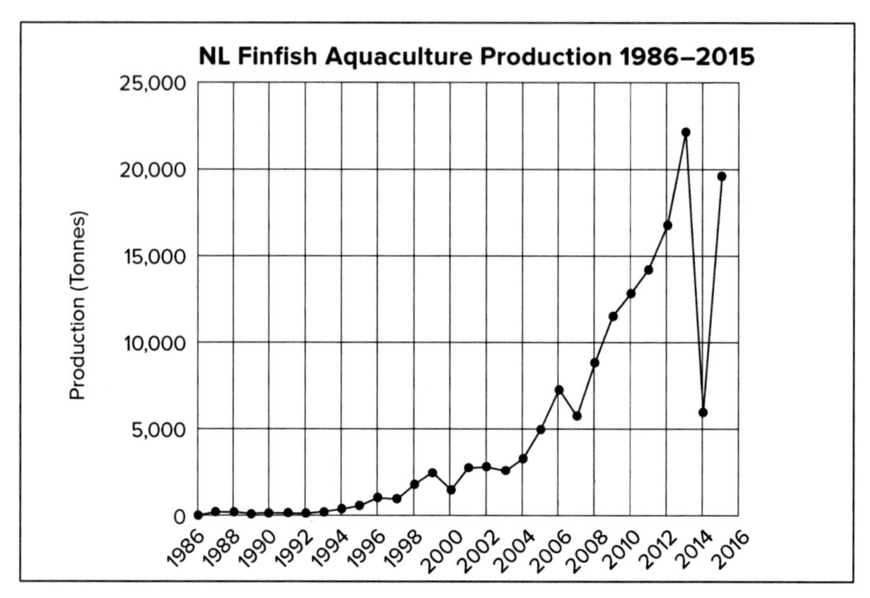

Figure 11.1. Finfish aquaculture production in Newfoundland and Labrador for the period 1986–2015. *Note: The significant decline from 2013 to 2014 is due to marine losses related to cold winter water temperatures and infectious salmon anemia (ISA). Sources: Canada (2014); Newfoundland and Labrador (2015).*

Hopeall

Construction of a rainbow trout hatchery and growout started in February 1976 at Hopeall through the efforts of the Upper Trinity South Regional Development Association (Jamieson, 1978). It was thought that the farm would produce pan-size rainbow trout for local and export markets and, in time, expand into Atlantic salmon smolt production (Aggett, 1985). The original intention was that trout would be imported from Ontario, but concerns that introduced disease could spread to local stocks led to the use of a local broodstock (Jamieson, 1978). Eventually, however, the superiority of an imported Ontario stock was proven when growth rates doubled those achieved with local fish (Newfoundland and Labrador, 1985).

In 1980, 15,000 lb (~6,800 kg) of 6–12 oz (170–340 g) rainbow trout were harvested from the Hopeall site (Aggett, 1985). Unfortunately, also in 1980, the viral disease infectious pancreatic necrosis (IPN) was found in fish both at the hatchery and in wild fish in the watershed above the hatchery (Aggett, 1985). This effectively meant that fish could not be sold into other watersheds for ongrowing.

Between the early 1980s and the early 2000s the Hopeall site was primarily a private operation for the production of rainbow trout for on-site fish-out sales.

In June 2012, Newhope Fish Farm Inc. registered an environmental assessment for the construction of a rainbow trout hatchery at Hopeall, Trinity Bay (Newhope Fish Farm, 2012). The project was released from the environmental assessment process in September 2012 (Newfoundland and Labrador, 2012). Newhope proposed to construct a land-based facility to produce 2.5 million fingerlings/smolt (60–80 g) per year and operations were to start in June 2013 (Newfoundland and Labrador, n.d.). Start-up of this operation has been delayed by aquaculture licensing processes.

Connaigre Peninsula/Coast of Bays Region

After the initial foray into commercial fish farming at Hopeall, people began to look at the Bay d'Espoir region as a potential location for expansion of the fledgling industry. Presently, a large portion of the province's finfish aquaculture production is focused within the Connaigre Peninsula/Coast of Bays region. Atlantic salmon ongrowing sites are concentrated in Facheux Bay west of Bay d'Espoir, Bay d'Espoir, Hermitage Bay, Connaigre Bay, Great Bay de l'Eau, and the western half of Fortune Bay. Steelhead trout ongrowing takes place on the inner, brackish part of Bay d'Espoir.

From the outset, MUN was involved with the development of aquaculture in the Bay d'Espoir region. The involvement included some of the earliest site selection work (1974) in the region, with growth and rearing trials (1977–79) at the hydroelectric facility (MUN, 1995). Growth and rearing studies examined the feasibility of using waste heat from the hydroelectric generating station for hatching and raising salmonids (Aggett, 1985). By 1980 the university was involved with the set-up of ongrowing cages in Roti Bay and the Conne River areas of Bay d'Espoir (MUN, 1995) and had completed a two volume report, "Bay d'Espoir Aquaculture Feasibility Study, MUN MSRL, 1980" (Apold et al., 1996). The report recommended monitoring of overwintering sites and construction of a hatchery at the hydroelectric dam (Apold et al., 1996).

Surveys conducted during the winters of 1981 and 1982 concluded that water below the ice in Bay d'Espoir was suitable for the overwintering of salmonids, and Roti Bay was seen as the best site for overwintering (Apold et al., 1996). Interestingly, lethally low temperatures were measured at all coastal locations sampled, making the sites unsuitable for year-round culture (Apold et al., 1996). Concurrent with the winter surveys, the potential hatchery site, adjacent to the hydroelectric facility, was evaluated (MUN, 1995).

Construction began on the hatchery in 1984, which was officially opened on 19 November 1985 by Premier Brian Peckford (Apold et al., 1996). The Bay d' Espoir Development Association built the hatchery and encouraged family-owned and -operated grow-outs of the juvenile fish produced at the hatchery. This led to the establishment of the Bay d'Espoir Salmon Growers Cooperative (Apold et al., 1996) for joint feeding, purchasing, processing, and marketing (Newfoundland and Labrador, 1998a).

Early on in the development of salmonid farming in the Bay d'Espoir region, the farmers were required, by federal regulation, to use local Newfoundland salmon stocks. Fisheries and Oceans Canada (DFO) conducted a great deal of the research on the identification of an appropriate Newfoundland stock (Pepper et al., 1996, 1999, 2001, 2003, 2004; Peterson et al., 1990). However, it was found that many of the local stocks were grilsing, meaning that a high proportion would sexually mature before reaching market size. Sexual maturation is not desirable among production fish in an aquaculture setting because feed energy that would otherwise go to somatic growth ends up contributing to the production of gametes and secondary sexual characteristics (i.e., distinctive external signs of sexual maturation).

In 1988, a significant bacterial infection (*Aeromonas salmonicida*, the causative agent of furunculosis) at the hatchery led to the destruction of the entire Newfoundland stock of salmon (Newfoundland and Labrador, 1998a). Perhaps fortuitously, this setback led to the lifting of the embargo on non-native salmon strains. From this point on, Bay d'Espoir salmon growers were permitted to bring in other Atlantic Canadian salmon stocks, with the Saint John River (New Brunswick) strain being the most widely used. The move to non-local strains of Atlantic salmon in the Bay d'Espoir region led to significant research on various reproductive technologies for the production of all female (typically later maturing than males) or sterile Atlantic salmon (Pepper et al., 2004). There was still concern that these non-local strains would negatively affect local stocks of Atlantic salmon should they escape into the wild. As per their mandate as stewards of wild fisheries resources, much of DFO's work with the salmonid farming industry, in recent years, has been centred on the interaction of the salmonid farming industry and the environment. This bidirectional interaction includes examining the industrial impact on the environment (Anderson et al., 2005; Hamoutene, 2014; Hamoutene et al., 2013, 2015a; Mabrouk et al., 2014; Mansour et al., 2008; Salvo et al., 2015; Tlusty et al., 1999, 2000, 2005) as well as the impact of environmental conditions on production (Tlusty et al., 1998; Burt et al., 2012, 2013; Hamoutene et al., 2015b).

In August 1989, 33 investors from the original 10 ongrowing farms restructured to form SCB Fisheries (Newfoundland and Labrador, 1998a). With the destruction of the salmon stock at the hatchery, SCB Fisheries estimated that it would take two years before they could bring the next crop of salmon to market. This led to a request to access steelhead trout (*Oncorhynchus mykiss*) that could reach market size in 15 to 19 months (Newfoundland and Labrador, 1998a). Permission was granted to bring in triploid (i.e., sterile) steelhead trout. In 2001, North Atlantic Sea Farms Inc. became the successor of SCB's farming assets (Newfoundland and Labrador, 2001). This firm operated until 2008 when it was bought out by Northern Harvest Sea Farms Inc. (Newfoundland and Labrador, n.d.), a company long established in the New Brunswick salmon farming industry. In 2011, Northern Harvest opened a smolt hatchery (Northern Harvest Smolt Inc.) in Stephenville to supply the company's marine operations in the Coast of Bays region. This facility has a total capacity of more than 4 million smolt, making it one of the largest land-based closed-containment facilities in North America (Newfoundland and Labrador, n.d.).

The Miawpukek First Nation of Conne River became involved in aquaculture in 1991 when development of a steelhead trout farm began (Newfoundland and Labrador, 1998a).

In 2006, Cooke Aquaculture Inc. (and its Newfoundland subsidiary, Cold Ocean Salmon), another well-established, family-owned salmon farming company from New Brunswick, started farming operations in Newfoundland and Labrador (CBC News, 2006). In order to supply its marine operations in the Coast of Bays region, the former Arctic char hatchery in Daniel's Harbour was purchased and expanded (Newfoundland and Labrador, n.d.). In 2010, Cooke began construction on a salmon nursery in the Swanger Cove area, near St. Alban's (Newfoundland and Labrador, n.d.). Operational by 2011, this nursery receives fry from the Daniel's Harbour hatchery and produces up to 3 million smolt annually (Newfoundland and Labrador, n.d.). Various aspects of contemporary salmon farming in the province are shown in Figure 11.2.

A
B
C
D
E
F
G

H
I
J

Figure 11.2. A view of modern salmon farming in Newfoundland and Labrador. (A) Salmon farmer at harvest time; (B) salmon egg stripping; (C) salmon egg and yolk-sac fry incubators; (D) vaccinating salmon smolt prior to saltwater entry; (E) modern recirculating aquaculture system (RAS) for salmon smolt production; (F) water treatment system on modern RAS closed-containment tanks; (G) finfish feed storage and distribution barge capable of feeding up to 1 million fish at a time; (H) modern feed control centre with surface and subsurface video monitoring to prevent feed wastage and optimize feeding in finfish; (I) large ocean containment net pen systems (150 m circumference) holding up to 100,000 fish per system and never exceeding 2 per cent rearing density (kg per cubic metre) of the net; (J) value-added fresh production of salmonid fillets, from farm to fork in 24–48 hours. (Photos courtesy of C. Couturier, K. Rideout, J. Westcott, T. Granter, Northern Harvest Sea Farms, and Cold Ocean Salmon)

Arctic Char Initiatives

Since the 1980s, there has been great interest in farming Arctic char in the province. The species' preferences for high rearing densities and cooler waters, particularly when compared to salmon and trout, make it an ideal candidate for culture. That said, variable growth rates among individuals of a cohort indicate that work must be done to produce an appropriate commercial stock.

Focus on Arctic char in this province began in 1988 (MUN, 1995). In 1989, eggs from the Fraser River in Labrador were imported to the Bay d'Espoir hatchery from New Brunswick and Manitoba (Sutterlin, 1991). Performance of the char was compared to all-female triploid (i.e., sterile) rainbow trout and the Saint John River strain of Atlantic salmon. Survival to 4 g was poorest in the char, compared to other species (28 per cent versus 70 and 75 per cent for the trout and salmon, respectively) (Sutterlin, 1991).

More recently, commercial production efforts for Arctic char in Newfoundland and Labrador have taken place in Port Rexton (land-based), the Main Dam area of Grand Lake near Deer Lake (cage-rearing), at Daniel's Harbour (land-based), and finally in the inner part of Bay d'Espoir (cage-rearing).

The Daniel's Harbour Arctic char hatchery and grow-out operated from 1991, using an 800-foot (~240-metre) decommissioned well from a former zinc mine. Initially, the operation was quite rudimentary, using temporary trailers, until a modern recirculating hatchery and grow-out was constructed in the mid-1990s. The intended production capacity was 200,000 lb (90 tonnes) (Newfoundland and Labrador, 1998b). It operated as a char facility until purchased by Cooke Aquaculture in the mid-2000s.

Marine Finfish Farming Initiatives

In addition to a significant interest in salmonid aquaculture in Newfoundland and Labrador, there recently has been significant work completed on a range of marine finfish species: Atlantic cod (*Gadus morhua*), Atlantic halibut (*Hippoglossus hippoglossus*), wolffish (*Anarhichas sp.*), yellowtail flounder (*Pleuronectes ferruginea*), winter flounder (*Pseudopleuronectes americanus*), witch flounder (*Glyptocephalus cynoglossus*), lumpfish (*Cyclopterus lumpus*), and ocean pout (*Macrozoarces americanus*). Following on the pioneering hatchery work of Adolph Nielsen with Atlantic cod in Dildo, Trinity Bay (1890–96), much of the more recent work has been related to achieving a better understanding of the reproductive biology and rearing requirements (e.g., rearing conditions, nutritional requirements, etc.) of these species (see Table 11.1). For various technical and/or economic reasons none of these species are currently commercially cultured within the province.

Sustainable Shellfish Production in Newfoundland and Labrador

Commencing in the 1960s and 1970s, efforts were undertaken to diversify capture fisheries production by investigating natural seed supply of several native shellfish species, including blue mussels (*Mytilus edulis*) and sea scallops (*Placopecten magellanicus*). Efforts were continued throughout the 1980s and 1990s to refine production methods for hatchery, nursery, and farming to suit the local environment. By the early 2000s, the mussel sector became a viable food-producing industry that produced several thousand tonnes. In contrast, the scallop industry — having accomplished many world "firsts" — eventually collapsed. This was attributed to lack of investor confidence, unexplained mortalities at the growing stage, and increased interest by producers in focusing on a more reliable species, the blue mussel.

Mussel Aquaculture in Newfoundland

Researchers began experimenting with wild seed collection of mussels in the late 1960s, at a time when mussel farming began to increase in Europe and Asia. This was seen as an opportunity to potentially diversify some areas in the province where capture fisheries were low or unpredictable. Experiments were conducted with growing mussels in the harsh winter and ice conditions prevalent along our coast. Soon, European farming methods were abandoned. These experimental growing methods continued on the south coast of Newfoundland in Garden Cove, Placentia Bay, until 1981, when the first pilot-scale longline farm, resistant to local drift ice, was turned over to a commercial interest for expansion and marketing. This new farming operation, Atlantic Ocean Farms, operated in the region for a few years before moving to Notre Dame Bay on the northeast coast, where ice-fast longline culture methods were developed along with year-round processing methods for live, fresh shellfish. The mussel R&D efforts up to that point are summarized in Sutterlin et al. (1981). Additional research using ice-fast longline methods on the northeast coast was undertaken in Bonavista Bay with the view of expanding the industry to more regions of the province (Couturier, 1983).

In 1987, a newly discovered, naturally occurring phytotoxin called domoic acid had poisoned several hundred people, with three fatalities in the developing Prince Edward Island mussel farming industry (Couturier, 1988). New toxin monitoring programs had to be developed by the food inspection agencies to assure consumers the products were safe, but not before consumer confidence in farmed mussels had been shaken across North America, including in regard to the burgeoning Newfoundland industry.

The mussel farming industry had just begun to stabilize and expand in the early 1990s when unusual summer conditions (low temperatures, low natural food levels) caused major reductions in natural mussel seed collection. These same conditions coincided with the now famous 1992 cod moratorium, and the failure of cod and mussel declines are likely linked in terms of lack of food and poor growing conditions for juveniles. In the cod analogy, the loss of capelin, cod's principal food source, is thought to have contributed to stock recruitment and collapses at this time (Rose, 2007).

University faculty were asked to develop an environmental monitoring and seed supply program for the mussel industry to try and avoid future collapses in seed supply. This program was highly successful in that dozens of new mussel farmers were trained in the art of mussel larval and environmental monitoring, and the protocols established in the early 1990s largely by the Marine Institute's faculty and staff continue to today, without a single failure in seed supply since that time. The program ceased in 2002 and has been internalized within the provincial Department of Fisheries and Aquaculture as a service to industry.

By the mid-1990s, there was variable mussel production and growth along the Newfoundland and Labrador coasts. In spite of increased interests by numerous investors, the mussel farming industry had not been able to expand beyond annual production figures of 500 tonnes, valued at about $1 million. Once again, faculty from Memorial were asked to develop a comprehensive mussel production and enhancement program for the sector. Led by the Marine Institute, research started in 1996 into new and more efficient production methods, mussel carrying capacity, mussel health assessment and stress assessment, stock performance, and mussel aquaculture physiology and ecology (Macneill et al., 1999; Clemens et al., 1999; Moret et al., 1999; Brown et al., 1999; Innes et al., 1999; Ibarra and Couturier, 1998; Ibarra et al., 2000; Mooney et al., 2002; Harding et al., 2004a, 2004b, 2007; Ross et al., 2007). These major efforts ceased in 2004 when grants from NSERC, AquaNet — Network Centers of Excellence, and provincial and federal agreements came to an end; however, the industry had grown from a $1 million to an $8 million per year industry, largely with the R&D support by the university's research units.

Newfoundland and Labrador mussel farming historical production and key inputs are shown in Figure 11.3. It is clear that efforts by MUN were critical in assisting the sector reach true commercial status, where fresh-farmed mussels are supplied locally and throughout North America 52 weeks of the year.

In 2015, the mussel industry is viable, worth between $10 million and $15 million annually to the provincial economy. Newfoundland and Labrador is now the second largest producing province of farmed mussels in Canada and has significant potential for more growth. Like salmon and trout farming outputs, it is one of the few foods available fresh, locally and across North America, 365 days of the year. Advances are still required in novel seafood formats, new site developments, deep-water cultivation, enhanced production technologies, and climate change adaptations for the mussel industry, and students and faculty are participating in these efforts on a regular basis (e.g., Gallardi et al., 2014).

Figure 11.4 illustrates some of the advances in farming technology and methods developed with the industry in rural Newfoundland and Labrador. These are the most cost-efficient and environmentally sustainable producers of farmed mussels in North America.

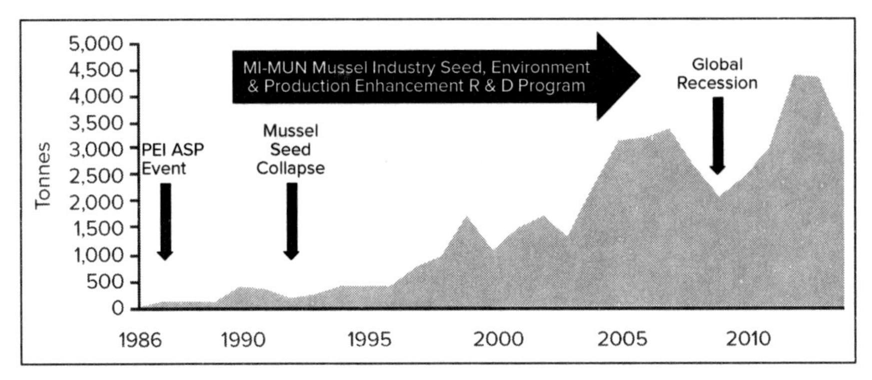

Figure 11.3. Mussel production by volume, 1986–2014. Significant events in the history of the industry are shown at the top of the graph. Memorial's involvement in applied research and development began in earnest in 1993 and continues to today. During this period the industry grew from about 50 employees producing $1 million per annum to today's 250 employees producing about $15 million in mussel products per annum. The R&D program focused initially on environmental and stock performance, genetics, production technology, carrying capacity, and mussel health. More recent efforts have focused on improving production efficiencies on the farm and in post-harvest processing. During this period, approximately 25 undergraduate, M.Sc., Ph.D., and post-doctoral students completed their studies on industry-related R&D on the farms.

Alternate Species of Shellfish

Research and development on alternate shellfish and seaweed production has been ongoing since the 1960s. Research on scallop farming, sea urchin culture, non-native oyster farming, clam culture and enhancement, and various seaweeds has ranged the gamut from seed supply to commercial production. A number of "firsts" were established as a result: the world's first commercial sea scallop farming operation in Placentia Bay (Couturier et al., 1995), the first scallop hatchery patent and the first commercial sea scallop hatchery (Parsons and Robinson, 2006). Researchers were instrumental in developing methods for broodstock control, wild seed collection, hatchery and nursery seed production, and at-sea farming methods for the native sea scallops (Couturier and Newkirk, 1991; Couturier, 1994a, 1994b; Couturier et al., 1995; Grecian et al., 2000, 2003; Parsons and Robinson, 2005). As noted above, efforts to farm scallops in the province ceased in 2002 owing to unexplained mass mortalities, lack of investor confidence, and market pressures. However, the methods developed for this species in Newfoundland and Labrador, or variations thereof, are now employed by sea scallop farmers along the northeast coast of North America from Quebec to Maine.

A
B
C
D
E
F
G
H
I
J

Figure 11.4. A view of modern mussel farming in Newfoundland and Labrador. (A) Mussel production farm; (B) mussel seed collectors after a few months' deployment; (C) continuous socking method for mussels; (D) mussel growing ropes with biodegradable cotton "socks"; (E) deployment of growing ropes on a farm; (F) harvestable organic mussels after three years from seed to market size; (G) harvest boat returning to shore; (H) wet (live) storage of mussels after harvesting; (I) washing and grading mussels before packaging for market; (J) packaging mussels for live, fresh market. (Photos courtesy of C. Couturier)

Research on sea urchin and seaweed cultivation in Newfoundland and Labrador was undertaken throughout the 1990s and early 2000s; however, to date there are no commercial producers in the province (Hooper at al., 1997; Hiemstra, 2001). Efforts to develop eastern oyster (*Crassostrea virginica*) farming and softshell clam (*Mya arenaria*) aquaculture in the province have been the subject of ongoing research since 2005, and commercialization of both species is expected soon.

SUSTAINABILITY AND CERTIFICATION

Sustainability of Aquaculture Practices

There are potential negative impacts of any form of food production. The question becomes: How does aquaculture mitigate or prevent impacts in order to achieve sustainable production? When considering the sustainability of aquaculture, one must include the three tenets of sustainability (World Bank, 2014):

1. Environmental sustainability — Aquaculture should not create significant disruption to the ecosystem, or cause the loss of biodiversity or substantial pollution impact.
2. Economic sustainability — Aquaculture must be a viable business with good long-term prospects.
3. Social and community sustainability — Aquaculture must be socially responsible and contribute to community well-being.

Presumably, aquaculture activities that are not subsidized through direct or indirect governmental processes and programs will either be economically sustainable or they will fail. Economically dubious aquaculture activities will not persist for long. Likewise, aquaculture activities that are not socially responsible or that do not contribute to the well-being of the local community will not be maintained unless there is an inequitable distribution of power in the community. Social licence is a must for any form of aquaculture or food production activity.

All aquaculture operations in Newfoundland and Labrador must be vetted for environmental impacts, social acceptability, and financial sustainability. No less than 17 federal and provincial regulatory agencies, with mandates to protect the public trust, public health, and the environment, must approve the plans of a fish farmer before he or she can even commence to put fish or shellfish in the water. It is one of the most highly regulated sectors of the economy anywhere in Canada.

When most people think of the sustainability of aquaculture practices, environmental sustainability is top of mind. In a 2011 study, *Blue Frontiers — Managing the Environmental Cost of Aquaculture* (Hall et al., 2011), the authors delineated the environmental impacts of aquaculture into six categories:

1. Eutrophication — release of macronutrients into the environment. Expressed as t PO4 equivalents.
2. Acidification — release of acidifying substances into the environment. Expressed in t SO_2 equivalents.
3. Climate change — based on the characterization model developed by the Intergovernmental Panel on Climate Change (IPCC). Expressed as climate change potential in tonnes of CO_2 equivalents.
4. Cumulative energy demand (CED) — the direct and indirect use of industrial energy throughout the production process. Expressed in Gj (giga Joules).
5. Land occupation — sum of direct and indirect land occupation. This analysis used equivalence factors adjusted for the relative levels of bioproductivity. The higher the bioproductivity of the land, the higher equivalence factor and therefore the land occupation. Expressed in ha equivalents.
6. Biotic depletion (fish) — the small, pelagic fish used to produce the cultured species.

For the life-cycle analyses (LCA: cradle-to-grave environmental costs, i.e., the cumulative energy demand indicator, one example of a life-cycle analysis), cultured species were placed into 13 groups, based on 2008 FAO statistics. Collectively, these 13 groups accounted for 82 per cent of global aquaculture production in that year.

When the LCA were completed for the 13 species groups, it was clear that relative to other cultured species, the greatest impact of salmonid aquaculture (the largest portion of the province's aquaculture industry production and value) is from the biotic depletion perspective. In fact, salmonid farming compares quite favourably to the other species groupings, with only bivalves and gastropods doing better than salmonids across the other five impact categories. Seaweeds and aquatic plants and other vertebrates perform better than salmonids from a eutrophication perspective, and other invertebrates score better on the land occupation category compared to salmonid culture.

On biotic depletion (i.e., fish consumption), only eel production does more poorly than salmonid farming. While significant improvements have been achieved in recent decades, work still needs to be done to reduce the amount of wild fish (in the form of fishmeal and fish oil) needed to feed farmed salmon and trout.

The conversion of wild fish to farmed fish is termed the Fish In: Fish Out (FIFO) ratio. FIFO is a way to describe precisely how much wild fish is needed to produce farmed fish, such as salmon. To determine how much wild fish is needed to produce farmed salmon, an understanding of a few fundamental relationships is necessary. First, one must assume a yield of fishmeal and oil from wild pelagic fish most commonly used for the production of these products. Based on the data of Jackson (2009) and Hicks (2011), the yield in meal and oil from wild pelagic fish sources is approximately 32 per cent. The meal and oil are used in the production of finished, pelleted or extruded diets that are then fed to the salmon. The amount of fishmeal and oil added to salmon diets has declined significantly in the last 20 years. On average, a salmon diet in 1995 contained 45 per cent fishmeal and 25 per cent fish oil (Tacon et al., 2011). By 2010 these values had been reduced to 22 per cent and 12 per cent for meal and oil, respectively (Tacon et al., 2011). The next important relationship is the conversion of the finished feed to salmon flesh. This is referred to as feed efficiency and is expressed, very often, as a feed conversion ratio (FCR) the ratio of feed fed to fish biomass produced.

With improved diets and feeding practices, aquaculturists have been able to reduce (i.e., improve) FCR over the years. In the 1970s FCRs in excess of 2:1 for salmon aquaculture were not uncommon. Today, FCRs in the 1.0–1.3:1 range are routine (Sarker et al., 2013).

These values, in aggregate, mean that today it takes approximately 1.3 kg of wild, pelagic fish to produce a kilogram of cultured salmon (FIFO = 1.3:1). While this value is still greater than 1:1 and means that more fish is being used to produce the salmon than is contributed by the salmon production, it is significantly more efficient than even two decades ago. In recent years, MUN researchers have played an important role in evaluating alternative diets for reducing fish meal and oil components in farmed fish diets even further, using molecular and nutrigenomics tools (e.g., Hixson et al., 2014).

Much of the research around the impacts of aquaculture in Newfoundland has centred on finfish farming, particularly the portion carried out in marine waters. This research has concentrated on two broad areas: the impacts associated with organic enrichment from farms (Anderson et al., 2005; George, 2011; Hamoutene, 2014; Hamoutene et al., 2013, 2015a; Mabrouk et al., 2014; Oldford, 2013; Power, 1999; Tlusty et al., 2000, 2005) and interactions of escapees from farms (Bridger, 2002; Hamoutene et al., 2015b; Lush et al., 2014; Rideout, 2006; Wringe et al., 2015; Zimmerman et al., 2012, 2013). The impacts of shellfish farming in Newfoundland and Labrador have also been examined, and these have been determined to be negligible with respect to the environment unless there is overstocking of small inlets (Cranford et al., 2006). In fact, quite a few recent studies show positive influences on the local environment from shellfish culture, including enhanced habitat and fisheries production near shellfish farms (e.g., McKindsey et al., 2012; Wang, 2014).

The organic enrichment work has focused on identifying appropriate sampling methods and indicator organisms of enrichment. Common sampling techniques employed include sediment sampling (Barkhouse, 2003) and video collection (Hamoutene et al., 2015a; Mabrouk et al., 2014; Oldford, 2013). Video collection is preferred in the province because the rocky and patchy nature of substrates makes it difficult to consistently collect sediment samples. Indicator organisms identified as having some correlation with organic inputs include the sulphur-oxidizing bacterium *Beggiatoa sp.* (Hamoutene, 2014; Hamoutene et al., 2013), nematodes (Power, 1999), and opportunistic polychaete worm complexes (OPC) (Hamoutene, 2014; Hamoutene et al., 2013; Salvo et al., 2015).

Research on escapement of farmed fish in Newfoundland has concentrated on reproductive success (Lush et al., 2014; Wringe et al., 2015), survival (Hamoutene et al., 2015b), or dispersal after escapement (Bridger, 2002; Zimmerman et al., 2013). Potential consequences of escaped fish include hybridization with wild conspecifics and reduced fitness or competition for natural resources. There is no evidence of competition for wild resources in escaped salmonids; the risk is small and most of the farmed, domesticated fish are preyed upon or do not feed well once they escape (C. Hendry, personal communication, DFO, 2015). Escaped fish have been relatively few in the past two decades owing to stringent containment protocols. However, at the time of publication a large, high-profile spill of farmed salmon from the state of Washington has renewed the public's interest in the potential interactions between farmed and wild fish, and invasive aquatic species. In this particular event, the large quantity (305,000 fish), and a relatively high number of sightings in BC, will likely be the subject of study in the years to come. For contextual purposes, between 2011 and 2017, there were only three reports of escaped farmed Atlantic salmon in BC, but directly following the August 2017 event, there were 40 reports as far north as 250 kilometres from the accident (CBC News, 2017). Research is ongoing to validate whether hybridization and reduced fitness in the hybrids have occurred in the province's natural systems, but based on the presently available science, no evidence suggests this will have a negative consequence on future populations.

Another potential impact of farmed fish is the transfer of pests or diseases to wild stocks. To date, all farmed fish are vaccinated against the major natural disease agents and must be certified by veterinary authorities as disease-free before being deployed in ongrowing cages. Farmed fish, however, are exposed to natural disease agents from wild fish and the cage environment. If infected, fish health treatments are prescribed and monitored by veterinarians, for animal welfare purposes and to prevent potential spread back to wild stocks. All of this is carefully monitored with approvals from the various regulatory agencies such as Environment Canada, Fisheries and Oceans, Health Canada, and the Canadian Food Inspection Agency, as well as provincial counterparts. Most fish are never subjected to pesticide or therapeutant treatments, and there is little evidence that disease outbreaks in farmed fish re-infect wild fish in Newfoundland waters, presumably due to the proactive fish health management strategies employed by the farmers.

The university has been engaged in R&D over the past 15 years to provide better fish health management tools for finfish farmers, including vaccines, immunostimulants, selective breeding for disease resistance, and alternative organic pest treatment options.

Certification of Finfish Farming Practices

Both Cooke Aquaculture Inc. and Northern Harvest Sea Farms Inc. carry the Best Aquaculture Practices — Four Star Rating for their farming operations (Global Aquaculture Alliance, n.d.). The Best Aquaculture Practices (BAP) certification standards were developed in 1997 by the non-profit NGO, Global Aquaculture Alliance (GAA), to encourage the use of responsible aquaculture practices for a variety of species, including shrimp, tilapia, catfish, and most recently, salmon. The BAP salmon standards are built to specifically address a full range of issues, including environmental and social responsibility, animal health and welfare, food safety, and traceability for each phase of the salmon production cycle. The standards promote the responsible use of resources including land, water, nutrients, and other resources.

These are science-based, independently verified standards, providing customers assurances that the production facilities meet the highest standards of environmental and social sustainability, food safety, and animal welfare. The standards were developed by conservation-minded, environmental non-governmental organizations (ENGOs), academics, governments, and industry personnel with a keen interest in sustainable aquaculture.

Certification of Mussel Farming

The Newfoundland and Labrador mussel farming industry was the first in the world to be certified to the Canadian Organic Aquaculture Standard (COAS; Standards Council of Canada, 2012) and the Best Aquaculture Practices Mussel standard (BAP Mussel, Global Aquaculture Alliance, n.d.). Both eco-labels offer third-party certification of adherence to international standards for environmental and social sustainability, animal welfare, and food safety. They are among the most rigorous, science-based standards available anywhere. Given their expertise, MUN Marine Institute faculty participated in the development of these science-based certification standards (COAS, BAP) as well as the Aquaculture Stewardship Council's mollusc standard (ASC Bivalve, Aquaculture Stewardship Council), also an increasingly recognized global ecolabel for consumers. Many retailers are now asking for one or more ecolabels of this sort to give increased assurances to the customers that their seafood is sustainable, safe, and of reputable origin.

Educational Programs

Education and training have gone hand in hand with the research associated with development of the Newfoundland and Labrador aquaculture industry. MUN and its research units, particularly the Marine Sciences Research Lab, the Ocean Sciences Centre, and the Fisheries and Marine Institute, have been the primary providers of aquaculture-related education. This programming has run the educational gamut from short-duration industry training through multi-year doctoral programming.

The province's pioneering aquaculture program was the Graduate Diploma in Aquaculture, established in 1987 at the Newfoundland and Labrador Institute of Fisheries and Marine Technology. In 1992, this institution became part of the larger MUN system and was renamed the Fisheries and Marine Institute of Memorial University (MI). This post-graduate program has undergone a few name changes over the years, and it continues today as the Advanced Diploma in Sustainable Aquaculture program.

In 1994, Memorial University established an interdisciplinary Master of Science program in Aquaculture that is jointly administered by the Ocean Sciences Centre and the Marine Institute, with representatives from the Ocean Sciences Department, Biology, Biochemistry, and the Marine Institute. In some cases, graduates of this program have gone on to aquaculture-specific doctoral studies within the Faculty of Science at Memorial (in Biology or Environmental Science).

The Marine Institute has been the primary provider of industrial training opportunities for individuals already working within the aquaculture industry as well as those looking for employment in this rapidly expanding field. This programming has consisted of stand-alone courses or collectives of courses packaged together into larger certificate programs. Examples include the Certificate in Aquaculture programs of the 1990s and, more recently, the Technical Certificate in Aquaculture programs that have been going since 2010 for the salmonid and shellfish segments of the provincial industry.

In 2008, the Marine Institute established a professional development program at distance entitled Master of Technology Management (Aquaculture).

This program caters to current professionals who aspire to upgrade their skills in managing aquaculture production and development.

Over the course of the past two decades, the Marine Institute has aided in the development and establishment of aquaculture diploma and degree programs in a variety of countries, including Vietnam, Cambodia, Mozambique, and Malawi. Within the province, several science credit courses with significant aquaculture components at the high school level have been developed with the aid of faculty at the Marine Institute, as well as teacher summer marine sciences courses consisting of aquaculture and fisheries components.

Many of the 400 graduates from the diploma and degree programs since the 1980s are now leaders in the aquaculture sector, in research, and in government positions across the province and Canada, and in more than 10 other countries. Well over 1,000 industry personnel have been trained in Newfoundland and Labrador since the early 1990s. Undergraduate and graduate training and research collaborations have extended over several continents, including Africa (Mozambique, Malawi), Asia (Indonesia, Cambodia, and Vietnam), South America (Brazil and Chile), and Europe (Iceland and Norway).

CONCLUSIONS

The three research questions posed in the introduction to this chapter were:

1. Can we produce sustainable aquaculture products in Newfoundland and Labrador?
2. What are the best practices in aquaculture?
3. What are the future prospects for sustainable aquaculture in the province?

Through a historical narrative we show that sustainable aquaculture is prevalent in our province. This sector is not only using best practices for responsible aquaculture, but in most cases it accedes to international sustainable aquaculture practices for environmental protection, animal welfare, and social acceptance. Today, Newfoundland and Labrador is the second largest producing province in Canada for sustainable aquaculture. Many of the bottlenecks to production have been solved by close partnership among researchers at MUN over the past two decades. The university will continue to assist with the advancement of this increasingly important, sustainable food industry.

Graduates from our certificate to post-graduate programs are leaders in industry, government, and academia across Canada and the globe. Existing and new programs are continually being developed to meet various segments of the sustainable aquaculture sector (farm technicians, farm managers, researchers, and government policy and management agencies).

MUN's research and development capacity, expertise, and collaborations in cold-water aquaculture are unmatched in Canada; we are recognized nationally and internationally for our efforts. We are increasingly involved in training and research on aquaculture in a variety of locales globally, from the countries in the Northeast Atlantic to South America, sub-Saharan Africa, and Southeast Asia.

The future of sustainable aquaculture in Newfoundland and Labrador is indeed promising. However, there will continue to be challenges faced by the sector with respect to continued sustainability. These challenges include climate change impacts on production, mitigating impacts on the environment, production of better-performing stocks, improved fish health management, and training the leaders of the future. The university is well poised with its complement of faculty and staff and world-class facilities to assist with the challenges and future prospects of our province's sustainable aquaculture sector.

ACKNOWLEDGEMENTS

The sustainable aquaculture initiatives presented in this chapter would not have been possible without the ongoing commitment of numerous funding agencies, granting councils, entrepreneurs, aquaculturists, First Nations, interested individuals, government departments, and environmental groups. There are too many to mention, and at the risk of forgetting someone, we can only give our thanks for their belief in aquaculture as a means to diversify our economy, sustain our communities, and provide some of the most nutritious, sustainable foods available.

NOTE

1. "Ongrowing" is the process whereby the small juvenile fish or shellfish are grown to market size.

REFERENCES

Aggett, D. 1985. *Aquaculture Activities in Newfoundland, 1970–1980.* Department of Fisheries, Industry Support Services Division, Data Report Series No. 19, Feb. St. John's, Government of Newfoundland.

Anderson, M. R., M. F. Tlusty, and V. A. Pepper. 2005. "Organic enrichment at cold water aquaculture sites — the case of coastal Newfoundland." In B. T. Hargrave, ed., *Handbook of Environmental Chemistry*, vol. 5, 99–113. Berlin: Springer.

Apold, W. O., A. Sutterlin, and C. Couturier, C. 1996. *South Coast Region Aquaculture Strategic Plan*. Halifax: Tavel Limited.

Aquaculture Stewardship Council. n.d. "Bivalve standards." At: http://www.asc-aqua.org/?act=tekst.item&iid=6&iids=290&lng=1.

Barker, D. E., D. K. Cone, and M. D. B. Burt. 2002. "*Trichodina murmanica* (Ciliophora) and *Gyrodactylus pleuronecti* (Monogenea) parasitizing hatchery-reared juvenile winter flounder (*Pleuronectes americanus*): Effects on host growth and assessment of parasite interaction." *Journal of Fish Diseases* 25: 81–89.

Barkhouse, C. 2003. "How many sediment samples are required for an environmental assessment of an existing or potential aquaculture site?" Master of Env. Sci. Report, Memorial University.

Benfey, T. J., and A. M. Sutterlin. 1984. Growth and gonadal development in triploid landlocked Atlantic salmon (*Salmo salar*). *Canadian Journal of Fisheries and Aquatic Sciences* 41, 9: 1387–92.

Best, K., C. McKenzie, and C. Couturier. 2014. "Investigating mitigation of juvenile European green crab *Carcinus maenas* from seed mussels to prevent transfer during Newfoundland mussel aquaculture operations." *Management of Biological Invasions* 5, 3: 255–62.

Bridger, C. J. 2002. "Movement and mitigation of domestic triploid steelhead trout (*Oncorhynchus mykiss*) escaped from aquaculture grow-out cages." M.Sc. thesis, Memorial University.

Brown, C., C. Couturier, T. Zokvic, and J. Parsons. 1999. "Towards best practices: A practical guideline for mussel aquaculture in Newfoundland." Proceedings of the Workshop on Mussel Production Capacity (Part 2) Held at Aquaculture Canada '98. *Bulletin of the Aquaculture Association of Canada* 99, 2: 38–41.

Brown, J. A., and P. W. Colgan. 1984. "The ontogeny of social behaviour in four species of centrarchid fish." *Behavioural Processes* 9, 4: 395–411.

Bruinsma, J., ed. 2003. *World Agriculture: Towards 2015/2030, an FAO Perspective.* London: Earthscan.

Burt, K., D. Hamoutene, G. Mabrouk, C. Lang, T. Puestow, D. Drover, and F. Page. 2012. "Environmental conditions and occurrence of hypoxia within production cages of Atlantic salmon on the south coast of Newfoundland." *Aquaculture Research* 43, 4: 607–20. doi:10.1111/j.1365-2109.2011.02867.x.

———, ———, J. Perez-Casanova, A. K. Gamperl, and H. Volkoff. 2013. "The effect of intermittent hypoxia on growth, appetite and some aspects of the immune response of Atlantic salmon (*Salmo salar*)." *Aquaculture Research* 45, 1: 124–37. doi:10.1111/j.1365-2109.2012.03211.x.

CBC News. 2006. "Cooke plans salmon bonanza for Newfoundland's south coast," 27 Oct. At: http://www.cbc.ca/news/canada/newfoundland-labrador/cooke-plans-salmon-bonanza-for-newfoundland-s-south-coast-1.575310.

———. 2017. "Escaped Atlantic salmon reported 250 km north of collapsed fish farm," 13 Sept. At: http://www.cbc.ca/news/canada/british-columbia/escaped-atlantic-salmon-reported-250-km-north-of-collapsed-fish-farm-1.4288626.

Cho, C. Y., and S. J. Kaushik. 1990. "Nutritional energetics in fish: Energy and protein utilization in rainbow trout (*Salmo gairdneri*)." *World Review of Nutrition and Dietetics* 61: 132–72.

Cochrane, K., C. De Young, D. Soto, and T. Bahri, eds. 2009. *Climate Change Implications for Fisheries and Aquaculture: Overview of Current Scientific Knowledge.* FAO Fisheries and Aquaculture Technical Paper No. 530. Rome: FAO.

Clemens, T., C. Couturier, G. J. Parsons, and P. Dabinett. 1999. "Newfoundland Aquaculture Industry Association environmental monitoring program of shellfish farms." Proceedings of the Workshop on Mussel Production Capacity (Part 2) Held at Aquaculture Canada '98. *Bulletin of the Aquaculture Association of Canada* 99, 2: 29–34.

Cooke Aquaculture. n.d. "Certification." At: http://cookeaqua.com/index.php/commitment-to-the-environment/certification-seafood-trust.

Costa-Pierce, B. A., D. M. Bartley, M. Hasan, F. Yusoff, S. J. Kaushik, K. Rana, . . . A. Yakupitiyage. 2012. "Expert Panel 1.1. Responsible use of resources for sustainable aquaculture." In R. P. Subasinghe et al., eds., *Farming the Waters for People and Food*, 113–48. Proceedings of the Global Conference on Aquaculture 2010, Phuket, Thailand, 22–25 Sept. 2010. Rome and Bangkok: FAO, NACA.

Couturier, C. 1983. "Mussel culture in Northeast Arm." Marine Sciences Research Laboratory Technical Report. St. John's: Memorial University of Newfoundland.

———. 1988. "Shellfish toxins aplenty." *Bulletin of the Aquaculture Association of Canada* 88, 2: 11–25.

———. 1994a. "Ovary, egg, and larval peptides in sea scallops, *Placopecten magellanicus*." In N. F. Bourne, B. L. Bunting, and L. D. Townsend, *Proceedings of the 9th International Pectinid Workshop, Nanaimo, B.C., Canada, April 22–27, 1993*, vol. 1, 128–37. Ottawa: Department of Fisheries and Oceans, Canadian Technical Report of Fisheries and Aquatic Sciences.

———. 1994b. "Spawning in sea scallops, *Placopecten magellanicus*." In N. F. Bourne, B. L. Bunting, and L. D. Townsend. *Proceedings of the 9th International Pectinid Workshop, Nanaimo, B.C., Canada, April 22–27, 1993*, vol. 1, 138–46. Ottawa: Department of Fisheries and Oceans, Canadian Technical Report of Fisheries and Aquatic Sciences.

———. 2014a. *Building Climate Change Resilient Food Systems: Integrating Reservoir and Rice-Fish Systems, Takeo Province, Cambodia. Report on Phase 1 and Phase 2 Technical Support Missions Adapting to Climate Change – Fisheries and Rice-Fish Farming Curriculum*. St. John's: Fisheries and Marine Institute of Memorial University.

———. 2014b. *Building Climate Change Resilient Food Systems: Integrating Reservoir and Rice-Fish Systems, Takeo Province, Cambodia. Report on Phase 3 Technical Mission – GAP Analyses, Surveys of Benefits to Farmers from the Project, PNCA Capacity Assessment, and Proposed Greenfield Water Management Options*. St. John's: Fisheries and Marine Institute of Memorial University.

———, P. Dabinett, and M. Lanteigne. 1995. "Scallop culture in Atlantic Canada." In A. D. Boghen, ed., *Cold-Water Aquaculture in Atlantic Canada*, 297–340. Sackville, NB: Tribute Press and Canadian Institute for Research on Regional Development.

——— and G. Newkirk. 1991. "Biochemical and gametogenic cycles in scallops, *Placopecten magellanicus*, (Gmelin 1791) held in suspension culture." In S. E. Shumway and P. A. Sandifer, eds., *An International Compendium of Scallop Biology and Culture*, 107–17. World Aquaculture Workshops No. 1. Baton Rouge, La.: World Aquaculture Society.

Cranford, P. J., R. Anderson, P. Archambault, T. Balch, S. S. Bates, G. Bugden, . . . P. Strain. 2006. *Indicators and Thresholds for Use in Assessing Shellfish Aquaculture Impacts on Fish Habitat*. Canadian Science Advisory Secretariat, Research Document 2006/034, 116.

Crim, L. W., D. M. Evans, and D. H. Vickery. 1983. "Manipulation of the seasonal reproductive cycle of the landlocked Atlantic salmon (*Salmo salar*) by LHRH analogues administered at various stages of gonadal development." *Canadian Journal of Fisheries and Aquatic Sciences* 40, 1: 61–67.

Dabinett, P. E. 1989. "Hatchery production and grow-out of the giant scallop *Placopecten magellanicus*." *Bulletin of the Aquaculture Association of Canada* 89, 3: 68–70.

Daniel, E. S., C. C. Parrish, D. C. Somerton, and J. A. Brown. 1993. "Lipids in eggs from first-time and repeat spawning Atlantic halibut, *Hippoglossus hippoglossus (L.)*." *Aquaculture Research* 24, 2: 187–91.

FAO (Food and Agriculture Organization of the United Nations). n.d. *CWP Handbook of Fishery Statistical Standards*, section J: Aquaculture. At: http://www.fao.org/fishery/cwp/handbook/j/en.

———. n.d. "FAO term portal." At: http://www.fao.org/faoterm/en/?defaultCollId=21.

———. 2014a. *State of World Fisheries and Aquaculture*. Rome: FAO.

———. 2014b. *FAO Yearbook of Fisheries and Aquaculture Statistics*. Rome: FAO. At: ftp://ftp.fao.org/FI/STAT/summary/default.htm.

———. 2015. "FAO Global Aquaculture Production database updated to 2013 — Summary information." Rome: FAO.

———. 2016. *The State of World Fisheries and Aquaculture 2016: Contributing to Food Security and Nutrition for All*. Rome: FAO.

Felt, L. 2010. "It all depends on the lens, b'y: Local ecological knowledge and institutional science in an expanding finfish aquaculture sector." In K. Culver and D. Castle, eds., *Aquaculture, Innovation and Social Transformation*, 167–90. The International Library of Environmental, Agricultural and Food Ethics. Dordrecht, Netherlands: Springer.

Fisheries and Oceans Canada. 2014. "Aquaculture production quantities and values." 18 Nov. At: http://www.dfo-mpo.gc.ca/stats/aqua/aqua-prod-eng.htm.

Fletcher, G., and P. L. Davies. 1991. "Transgenic fish for aquaculture." *Genetic Engineering* 13: 331–70.

Fleming, I. A., S. Einum, B. Jonsson, and N. Jonsson. 2003. "Comment on 'Rapid evolution of egg size in captive salmon'." *Science* 302: 59b.

Gallardi, D., and C. Couturier. 2007. "Stress in eastern oyster (*Crassostrea virginica*)." *AAC Special Publication* 13: 23–25.

———, K. Hobbs, T. Mills, C. Couturier, C. C. Parrish, and H. M. Murray. 2014. "Effects of extended ambient live holding on cultured blue mussels (*Mytilus edulis L.*) with reference to condition index, lipid profile, glycogen content and organoleptic testing." *Aquaculture* 430: 149–58.

Gamperl, A. K., H. A. Faust, B. Dougher, and K. J. Rodnick. 2004. "Hypoxia tolerance and preconditioning are not additive in the trout (*Oncorhynchus mykiss*) heart." *Journal of Experimental Biology* 207: 2497–2505.

George, E. M. 2011. "Organic footprint and composition of particles from marine finfish aquaculture operations." M.Sc. thesis, Memorial University.

Global Aquaculture Alliance. n.d. "Best aquaculture practices certification — The responsible seafood choice." At: http://bap.gaalliance.org/find-certified-facilities/.

Goddard, S. 1995. *Feed Management in Intensive Aquaculture*. New York: Springer Science and Business Media.

Grecian, L. A., G. J. Parsons, P. Dabinett, and C. Couturier. 2000. "Influence of season, initial size, depth, gear type, and stocking density on the growth rates and recovery of sea scallop, *Placopecten magellanicus*, on a farm-based nursery." *Aquaculture International* 8: 183–206.

———, ———, ———, and ———. 2003. "Effect of deployment date and environmental conditions on growth rate and retrieval of hatchery-reared sea scallops, *Placopecten magellanicus* (Gmelin, 1791), at a sea-based nursery." *Journal of Shellfish Research* 22, 1: 101–09.

Halfyard, L. C. 2009. *Sustainable Rice Fish Integration (SRFI) Project: Summary Report*. St. John's: Fisheries and Marine Institute of Memorial University.

———, C. C. Parrish, and K. Jauncey. 2000. "Nutritional requirements of Atlantic wolffish, *Anarchichas lupus L.* and *A. minor Olafsen*, two new coldwater aquaculture species." In F. Shahidi, ed., *Seafood in Health and Nutrition – Transformation in Fisheries and Aquaculture: Global Perspectives*, 449–76. St. John's: ScienceTech Publishing Company.

Hall, S. J., A. Delaporte, M. J. Phillips, M. Beveridge, and M. O'Keefe. 2011. *Blue Frontiers: Managing the Environmental Costs of Aquaculture*. Penang, Malaysia: The WorldFish Center. At: http://pubs.iclarm.net/resource_centre/WF_2818.pdf.

Hamoutene, D. 2014. "Sediment sulphides and redox potential associated with spatial coverage of spp. at finfish aquaculture sites in Newfoundland, Canada." *ICES Journal of Marine Science* 71, 5: 1153–57. doi:10.1093/icesjms/fst223.

———, I. Costa, K. Burt, L. Lush, and J. Caines. 2015b. "Survival of farmed, wild and first generation hybrid Atlantic salmon (*Salmo salar* Linnaeus, 1758) to low temperatures following seawater transfer." *Journal of Applied Ichthyology* 31, 2: 333–36. doi:10.1111/jai.12694.

———, G. Mabrouk, L. Sheppard, C. MacSween, E. Coughlan, and C. Grant. 2013. "Validating the use of *Beggiatoa* sp. and opportunistic polychaete worm complex (OPC) as indicators of benthic habitat condition at finfish aquaculture sites in Newfoundland." Can. Tech. Rep. Fish. Aquat. Sci. 3028. St. John's: Fisheries and Oceans Canada. At: http://publications.gc.ca/collections/collection_2013/mpo-dfo/Fs97-6-3028-eng.pdf.

———, F. Salvo, T. Bungay, G. Mabrouk, C. Couturier, A. Ratsimandresy, and S. C. Dufour. 2015a. "Assessment of finfish aquaculture effect on Newfoundland epibenthic communities through video monitoring." *North American Journal of Aquaculture* 77, 2: 117–27. doi:10.1080/15222055.2014.976681.

Harding, J. M., C. Couturier, G. J. Parsons, and N. W. Ross. 2004a. "Evaluation of the neutral red assay as a stress response indicator in cultivated mussels (*Mytilus spp.*) in relation to post-harvest processing activities and storage conditions." *Aquaculture* 231: 315–26.

———, ———, ———, and ———. 2004b. "Evaluation of the neutral red retention assay as a stress response indicator in cultivated mussels (*Mytilus spp.*) in relation to seasonal and environmental conditions." *Journal of Shellfish Research* 23, 3: 745–51.

———, ———, ———, and ———. 2007. "Evaluation of short-term transportation conditions on stress response of cultivated blue mussels (*Mytilus spp.*)." In R. Tremblay, B. Myrand, and D. Proulx, eds., *Atelier de travail "Indicateurs de stress chez les mollusques"*, 14–15. Les Publications de la Direction de l'Innovation et des Technologies, No. 20. Québec: Ministère de l'agriculture et de l'alimentation du Québec.

Hickling, C. F. 1962. *Fish Culture*. London: Faber and Faber.

Hicks, B. 2011. "What Is FIFO?" *Aquaculture North America* (Jan./Feb.).

Hiemstra, L. 2001. Proceedings of the Sea Urchin Culture Workshop, 24–25 Sept., Malaspina University College, Nanaimo, BC.

Hixson, S. M., C. C. Parrish, and D. M. Anderson. 2014. "Use of camelina oil to replace fish oil in diets for farmed salmonids and Atlantic cod." *Aquaculture* 431: 44–52.

Hooper, R. G., F. M. Cuthbert, and T. McKeever. 1997. "Feasibility of sea urchin aquaculture using natural feeds." *Bulletin of the Aquaculture Association of Canada* 97, 1: 5–7.

Hustins, D. 2007. *Brown Trout and Rainbow Trout: A Journey into Newfoundland Waters*. St. John's: Tight Lines Publishers.

Ibarra, D., and C. Couturier. 1998. "Factors influencing cultured mussel meat yields and recommendations for a standard method." *Bulletin of the Aquaculture Association of Canada* 98, 2: 59–61.

———, ———, and T. Mills. 2000. "Calculating meat yields by mussel growers." In F. Shahidi, ed., *Seafood in Health and Nutrition*, 261–69. St. John's: ScienceTech Publishing Company.

Innes, D. J., A. S. Comesaña, J. E. Toro, and R. J. Thompson. 1999. "The distribution of *Mytilus edulis* and *M. edulis* at spat collection sites." Proceedings of the Workshop on Mussel Production Capacity (Part 2) Held at Aquaculture Canada '98. *Bulletin of the Aquaculture Association of Canada* 99, 2: 22–23.

Intergovernmental Panel on Climate Change (IPCC). (2007). *IPCC Fourth Assessment Report: Climate Change 2007*. Geneva: IPCC.

Jackson, A. 2009. "Fish in-Fish out (FIFO) ratios explained." International Fishmeal and Fish Oil Organization. At: http://www.iffo.net/cn/system/files/100.pdf.

Jamieson, A. 1978. *A Report on the Establishment and Operation (18 months) of a Rainbow Trout Hatchery/Farm at Hopeall, Trinity Bay, Newfoundland.* St. John's: Research and Resource Services, Fisheries and Marine Service, Newfoundland Region.

Larsen, J., and M. Roney. 2013. "Farmed fish production overtakes beef." Earth Policy Institute, Rutgers University, Rutgers, NJ. At: http://www.earth-policy.org/plan_b_updates/2013/update114.

Le Vallée, J. C., and M. Grant. 2015. *Report Card on Food: Performance and Potential. Comparing Canada's Food Performance Internationally*. Ottawa: Centre for Food in Canada, Conference Board of Canada (draft Oct. 2015).

Lloyd, L. E., B. E. McDonald, and E. W. Crampton. 1978. *Fundamentals of Nutrition*, 2nd ed. San Francisco: W. H. Freeman and Company.

Lush, L., K. Burt, D. Hamoutene, N. Camarillo-Sepulveda, J. Perez-Casanova, S. Kenny, and C. Collier. 2014. "Size and ATP content of unfertilized eggs from farmed and wild Atlantic salmon in Newfoundland." *North American Journal of Aquaculture* 76, 2: 138–42. doi:10.1080/15222055.2014.886648.

Mabrouk, G., T. R. Bungay, D. Drover, and D. Hamoutene. 2014. "Use of remote video survey methodology in monitoring benthic impacts from finfish aquaculture on the south coast of Newfoundland (Canada)." Research document (Canadian Science Advisory Secretariat); 2014/039. At: http://publications.gc.ca/collections/collection_2014/mpo-dfo/Fs70-5-2014-039-eng.pdf.

MacNeill, S., M. Pryor, C. Couturier, and G. J. Parsons. 1999. "Increasing spat collection for mussel culture: Newfoundland Aquaculture Industry Association larval and spatfall monitoring program." Proceedings of the Workshop on Mussel Production Capacity (Part 2) Held at Aquaculture Canada '98. *Bulletin of the Aquaculture Association of Canada* 99, 2: 24–28.

Mansour, A., D. Hamoutene, G. Mabrouk, T. Puestow, and E. Barlow. 2008. "Evaluation of some environmental parameters for salmon aquaculture cage sites in Fortune Bay, Newfoundland: Emphasis on the occurrence of hypoxic conditions. Can. Tech. Rep. Fish. Aquat. Sci. 2814. At: http://www.dfo-mpo.gc.ca/Library/341539.pdf.

McKindsey, C. W., P. Archambault, and N. Simard. 2012. "Spatial variation of benthic infaunal communities in baie de Gaspé (eastern Canada): Influence of mussel aquaculture." *Aquaculture* 356/357: 48–54.

Memorial University of Newfoundland (MUN). 1995. *Aquaculture at Memorial University of Newfoundland*. St. John's: MUN.

Mercier, A., R. H. Ycaza, R. Espinoza, V. M. A. Haro, and J.-F. Hamel. 2012. "Hatchery experience and useful lessons from *Isostichopus fuscus* in Ecuador and Mexico." In C. A. Hair, T. D. Pickering, and D. J. Mills, eds., *ACIAR Proceedings*, vol. 136, 79–90. Canberra: Australian Centre for International Agricultural Research.

Mooney, M., G. J. Parsons, and C. Couturier. 2002. "A comparison of feeding physiology of wild and cultured *Mytilus edulis* and *M. trossulus*." In B. Myrand and F. Coulombe, eds., *Compte rendu no. 12. Atelier de travail sur la problématique Mytilus edulis et M. trossulus*, 91–98. Gaspé, 18–20 octobre 2000. Québec: Gouvernement du Québec.

Moret, K., K. Williams, C. Couturier, and G. J. Parsons. 1999. "Newfoundland culture mussel (*Mytilus edulis*) industry 1997 health survey." Proceedings of the Workshop on Mussel Production Capacity (Part 2) Held at Aquaculture Canada '98. *Bulletin of the Aquaculture Association of Canada* 99, 2: 35–37.

——— and L. C. Halfyard. 2005. "Promoting aquaculture awareness and education through Canadian and overseas linkages." *Bulletin of the Aquaculture Association of Canada* 105, 1: 26–31.

Morgan, B. 1974. "The Marine Sciences Research Laboratory." *MUN Gazette* 6, 40: 3.

National Research Council of the National Academies. 2011. *Nutrient Requirements of Fish and Shrimp*. Washington: National Academies Press.

Neill, R. 2008. "Aquaculture property rights in Canada." In N. Scheider, ed., *A Breath of Fresh Air: The State of Environmental Policy in Canada*, ch. 11. Vancouver: Fraser Institute.

Newfoundland and Labrador. 1985. *Aquaculture Activities in Newfoundland, 1982*. Department of Fisheries, Industry Support Services Division, Data Report Series No. 20. St. John's: Government of Newfoundland and Labrador, Oct.

Newfoundland and Labrador, Department of Environment and Conservation. n.d. "Summary of environmental assessment process. Project description. Hopeall fish farm hatchery. Proponent: Newhope Fish Farm Inc." At: http://www.env.gov.nl.ca/env/env_assessment/projects/Y2012/1656/index.html.

———. 2012. "Environmental assessment bulletin," 6 Sept. At: http://www.env.gov.nl.ca/env/env_assessment/bulletins/Y2012/20120906.pdf.

Newfoundland and Labrador, Department of Fisheries and Aquaculture. n.d. "The development of salmon hatcheries in Newfoundland and Labrador." At: http://www.fishaq.gov.nl.ca/education/pdf/The%20Development%20of%20Salmon%20Hatcheries.pdf.

———. 1998a. "Bay d'Espoir area leads the way in aquaculture production." *Newfoundland Aquaculture Review*: 25–27.

———. 1998b. "Newfoundland charr culture on-going since 1991." *Newfoundland Aquaculture Review*: 30–31.

———. 2015. "Statistics." 21 Oct. At: http://www.fishaq.gov.nl.ca/stats/index.html.

Newfoundland and Labrador, Office of the Auditor General. 2001. *Report on Reviews of Departments and Crown Agencies*, 152–57 (Section 3.17, Aquaculture Program). At: http://www.ag.gov.nl.ca/ag/annualReports/2001AnnualReport/Aquaculture_SCB.pdf.

Newhope Fish Farm Inc. 2012. "Construction of a salmonid hatchery at Hopeall, Trinity Bay. Environmental assessment project registration." June. Submitted to Minister of Environment and Conservation, Newfoundland and Labrador. At: http://www.env.gov.nl.ca/env/env_assessment/projects/Y2012/1656/1656_Registration.pdf.

Oldford, V. G. 2013. "Spatial extent of visual indicators at finfish aquaculture sites after the first year of production — video sampling evaluation." Master of Env. Sci. Report, Memorial University.

O'Neill, S. M., A. M. Sutterlin, and D. Aggett. 1984. "The effects of size-selective feeding by starfish (*Asterias vulgaris*) on the production of mussels (*Mytilus edulis*) cultured on nets." *Aquaculture* 35: 211–20.

Parsons, G. J., and S. M. C. Robinson. 2006. "Sea scallop aquaculture in the Northwest Atlantic." In S. E. Shumway and G. J. Parsons, eds., *Scallops: Biology, Ecology and Aquaculture*, 907–44. Developments in Aquaculture and Fisheries Science 35. New York: Elsevier.

Penney, R. W., ed. 1991. *Report of the Arctic Charr Aquaculture Workshop March 12, 1991, St. John's, Newfoundland.* Canadian Industry Report of Fisheries and Aquatic Sciences, No. 212. St. John's: Newfoundland Region, Department of Fisheries and Oceans.

Pepper, V. A., E. Barlow, C. Collier, and T. Nicholls. 2001. *Quantitative Performance Measurement of Alternative North American Salmonid Strains for Newfoundland Aquaculture, 2000–2001. Aquaculture Component – Canada–Newfoundland Agreement on Economic Renewal (ACERA), Annual Report for 2000/01.* St. John's.

———, T. Nicholls, and C. Collier. 1996. *Marine Growth Performance of a Newfoundland Salmon Stock for Aquaculture.* St. John's: Department of Fisheries and Oceans.

———, ———, and ———. 1999. *Project Report: Broodstock Program for the Bay d'Espoir Region.* St. John's: Department of Fisheries and Oceans.

———, ———, and ———. 2004. *Reproductive Technologies Applied to Newfoundland Salmonid Aquaculture to Enhance Commercial Production.* Can. Tech. Rep. Fish. Aquat. Sci. 2541. At: http://www.dfo-mpo.gc.ca/Library/281192.pdf.

———, ———, ———, V. Watkins, E. Barlow, and M. F. Tlusty. 2003. *Quantitative Performance Measurement of Alternative North American Salmonid Strains for Newfoundland Aquaculture.* Can. Tech Rep. Fish. Aquat. Sci. 2502. At: http://www.dfo-mpo.gc.ca/Library/277174.pdf.

———, R. Withler, T. Nicholls, and C. Collier. 2004. *Quantitative Marine-Performance Evaluation of a Newfoundland Atlantic Salmon Strain for Bay d'Espoir Aquaculture.* Can. Tech. Rep. Fish. Aquat. Sci. 2540. At: http://www.dfo-mpo.gc.ca/Library/281193.pdf.

Peterson, R. G., L. K. Dunn, and G. A. Hunter. 1990. *Proposed Newfoundland Atlantic Salmon Broodstock Program.* St. John's: Department of Fisheries and Oceans, Science Branch, Newfoundland Region.

Power, V. 1999. "Nematodes as indicators of the environmental effects of salmonid aquaculture in Bay d'Espoir, Newfoundland." Master of Env. Sci. Report, Memorial University.

Purchase, C., D. Boyce, and J. A. Brown. 2000. "Growth and survival of juvenile yellowtail flounder *Pleuronectes ferrugineus* (Storer) under different photoperiods." *Aquaculture Research* 31, 6: 547–52.

Rana, K. J. 1997. "Guidelines on the collection of structural aquaculture statistics." Rome: FAO Statistical Development Series 5b.

Reece, J. B., M. R. Taylor, E. J. Simon, J. L. Dickey, and K. Scott. 2014. *Campbell Biology Concepts and Connections.* Toronto: Pearson.

Reid, V., C. McKenzie, K. Matheson, T. Wells, and C. Couturier. 2016. "Post-metamorphic attachment by solitary ascidian *Ciona intestinalis* (Linnaeus, 1767) juveniles from Newfoundland and Labrador, Canada." *Management of Biological Invasions* 7: 67–76, http://dx.doi.org/10.3391/mbi.2016.7.1.09.

Rideout, K. 2006. "The ecological and genetic impacts of escaped farmed salmon on wild salmon with recommended management measures for southern Newfoundland." Master of Marine Studies thesis, Memorial University.

Rise, M. L., G. W. Nash, J. R. Hall, M. Booman, T. S. Hori, E. A. Trippel, and A. K. Gamperl. 2014. "Variation in embryonic mortality and maternal transcript expression among Atlantic cod (*Gadus morhua*) broodstock: A functional genomics study." *Marine Genomics* 18: 3–20.

Rose, G. A. 2007. *Cod: An Ecological History of the North Atlantic Fisheries*. St. John's: Breakwater Books.

Ross, N. W., E. Egbosimba, N. T. Brun, V. M. Bricelj, T. H. MacRae, J. M. Harding, C. Couturier, and G. J. Parsons. 2007. "Development of biochemical indicators of stress response for bivalves: Recent studies on heat shock proteins and proteases." In R. Tremblay, B. Myrand, and D. Proulx, eds., *Atelier de travail "Indicateurs de stress chez les mollusques"*, 21–22. Les Publications de la Direction de l'Innovation et des Technologies, No. 20. Québec: Ministère de l'agriculture et de l'alimentation du Québec.

Salvo, F., D. Hamoutene, and S. C. Dufour. 2015. "Trophic analyses of opportunistic polychaetes (*Ophryotrocha cyclops*) at salmonid aquaculture sites." *Journal of the Marine Biological Association of the United Kingdom* 95, 4: 713–22. doi:10.1017/S0025315414002070.

Sarker, P. K., D. P. Bureau, K. Hua, M. D. Drew, I. Forster, K. Were, B. Hicks, and G. W. Vandenberg. 2013. "Sustainability issues related to feeding salmonids: A Canadian perspective." *Reviews in Aquaculture* 5: 199–219.

Scott, W. B., and E. J. Crossman. 1964. *Fishes Occurring in the Fresh Waters of Insular Newfoundland*. Life Sciences Contributions no. 58. Ottawa: Department of Fisheries.

——— and ———. 1973. *Freshwater Fishes of Canada*. Bulletin 184. Ottawa: Fisheries Research Board of Canada.

Shelton, C. 2014. "Climate change adaptation in fisheries and aquaculture — compilation of initial examples." Rome: FAO Fisheries and Aquaculture Circular No. 1088.

Smith, P., D. Martino, Z. Cai, D. Gwary, H. Janzen, P. Kumar, . . . J. Smith. 2008. "Greenhouse gas mitigation in agriculture." *Philosophical Transactions of the Royal Society B* 363: 789–813.

Soko, C. K., and D. E. Barker. 2005. "Efficacy of crushed garlic and lemon juice as bio-product treatments for *Ichthyophthirius multifiliis* ('ich') infections among cultured, juvenile Nile tilapia, *Oreochromis niloticus*." *Aquaculture Association of Canada Special Publication 9*: 108–10.

South, G. R. 1970. "Experimental culture of *Alaria* in a sub-arctic, free-flowing sea water system." *Helgoländer wissenschaftliche Meeresuntersuchungen* 20: 1: 216–28.

Standards Council of Canada. 2012. *Canadian Organic Aquaculture Standards.* CAN/CGSB-32.312-2012. Ottawa: Standards Council of Canada.

Statistics Canada. 2014. *Aquaculture Statistics 2013.* Catalogue no. 23-222-X. Ottawa: Ministry of Industry.

Sutterlin, A. M. 1991. "Some experience with Arctic charr compared with Atlantic salmon and rainbow trout at the Bay d'Espoir salmon hatchery, Newfoundland." In R. W. Penney, ed., *Report of the Arctic Charr Aquaculture Workshop, March 12, 1991*, 23–24. St. John's: Canadian Industry Report of Fisheries and Aquatic Sciences No. 212.

———, D. Aggett, C. Couturier, R. Scaplen, and D. Idler. 1981. *Mussel Culture in Newfoundland Waters.* Marine Sciences Research Laboratory Technical Report No. 23. St. John's: Memorial University.

Tacon, A. G. J., M. R. Hasan, and M. Metian. 2011. *Demand and Supply of Feed Ingredients for Farmed Fish and Crustaceans: Trends and Prospects.* Rome: FAO Fisheries and Aquaculture Technical Paper No. 564.

Tlusty, M. F., V. A. Pepper, and M. R. Anderson. 1999. *Environmental Monitoring of Finfish Aquaculture Sites in Bay d'Espoir Newfoundland during the Winter of 1997.* Can. Tech. Rep. Fish. Aquat. Sci. 2273. At: http://www.dfo-mpo.gc.ca/Library/237676.pdf.

———, ———, and ———. 2005. "Reconciling aquaculture's influence on the water column and benthos of an estuarine fjord: A case study from Bay d'Espoir, Newfoundland." In B. T. Hargrave, ed., *Handbook of Environmental Chemistry*, vol. 5, 115–28. Berlin: Springer.

———, ———, and J. A. Helbig. 1998. *Project Report: Overwintering Holding Capacity for Bay d'Espoir Salmonids, 1998/99. Aquaculture Component – Canada–Newfoundland Agreement on Economic Renewal (ACERA).* Annual Report for 1998/99. St. John's.

———, K. Snook, V. A. Pepper, and M. R. Anderson. 2000. "The potential for soluble and transport loss of particulate aquaculture wastes." *Aquaculture Research* 31, 10: 745–55. doi:10.1046/j.1365-2109.2000.00497.x.

UNEP. 2008. *Vital Water Graphics – An Overview of the State of the World's Fresh and Marine Waters*, 2nd ed. Nairobi, Kenya: UNEP.

UNFCCC. 2015. "Draft Paris Outcome. COP 21 agenda item 4 (b) Durban Platform for Enhanced Action (decision 1/CP.17) Adoption of a protocol, another legal instrument, or an agreed outcome with legal force under the Convention applicable to all Parties Version 1 of 9 December 2015." At: http://unfccc.int/resource/docs/2015/cop21/eng/da01.pdf.

Van Zyll de Jong, M., R. J. Gibson, and I. G. Cowx. 2004. "Impacts of stocking and introductions on freshwater fisheries of Newfoundland and Labrador, Canada." *Fisheries Management and Ecology* 11, 3: 183–93.

Vickerson, A., C. Couturier, and C. McKenzie. 2007. "Managing mussel, *Mytilus spp.* seed health: The effects of brine, lime and acetic acid antifouling treatments and transport on mussel seed performance." *AAC Special Publication* 13: 348–50.

Wang, G. 2014. "Use of inshore benthic cages for storage and outgrowth of adult lobsters *Homarus americanus*." Master of Science (Aquaculture) thesis, Memorial University of Newfoundland.

World Bank. 2013. *Fish to 2030. Prospects for Fisheries and Aquaculture.* Washington: Agriculture and Environmental Services Discussion Paper 03.

———. 2014. "Sustainable aquaculture." 5 Feb. At: http://www.worldbank.org/en/topic/environment/brief/sustainable-aquaculture.

———. 2015. "The World Bank Open Data Indicators." At: http://data.worldbank.org/indicator/AG.LND.ARBL.ZS.

Wringe, B., I. Fleming, and C. Purchase. 2015. "Spawning success of cultured and wild male Atlantic cod *Gadus morhua* does not differ during paired contests." *Marine Ecology Progress Series* 535: 197–211. doi:10.3354/meps11406.

Zimmermann, E., C. Purchase, I. Fleming, and J. Brattey. 2013. "Dispersal of wild and escapee farmed Atlantic cod (*Gadus morhua*) in Newfoundland." *Canadian Journal of Fisheries and Aquatic Sciences* 70, 5: 747–55. doi:10.1139/cjfas-2012-0428.

12

A Hive of Possibility: The Sustainability of Honey Bees and Apiculture in Newfoundland

Stephan Walke & Jianghua Wu

INTRODUCTION

Pollinator declines across the globe have become a hot topic among scientists, politicians, and the general public. This is not surprising, considering that approximately 35 per cent of human food sources (including 87 of the leading world food crops) are reliant on animal pollination (FAO, 2008). The western honeybee, *Apis mellifera*, is only one of more than 100,000 pollinating species (FAO, 2008). However, the honeybees' domestication, abundance, honey production, easy management, and long history with humans make them perhaps the most useful and most understood insect pollinator in the world (Kevan, 1999; vanEngelsdorp and Meixner, 2010). While many non-*Apis* species of pollinator have remained poorly understood and documented, the honeybee has allowed for more consistent and in-depth monitoring of declining bee health and abundance (Blackburn, 2012; FAO, 2008; OPERA, 2013).

In Canada, the 2013–14 winter season saw 25 per cent winter losses in honeybee colonies, with Ontario losing 58 per cent of its hives, far above the 15 per cent losses typically expected (Kozak et al., 2014). Canadian colonies fared considerably better in the 2014–15 winter season with an average loss of 16.4 per cent across the country, with Ontario's rate of loss dropping to 37.8 per cent (Leboeuf et al., 2015). While this is among the lowest national overwintering losses between 2006 and 2015, it should be noted that overwintering losses vary considerably on a regional scale and the stressors that contribute to colony losses are also highly variable (Leboeuf et al., 2015; Currie, Pernal, and Guzmàn-Novoa, 2010).

A number of interacting factors have contributed to increased losses around the world and to the recently observed colony collapse disorder (CCD) in the United States (Melhim, Weersink, Daly, and Bennet, 2010; vanEngelsdorp et al., 2009). Briefly, some of these factors include diseases and pests, pesticide exposure, habitat and biodiversity reduction, weather and climate, and socio-political factors. Potts et al. (2010) argue that these negative factors can act synergistically and compound with the effects of harsh winters and unfavourable forage seasons on honeybee colonies.

Given this global pollinator crisis, the island of Newfoundland currently finds itself in a very unique position. Some of the major stressors partially attributed to colony losses on the rest of the continent are noticeably absent. In the Atlantic provinces, for example, large-scale monocultural farming practices, intensive management, and movement of hives for pollination purposes have been attributed to ~30 per cent colony reductions between 2007 and 2010 (AMEC, 2010). These stressors are likely of little significance in Newfoundland at present due to the island's isolation and relatively dispersed use of pesticides arising from low levels of agricultural production.

Newfoundland boasts a very healthy honeybee population when contrasted with many colonies across the world that suffer from the deleterious effects of invasive pests and diseases, especially the *Varroa destructor* mite and *Nosema spp.* (Currie et al., 2010; Kozak et al., 2014; OPERA, 2013; vanEngelsdorp, 2009; Williams, Head, Burgher-MacLellan, Rogers, and Shutler, 2010). To date, Newfoundland hosts a honeybee population still unaffected by *V. destructor*, tracheal mites (*Acarpis woodi*), Israeli acute paralysis virus, Kashmir bee virus, and other significant invasive organisms (Shutler et al., 2014). A consequence of its isolation, therefore, is that Newfoundland has been endowed with substantial possibility as a potential supplier of disease-free bees, organic hive products, and significant scientific research opportunities (Fletcher, 2015; Shutler et al., 2014; Williams, 2010; Williams et al., 2010). It should be noted that while Newfoundland falls within the jurisdiction of the province of Newfoundland and Labrador and is often referred to in provincial terms in this paper, the geographic scope of this project is limited to the island of Newfoundland. There are no known beekeeping operations in Labrador at this time (Hicks, 2014; Williams, 2010). In addition, Newfoundland is separated from the mainland by approximately 15 km at its closest point, which is the primary reason why it remains free of a number of widespread honeybee pests and diseases to date.

With only about 500 honeybee colonies in the province being managed by approximately 50 beekeepers, beekeeping on the island remains relatively underdeveloped but promising (Hicks, 2014; Canadian Association of Professional Apiculturists, 2017; Newfoundland and Labrador Beekeeping Association, 2017a). However, the province also faces a number of biological and social challenges that could threaten the long-term sustainability of apiculture. There are biological risks like high winter mortality, limited genetic diversity, and potential for increased pesticide use from expanded agricultural production. Social and human risks also are associated with the capacity of government and industry to support an emerging industry. At this writing, the province requires no hive registration and the apiary technical support programs at the Agrifoods Development Branch of the Department of Natural Resources only began recently. Nevertheless, the Newfoundland and Labrador Beekeeping Association (NLBKA), formed at the end of 2014, represents a significant step towards formalizing management efforts among Newfoundland beekeepers. The aims of the organization are comprehensive and have relevance to beekeepers in the province as well as to government officials, policy-makers, researchers, and the general public (NLBKA, 2017b).

In light of the recent global issues facing honeybee populations, it is clearly important to develop policies and management plans to protect Newfoundland's unique, disease-free honeybee population, as well as to sustainably grow the apicultural industry. However, the paucity of information about the province's honeybee population poses a significant problem for policy-makers. Without sufficient understanding of honeybee abundance, distribution, forage use, health, genetic characteristics, and current management in Newfoundland, there are no scientific grounds on which to base policies affecting bee health and apicultural development.

This research is a stepping-stone towards guiding strategic conversations about the sustainability of apiculture in Newfoundland. Three questions were explored: (1) What are the key factors to consider with respect to sustainable apiculture in Newfoundland and what are the implications? (2) What are some major challenges and opportunities facing beekeepers in Newfoundland? (3) What considerations may be most relevant to policy-makers and managers regarding the sustenance of pollinators, honeybee health, and the apicultural industry in Newfoundland?

METHODS

We used a mixed methods approach to obtain both quantitative information (population size, distribution, and source stock) and qualitative information (opportunities, challenges, and influencing factors). This approach allowed us to create an industry profile of apiculture on the island of Newfoundland and to contextualize this profile by identifying factors of concern as they relate to challenges and opportunities facing the industry. We gathered information in a "concurrent embedded" strategy (Creswell, 2009; Driscoll, 2007) by way of a questionnaire sent to all members of the NLBKA in February 2015. Data were gathered between February and April 2015. Approximately seven hours of in-person and phone interviews as well as e-mail correspondences also were conducted between February and April 2015, with follow-up correspondences between April and May 2016. These communications involved beekeepers in Newfoundland, one entomology researcher in the province, two honeybee researchers in Ontario, representatives of the NLBKA, and one representative of the Agrifoods Development Branch of the Newfoundland Forestry and Agrifoods Agency.

Supplementing survey data with in-depth interviews and casual conversations has been used in researching small apiculture sectors in previous studies (Chemàs and Rico-Gray, 1991). This approach is applicable for contextualizing a small, emerging industry of natural resource products that reflects a unique network of biophysical, political, and economic factors (Keske, 2008). The number of questionnaire responses was inadequate to make accurate predictions with regard to a quantified industry profile (exact size, distribution, and genetic stock profile). However, member responses provided valuable information regarding potential challenges and opportunities facing the sustainability of individual operations and the emerging Newfoundland industry. Responses were then used to guide further interviews and correspondences to expand on key factors and interactions. We supplemented our primary data with a literature review, with an emphasis on Atlantic Canada and North America. Our results provide a unique contribution to the literature upon which future social and biological research can build.

RESULTS

A total of seven respondents replied to the questionnaire either by e-mail or through phone conversation. This represents about 13 per cent of NLBKA members at the time the survey was administered; not all NLBKA members are beekeepers or manage hives. The number of respondents represented approximately 23 per cent of beekeepers in the province operating about 59 per cent of the hives in 2015 (based on estimates from Hicks, 2014). Although the sample was small, it represented a diversity of operations and more than half of the verifiable hives. Three were hobby operations, two were potential/developing businesses, and two were established commercial operations. One respondent practised an unconventional method of beekeeping with the use of Kenyan top bar hives. Five apiaries were operated by newcomers to beekeeping in Newfoundland (started within the past five years), and two had practised apiculture for at least 30 years in Newfoundland. Both experienced beekeeper respondents had also trained or participated in apiculture outside the province. Three of the respondents completed the questionnaire by phone, which allowed for more elaboration on themes in an ethnographic manner, as compared to written e-mail responses. To date, no apiary operates more than 100 hives. In addition, no beekeeper in Newfoundland currently acquires his/her primary income from beekeeping.

Eight key factors influencing the sustainability of apiculture in Newfoundland were identified based on recurring themes in questionnaire responses. These are listed in Figure 12.1: industry profile, regulations and enforcement, genetic diversity, weather conditions, diseases and pests, agriculture and pollination, forage availability/quality, and education.

Each key factor listed in Figure 12.1 corresponds by graphic to its associated point in the "related primary factors" columns. Only primary factors are listed. The following discussion will elaborate on these factors and demonstrate how most are interconnected with one another, often on multiple levels. Education will be discussed separately within the final, recommendations section.

Industry Profile	Regulations & Enforcement	Genetic Diversity	Weather Conditions/ Overwintering	Diseases & Pests	Agriculture/Pollination	Forage Availability/Quality	Education

Opportunities	Related Primary Factors	Challenges	Related Primary Factors
out-of-province honeybee sales/market for research and commercial use	-maintenance of disease/pest-free bees -stable and sustaining honeybee population	potential reduction of genetic diversity	-small and unbalanced scale of apiculture industry -restrictions on honeybee imports
specialty/organic hive products sale	-maintenance of disease/pest-free bees -stable and sustaining honeybee population -healthy, abundant and diverse forage -sustainable agriculture and berry crop development	high winter mortality	-small and unbalanced scale of apiculture industry -harsh weather conditions
pollination service provision for local berry and fruit crops	-sustainable agriculture and berry crop development -stable and sustaining honeybee population -favorable support and regulations which favour domestic and native pollinators over imports	risk of disease/pest infestation	-regulatory framework and enforcement -education (beekeepers, crop producers, and public) -small and unbalanced scale of apiculture industry
research opportunities	-stable and sustaining honeybee population -maintenance of disease/pest-free bees -attention to ecological interactions	lack of capacity (regulatory, enforcement, scientific, financial, and industry)	-small and unbalanced scale of apiculture industry -small agricultural sector

Figure 12.1. Key factors identified from questionnaire responses. *Note: Each key factor listed in Figure 12.1 corresponds by graphic to its associated point in the "related primary factors" column. Only primary factors are listed. The following discussion will elaborate on these factors and demonstrate how most are interconnected with one another, often on multiple levels. Education will be discussed separately within the final, recommendations section.*

DISCUSSION: OPPORTUNITIES AND CHALLENGES WITHIN THE UNIQUE NEWFOUNDLAND CONTEXT

The following section expands on the major factors identified by Newfoundland apiarists as influencing the sustainability of apiculture in the province. Opportunities and challenges facing beekeepers on the island are also presented in the context of these key themes. The order of presentation does not represent a hierarchy of importance in any way.

Industry Profile

Across Canada, there has been a growing shift away from hobby beekeeping and towards larger-scale commercial operations. Despite increases of production since the 1960s to about 34,000 tonnes of honey annually, there has been an overall decrease in the number of beekeepers in Canada (Melhim et al., 2010). Colony numbers in Canada have remained relatively similar between 1945 and 2009, but the number of beekeepers in 2009 was at about 16 per cent of 1945 numbers. Essentially, honeybee colonies are becoming concentrated into more intensive operations.

Similar trends are also being seen across the Western world. In Europe, colony losses in recent years were most prominently observed in hobby operations (OPERA, 2013). This phenomenon was partly explained by hobbyists' lack of experience and resources. Treatment of the *Varroa* mite and other invasive pests presents additional major costs in an already very costly hobby, which can be overlooked by small-scale apiaries and entry-level beekeepers (OPERA, 2013; Roche, 2014).

When this research was being conducted in spring 2015 there was no verifiable number of beekeepers or hives in Newfoundland. Hicks (2014) provided the most recent research; therefore, his estimate of 300 hives operated by between 25–30 individuals was used as the guideline for this chapter. The 2015 season saw some growth in Newfoundland's apiaries. Five operations are considered commercial in 2016, two of which were included in this study. About 300 hives are estimated to operate commercially at this time with total hive numbers at about 500 on the island. To provide some context, 500 hives represents about 0.07 per cent of all the hives across Canada (Leboeuf et al., 2015). With neighbouring Nova Scotia hosting about 25,000 hives under the management of around 400 beekeepers, the scale of Newfoundland's beekeeping industry is comparatively very small (Nova Scotia Beekeepers, 2016).

Newfoundland's apicultural industry profile follows similar trends with the rest of the Western world. One operation on the island, the Newfoundland Bee Company, contains about 100 hives and is a major supplier of queens and nucleus colonies (nucs) for other operations across Newfoundland. In 2015, only about four individuals in the province operated more than 10 hives each (Hicks, 2014). Due to the scale of beekeeping in Newfoundland, this imbalance between hobby and commercial operations poses key challenges related to other factors such as genetic diversity, weather conditions, and diseases/pests. In addition, the small size of most operations (including commercial ones) could make the industry much more susceptible to the challenges posed by these factors. In particular, high winter mortality in smaller operations can be much harder to rebound from.

The ambitions and development goals of hobby and entry-level beekeepers are another important aspect of the industry profile. Many of these small apiaries may be operated by retired individuals or hobbyists who have no aspirations for growing their operation. One respondent identified a lack of young and enthusiastic beekeepers as a challenge to the long-term sustainability of beekeeping on the island. It is clear that current industry size must be evaluated in combination with existing levels of ambition and desired development. While results from the questionnaire cannot be used to make numerical estimates about hobby operations on the island or their distribution, operation size, demographics, and goals are all important considerations.

Regulations/Enforcement

In this section, regulations pertaining specifically to honeybees in the province will be discussed, with particular emphasis on biosecurity. In general, "biosecurity" is a set of measures intended for the protection of an animal population from transmissible infectious agents (AHBIC, 2003), and vanEngelsdorp and Meixner (2010) identify global economic liberalization and increasingly lax import regulations as contributing significantly to the global spread of invasive diseases and pests. While cross-border disease transmission occurs via legal importation of bees, illegal importation is also an important factor (AHBIC, 2003; vanEngelsdorp and Meixner, 2010).

Newfoundland and Labrador has room to establish regulatory policies more consistent with other provinces, but there are loopholes that point to potential vulnerabilities. Unlike many other provinces, including Ontario,

British Columbia, Alberta, Manitoba, and Nova Scotia, Newfoundland does not possess explicit legislation related to bees or apiculture (Canadian Honey Council, 2016a). The only specific mention of honeybees in provincial legislation exists in a section within Animal Health Regulations under the Animal Health and Protection Act (O.C. 2012-106) (NL Reg., 2012). The legislation prohibits the importation of live bees from outside the province unless a permit has been issued along with an accompanying veterinary certificate from the place of origin indicating the bees have been inspected and are free from any pests and diseases currently not found in Newfoundland (sections 6 and 7).

Unlike Prince Edward Island's Animal Health Act, however, Newfoundland's regulations do not necessitate vehicles transporting honeybees or beekeeping equipment to stop for inspection at weigh stations (PEI, 2014). In addition, the importation of non-*Apis* bee species, such as the common eastern bumblebee, does not fall within the Newfoundland and Labrador Animal Health and Protection Act despite some of these species being vectors for significant diseases (Fletcher, 2015).

In addition to the shortfalls of Newfoundland's regulation, the Agrifoods Branch currently has no means of monitoring and inspecting any imports – the regulation is only enforceable based on an honour system of notification in the event of suspicion. As a compounding factor, unlike most other regions Newfoundland does not require mandatory registration of hives. However, in the past year, a voluntary bee registry has been piloted by the Forestry and Agrifoods Agency and distributed to all NLBKA members. At this time, the registry is a one-time registration of beekeepers as opposed to an annually updated hive registry. The provinces of British Columbia, Alberta, Saskatchewan, Manitoba, Ontario, and Nova Scotia all require official notice of honeybee sales within the province, which effectively acts as a means of tracing distribution in the event of disease outbreak (Fletcher, 2015). Such a check, in combination with mandatory hive registration, could enhance biosecurity for Newfoundland's honeybee population.

Stringent import regulations and enforcement are paramount for protecting the Newfoundland honeybee population from disease and pest infestation. At the same time, import restrictions also inhibit the possibility of increasing honeybee stock numbers and genetic diversity (implications will be discussed in the following section), although at this writing, imported queens from other rare disease-free areas of the world are being encouraged. This issue is recognized on a national scale as well. The Canadian Honey Council lists "reliance on imported packages and queens" as one of the seven major industry concerns (Canadian Honey Council, 2016b), while beekeepers in Manitoba, Alberta, and blueberry farmers in eastern Canada maintain that imported honeybee packages are necessary to boost hive numbers in spring due to harsh winters (Fletcher, 2015).

Compounding influences pose challenges for effective control of apicultural activities in Newfoundland. These factors include: (1) absence of mandatory apiary registry; (2) lack of capacity to carry out monitoring and enforcement; and (3) lack of comprehensive regulation, which includes consideration of other bee species and relevant products and equipment. In 2014, there was no provincial apiarist; however, the title was recently appointed by the Forestry and Agrifoods Agency to a responsible crop development officer, which is commensurate with other Atlantic provinces. While regulations pertaining to land use and agricultural practices have not been explored explicitly, it is acknowledged that existing relevant regulations should be reviewed and assessed for adherence to common goals and for their implications on honeybee health and the sustainability of apiculture.

Genetic Diversity

Genetic diversity can be viewed as valuable "biological capital" (Büchler, 2013) because it contributes to overall fitness, evolutionary resilience, and adaptability within a population (Delaney, Meixner, Schiff, and Sheppard, 2009; Lacy, 1987; Le Conte et al., 2012; Sammataro, 2012). The genetic diversity of Newfoundland's honeybee population may pose a significant challenge to the sustainability of apiculture on the island. While this section will not explore the complex field of honeybee genetics, it will outline some important facets of the genetic diversity issue in managed honeybee populations with implications for the Newfoundland context.

Genetic diversity of managed honeybee colonies in North America and Europe is already argued to be a major issue of concern, especially when honeybees are threatened by multiple stressors (Le Conte et al., 2012). Coby et al. (2012) argue that three distinct genetic "bottlenecks" occurred in North America, which resulted in decreased genetic diversity of the entire North

American stock. The first was a sampling bottleneck where about one-third of the subspecies of *A. mellifera* were introduced to North America and these were only represented by a few tens to hundreds of queens from each subspecies. The second bottleneck involved the widespread decimation of feral honeybee colonies due to *Varroa* mite infestation. The third bottleneck involves the nature of selective honeybee queen breeding in the US, which is concentrated in two distinct regions that produce around one million queens in a year from less than 600 mothers.

The need to maintain stock diversity is recognized on a national scale. For example, queens are regularly imported to Canada from the US for the purpose of infusing new genetic material into Canada's honeybee stocks, while US bee package importation remains forbidden (CFIA, 2013). Queens are more easily and effectively inspected than packaged bees so they are not included in the import restrictions. A report by the Standing Committee on Forestry and Agriculture in May 2015 controversially recommended that *Honeybee Importation Prohibition Regulations, 2004* be amended to once again facilitate movement of honeybee packages across the US–Canada border (Fletcher, 2015). International trade of honeybees allows for a great deal of genetic migration between colonies. The export value of queen honeybees from New Zealand alone was estimated to be around NZ$4.4 million in 2013 (Roche, 2014). This movement of bees across borders is thought to benefit the genetic diversity of populations, but it puts the importing country at greater risk of losing its disease-free status.

In order to actually consider the risks associated with low genetic diversity and population isolation, some aspects of honeybee biology must be understood. Since every colony possesses a single queen who rears all workers (females) and drones (males), every bee in a honeybee colony is related by varying degrees. For this reason, a honeybee colony is considered an "individual" within a population regardless of the number of bees residing in each colony (Cauia et al., 2010; Jaffé et al., 2009). Despite this fact, Newfoundland's honeybee population must not be seen merely as a small, disperse village of about 300 individuals. The actual scenario is more complex and involves particular honeybee traits and behaviours.

Honeybees possess a number of traits that act to maximize genetic diversity within colonies (Büchler et al., 2013). Multiple mating is one major source of genetic diversity, which results in multiple sub-families within a colony. Estimates of the number of drones that may mate with a queen range up to 40; however, between 5 and 20 is a more common approximation (Harpur, Minaei, Kent, and Zayed, 2012; Oldroyd, Rinderer, Harbo, and Buco, 1992). Drones may fly up to 15 km to drone congregation areas for mating (Jaffé et al., 2009). In Newfoundland, however, there are no feral colonies and many small apiaries exist outside a 15–20 km radius from other apiaries (Hicks, 2011; personal communication). Generally, queens will need to be replaced every two years on average (Büchler et al., 2013). The sperm that inseminates a queen on her mating flight serves to fertilize eggs for her entire lifespan: usually about two years of viability (Laidlaw and Page, 1997). It is therefore plausible that the genetic diversity of a colony in isolation may decrease significantly in a few generations (Cauia et al., 2010).

The introduction of a relatively small sample of honeybees to the island of Newfoundland and the subsequently low levels of genetic migration from outside the province could be considered a form of translocation. Thrimawithana, Ortiz-Catedral, Rodrigo, and Hauber (2013) observe that translocated groups tend to have lower genetic diversity than their larger source group. In addition, smaller populations are prone to loss of genetic diversity much faster than larger populations as a result of genetic drift (random sampling of genes that progress to the next generation) (Lacy, 1987).

It may be quite difficult to assess the actual risk of low genetic diversity in Newfoundland's honeybee population. New genetic stock has been brought into Newfoundland legally a number of times since the import restrictions were put in place. Such imports are allowed with a permit because they involve fertilized eggs as opposed to live bees. These imports introduced Russian and Buckfast strains to the existing Italian, Caucasian, and Carniolan lines in the province. Given that frequent immigration is the most effective way to counter loss of genetic diversity (Lacy, 1987), the genetic diversity issue may appear to be significantly curbed. For this very reason, one respondent believed genetic diversity was not a significant issue for Newfoundland's honeybees.

On the other hand, a majority of Newfoundland's apiaries are not only very small, but also they rely on just a couple queen and nuc sources (commercial and semi-commercial operations) for what is often yearly replacement. Perhaps genetic diversity within the entire island population is a separate issue from genetic diversity within individual apiaries. If enough genetic diversity exists on the island but is not accessible to all beekeeping operations for logistical reasons, the "biological capital" cannot fully be utilized and some operations may suffer.

Due to the complexities of honeybee genetics and the multiple factors influencing genetic flow among and between colonies, it may be impossible to definitively prescribe requirements regarding "safe" levels of genetic diversity. What can be ascertained is that the Newfoundland reality involves many apiaries of less than 10 hives, often existing in isolation. These unique conditions are certainly not favourable for enhancing genetic diversity on their own.

While three respondents claimed that they actively pursue some form of bee breeding in their operations, many small apiaries in Newfoundland perform neither breeding nor queen rearing. This may be attributed to a lack of expertise or it may also be an issue of time and resources to perform these involved tasks. If few apiaries actively increase their stock, ensure self-sufficiency of viable queens, or breed for desired traits such as overwintering ability, a large demand pressure will persist for the very few commercial operations to replace the stock of small apiaries. The problem may be compounded by isolation from other hives and variations in environmental conditions and stressors. Essentially, some operations may experience no problems relating to genetic diversity while others (especially very small numbers of colonies in sustained breeding isolation) may face potential inbreeding depression or some consequences of external risks on inbred colonies.

Maintaining immigration of new genetic material may be the most effective method of mitigating inbreeding in small isolated populations; however, subdividing the population also proves to be beneficial (Lacy, 1987). Such a management technique involves splitting a population into distinct units that cannot interbreed and using these units as supplemental genetic sources for one another in a planned scenario. Such a technique was used for the Russian honeybee breeding program in the United States beginning in 1997 (Rinderer et al., 2000).

No genetic profile of Newfoundland's honeybee population has been conducted to date. It is therefore difficult to assess any actual long-term or short-term risk of inbreeding within Newfoundland's colonies. However, it is clear that some beekeeping operations stand at a disadvantage when faced with very low colony numbers, geographic isolation, high winter mortality, and the absence of a guaranteed supply of bees from within the island. As a result, co-operation among Newfoundland's beekeepers in the design of a breeding scheme or breeding program may be highly beneficial in order to ensure the sustainability of individual operations. At least three operations on the island are already practising honeybee breeding, but breeding programs can be extremely labour-, knowledge-, and resource-intensive. In the event of pathogen infestation, increased pesticide exposure, or other stressors, decreased genetic diversity could be a significant, negative compounding factor (Sammataro, 2012). Ensuring that all beekeepers have access to sufficient genetic material is vital for the overall sustainability of the industry in Newfoundland. Further research into the exact number and location of operations/potential operations, their current sources of queens/nucs, and some of the logistical challenges these operations face may be beneficial in planning the goals and design of any breeding program or scheme.

Weather Conditions and Overwintering

Newfoundland's harsh climate is perhaps the most obvious and substantial challenge faced by beekeepers on the island. In Canada alone, long and harsh winters are considered a major challenge to beekeepers (Currie et al., 2010; Kozak et al., 2014). In particular, wet, cold spring conditions are a significant obstacle for spring build-up of colonies (Kozak et al., 2014). Considering honeybees will not forage during inclement weather conditions (Javorek, Mackenzie, and Vander Kloet, 2002), long winters with harsh spring conditions and freeze–thaw fluctuations increase the challenge for many Newfoundland beekeepers. Until the import ban on live honeybees was implemented in the 1980s, most beekeeping operations on the island purchased new bees every season due to high winter mortality (Hicks, 2014). Since the import ban, however, the independence and security of apiculture in Newfoundland faces considerable instability.

Five of the seven respondents mentioned weather or harsh winters as a major factor affecting their operations (including those respondents with some of the longest beekeeping experience on the island). During interviews, multiple concerns with winter protection of hives were discussed. Many adaptive techniques are being implemented to mitigate local weather conditions and harsh winters. Co-operation and communication among the beekeeping community, in combination with support for training workshops and effective educational material, may help beekeepers (especially inexperienced hobbyists and newcomers) to cope with the island's often severe climatic conditions.

Severe weather also interacts indirectly with the issues of genetic diversity, disease/pest risk, and regulation enforcement. Two respondents mentioned unfilled queen bee orders placed during the past few years due to high demand. High overwintering mortality has been linked to a high demand and low supply for honeybees on the island, which exacerbates concerns about shortages. High yearly demand for new honeybee stock is not specific to Newfoundland; it is a widespread reality made more precarious by higher colony losses in recent years. In the US alone, yearly demand for the replacement/restocking of honeybees is estimated to be about 2.4 million colonies (Coby et al., 2012). However, Newfoundland experiences this issue on a small scale resulting in the entire province's demand being met by a couple major suppliers. If these operations experience a significant loss due to severe weather or any stochastic event, the entire industry could be in jeopardy.

This insecurity within the province's honeybee population not only poses a challenge to the growth of apiculture, it also increases the potential problems relating to decreased genetic diversity in individual, small operations. In addition, the risk of illegal bee importation may increase when demand for honeybees cannot be met. Increasing education and training measures for beekeepers on regionally focused management strategies, increasing awareness about the importance of the import restrictions, and considering genetic diversity in the strategic growth and diversification of apiculture on the island may all aid in minimizing the risks incurred by high winter mortality and harsh weather on the island.

Diseases and Pests

Currie et al. (2010) recognize acaricide resistance and failure to control *Varroa* mites as one of the most important factors related to colony losses in Canada. As mentioned previously, Newfoundland stands at a considerable advantage to mainland North America as its honeybee population remains unaffected by *V. destructor*, tracheal mites (*Acarpis woodi*), Israeli acute paralysis virus, Kashmir bee virus, and significant other invasive organisms (Shutler et al., 2014). With these threats persisting in most other populations around the world (including neighbouring Nova Scotia), the risk of disease/

pest infestation in Newfoundland remains relatively high. It is therefore important to consider the probable introduction of these pests/diseases onto the island as well as their implications.

Three of the questionnaire respondents included introduced pests/ diseases or *Varroa* mite infestation as major challenges facing either their individual operations or apiculture in Newfoundland as a whole. Two respondents listed the risks from illegal importation of bees as a major challenge that would jeopardize the biosecurity from pests and disease. One respondent/interviewee considered the infestation of pests and diseases an imminent risk.

Indeed, other island honeybee populations have experienced a delayed exposure to some of these pests. Hawaii was mite-free until the discovery of a *Varroa*-infested colony on Oahu in 2007 (State of Hawaii, 2016). New Zealand was another isolated location that remained unaffected by *Varroa* mites until the pests' detection on the North Island in 2000 and the South Island in 2006 (Roche, 2014). In the words of one interviewee, infestation may "not [be] a matter of if, but when." Preparedness for mite and disease infestation is therefore paramount.

Some believe that the small number of widely dispersed hives present on the island offers an advantageous buffer to the possible spread of diseases. With no feral honeybee populations in Newfoundland and often large distances between hives, mitigating the transfer of pests and diseases may be more easily manageable than in a mainland scenario. Nevertheless, the supply chain of bees within the province must be considered. If any one of the major suppliers of honeybees in the province were affected by disease/ pest infestation, the entire island population would be in danger. Pathogen transmission via the transportation of bees, hive products, and equipment between apiaries within the province is a notable risk.

The issue of infestation is not necessarily restricted to honeybee-to-honeybee transfer. "Pathogen spillover" (transfer of infection usually between wild and managed populations) has been shown to occur between managed and wild bee populations through shared flower use (Colla, Otterstatter, Gegear, and Thomson, 2006; Fürst, McMahon, Osborne, Paxton, and Brown, 2014). It is rumoured that bumblebees are brought to Newfoundland from the mainland for berry pollination purposes (often after they have already serviced crops in Nova Scotia). This should be considered a major risk for both native pollinators and commercial honeybee populations. Graystock et al. (2013) not only demonstrated pathogen spread from bumblebees to honeybees, but also noted that 77 per cent of commercially produced bumblebee colonies used in an experiment carried microbial parasites despite being advertised as parasite-free. Pollinators, even when commercially produced, do not exist in isolation and their distribution cannot be truly controlled. Continued importation of non-*Apis* bee species puts Newfoundland's native and managed bees at risk.

If pests such as the *Varroa* and tracheal mites were to infect Newfoundland's honeybee colonies, the effects could be detrimental. The *Varroa* can act as a vector for significant other pests and diseases (Le Conte et al., 2012; Shutler et al, 2014; vanEngelsdorp and Meixner, 2010). The cumulative or synergistic impact of these potential inhibitions, along with already harsh climatic conditions in Newfoundland, could be severe.

As it stands now, Newfoundland's disease-/mite-free honeybee population is regarded as a significant opportunity. Beekeepers are given the advantage of not having to deal with the compounding stresses of infected colonies. Without the requirement of miticides and other chemical treatments, truly organic hive products, with proper market development, could be sold as specialty items (Williams et al., 2010). In addition, significant potential exists to provide disease-free, chemical-free bees for research purposes (Shutler et al., 2014; Fletcher, 2015).

Given that Newfoundland's honeybees have not been exposed to many of the stresses on the mainland, one respondent identified the sale and use of honeybees for research purposes as a more important opportunity than commercial sale off the island. Here we encounter the issue of honeybee genetics interacting with disease resistance and therefore market potential.

Significant attention is being given to the use of genetic research and honeybee breeding in order to increase mite resistance (OPERA, 2013; Rinderer et al., 2000; University of Guelph, 2016). Breeding traditionally has been focused on maximizing commercially significant traits such as honey production, temperament, and colony growth (Delaney et al., 2009). Breeding for mite resistance in honeybees can involve a number of behavioural traits (Sammataro, 2012) or even targets of mite growth rate (Fries, 2012). However, breeding for resistance requires exposure to infestation pressure (Cauia, 2010; OPERA, 2013). Therefore, the sale of mite/disease-free honeybees to infested commercial operations outside the province will not be viable unless collaboration with mainland breeding programs is maintained and a focus on producing mite-resistant honeybee strains is upheld in Newfoundland. It would be wise to prioritize production goals and assess market feasibility for the potential sale of honeybees outside the province. Marketing honeybees for research purposes and to provide other breeding programs and mite-free locations may prove to be a more lucrative development direction.

Agriculture, Pollination, and Industry Development

Plant pollination occurs through the transfer of pollen between flowers while honeybees forage for energy resources. In this way, honeybees not only collect the nectar with which honey is produced, but also perform the invaluable pollination service necessary for so many plant species and economically significant crops. Apiculture and agriculture can thus be considered complementary industries. The inherent tie between beekeeping and agriculture is tainted, however, as unsustainable agricultural development can prove detrimental to bee health. Developing these industries with a mutual knowledge base and congruent goals may aid in improving the sustainability of both apiculture and agriculture.

The European Pollinators Support Farm Productivity (STEP project) report published in 2011 noted the most important factors associated with recent pollinator declines are linked to land-use changes that occurred in the agricultural landscape after World War II (OPERA, 2013). Increased intensity of agriculture can involve destructive practices that reduce pollinator habitat and forage availability/quality. Such practices include large-scale monoculture ecosystems, reduced hedgerows and marginal habitats, and increased use of chemical inputs such as pesticides and herbicides (Allen-Wardell et al., 1998; FAO, 2008; Le Conte et al., 2012; OPERA, 2013). In addition, low-diversity agro-ecosystems, which cannot support sufficient pollinators naturally, necessitate the rental of large numbers of managed pollinators to provide this ecosystem service. As an example, some colonies might travel up to 40,000 miles over one year to pollinate four or more different crops (vanEngelsdorp and Meixner, 2010).

Newfoundland's agricultural output is extremely low compared to its Atlantic Canada counterparts. In 2009, Newfoundland had less than half the farms present in Prince Edward Island and only about 5 per cent of the cropland. Currently, fruit producers comprise a demand for pollination. In particular, blueberry and cranberry crops require insect pollination for successful fruit set. There is some degree of dispute as to the efficacy of honeybees in pollinating these crops (Aras, de Oliveira, and Savoie, 1996; Hicks, 2011; Javorek, Mackenzie, and Vander Kloet, 2002).

Non-*Apis* pollinators such as *Augochlora, Augochlorella, Andrena, Bombus, Halictus, Agapostomon,* and *Lasioglossum* have all been shown to demonstrate greater pollination efficiency compared to honeybees through sonication of flowers, or buzz pollination (Javorek, Mackenzie, and Vander Kloet, 2002). These species also exhibit higher degrees of tolerance for foraging during marginal weather conditions than honeybees. It has been shown that it is feasible for honeybees to be supplied in abundance and supplement low native bee numbers to successfully increase blueberry pollination (Eaton and Nams, 2012). Stakeholders in Atlantic Canada established that honeybees can provide the best and most easily managed method of crop pollination (AMEC, 2010). However, Hicks (2011) found stocking blueberry fields with imported *Bombus impatiens* and *Apis mellifera* to be ineffective at increasing fruit set in eastern Newfoundland. Globally, crop yields have also been shown to respond more positively to higher densities of wild pollinators than honeybees (Rose et al., 2015). Honeybees are doubtless an important pollinating species and can be used in Newfoundland. However, in consideration of this ongoing debate about pollination effectiveness, it is essential that pollination strategies and pollinator health take into consideration regional and local ecological factors and interactions beyond managed species. Such considerations should be guided by more pollination research within Newfoundland.

Unlike other more agriculturally productive provinces, Newfoundland's honeybees are not used on a large scale for pollination and migratory beekeeping. However, berry crop producers sometimes import other bee species for pollination purposes due to low numbers of native pollinators (Hicks, 2011). Anecdotal reports are that farmers import quads of bumblebees from Nova Scotia, often in the back of pickup trucks. One interviewee claimed of knowing of berry producers who would refrain from importing bumblebees if they could be guaranteed a supply of local honeybees to serve the purpose. Three respondents included the provision of pollination services as a future goal in the development of their operations. It is clear that Newfoundland's pollination capacity is not matched to its agricultural productivity, either in terms of pollinator numbers or the logistics of their rental, distribution, or transport. Nevertheless, mutual interest from both beekeepers and crop producers has been identified. Therefore, boosting honeybee populations on the island and co-ordinating communication and co-operation between fruit producers and beekeepers could aid in increasing the sustainability of these industries as well as their provincial independence.

While Newfoundland apiculture, just like agriculture, can be considered relatively underdeveloped, it also places the province at a significant advantage. The apicultural and agricultural practices linked to colony losses in other parts of the Western world are not observed on the island. A relatively small portion of the province has been converted to agricultural land, and none of that land is managed on a scale comparable to large, mainland monocultures. Honeybee exposure to pesticides may therefore be comparatively small. In addition, migratory apiculture (which incurs some of the highest cost of the industry) is not practised to any large degree in Newfoundland.

The scale of agriculture and apiculture in Newfoundland therefore affords the province an advantage in terms of pollinator health. However, the growth and development of both of these sectors will necessitate careful planning to mitigate the impacts observed in more agriculturally intense regions. One respondent identified agricultural development in the absence of pollinator knowledge and consideration as a major concern both to the sustainability of an individual operation and to apiculture in Newfoundland as a whole.

One area of concern related to agricultural development is the potential increase in pesticide use, which could accompany increased productivity. Honeybees are already noted for their lack of detoxification enzymes associated with moderate levels of pesticide resistance (vanEngelsdorp et al., 2009). Pesticide use and misuse have been linked to pollinator declines (AMEC, 2010; FAO, 2008; Health Canada, 2014; Hopwood et al., 2012; Le Conte et al., 2012; Melhim et al., 2010; OPERA, 2013). In particular, the group of pesticides known as neonicotinoids has become a major concern for beekeepers around the world. The European Commission restricted the use of three neonicotinoid pesticides in 2013 after they were found to cause "high acute risks" for bees (EC, 2013). These pesticides are still legally used across Canada; however, Health Canada's Pest Management Regulatory Agency has recognized the link between neonicotinoid use/misuse and declining bee health (Health Canada, 2014) and Ontario has imposed progressive restrictions (Ontario, 2014). It would be wise for Newfoundland to implement pre-emptive measures in the form of pesticide regulations that reflect the most recent research on pesticide use and pollinator health.

In order to avoid destructive agricultural and apicultural practices, clear goals for pollinator health and apiculture must be integrated into agricultural land-use regulations, farm best management practices, farm support programs, and other farm-related policies and management initiatives. The implications of unsustainable agricultural practices have been shown to negatively affect bee health in myriad ways. At this point, the relatively underdeveloped nature of both apiculture and agriculture in Newfoundland can be considered a blessing. These sectors are provided somewhat of a "blank slate" and the opportunity to develop with harmonized goals and management practices oriented towards the mutual sustainability of both industries.

Forage Availability, Quality, and the Broader Ecological Context

As an extension of the pollination theme comes reflection on non-agricultural forage sources for honeybees. It is necessary also to recognize the impacts that other industries and sectors have on pollinator habitat and forage sources. Newfoundland contains a diverse range of ecosystems and habitats and it is vitally important that consideration be given to the abundance, diversity, type, and quality of forage available for honeybees in specific regions, as well as for the broader group of pollinators and their ecological importance.

In response to a question regarding the largest perceived challenges to the sustainability of individual operations, one respondent listed the need for adequate land base and floral sources as a major factor. Another respondent mentioned carrying capacity (maximum population sustainable given the available food resources) of their region as a concern. Determining carrying capacity estimates for pollinators and minimum plant pollination requirements can be very difficult (Allen-Wardell et al., 1998). It would be advantageous to conduct studies on the floral abundance and diversity within ecoregions in Newfoundland in order to better understand target locations for potential apicultural development, as well as to assess the carrying capacity of different areas on the island. That being said, the NLBKA has made some headway in this regard. Their website contains an ongoing photo inventory of honeybee forage species submitted by volunteers. The list currently identified 28 separate forage species by common and scientific name along with clear evidence of their use by bees (NLBKA, 2017c).

It is important not only to assess available forage sources for pollinators, but also to create measures to protect the health, abundance, and diversity of these sources. Forestry and public land development are two areas where pollinator protection measures can be very beneficial for the preservation of *Apis* and non-*Apis* pollinator health in Newfoundland. One area of concern is the use of non-agricultural pesticides. Agricultural pesticide use was discussed in the previous section; however, pesticide use for cosmetic purposes (private and public aesthetic applications), public land management (such as roadside spraying), and forestry management are all additional areas of concern.

Since honeybees and other pollinators do not adhere to property or jurisdictional boundaries in their flight range, all private and public land-use changes and practices are relevant to pollinator health. In Australia, beekeeping is incorporated into the public planning process (Victorian State Department, 2013). This includes regional management plans, operational and management plans, management prescriptions, and forest zoning. Part of this process involves the designation of bee sites allocated under a licensing system. In 2012, there were 3,637 such bee sites on 7.6 million ha of forests, parks, and conservation reserves in Victoria. Of course, vast differences between Newfoundland's and Australia's climate, ecology, and floral abundance/diversity must be recognized. However, this example illustrates how apiculture development and sustainability can be incorporated within public and private management and sustainability plans and policies.

When speaking about a wider ecological context, we must be sure to include non-*Apis* pollinators as well and recognize the clear link between biodiversity and pollination (FAO, 2008). Canada currently has no provincial or federal legislation with explicit mention of or attention to native pollinators (Byrne and Fitzpatrick, 2009; Tang, Wice, Thomas, and Kevan, 2007). Blackburn (2012) criticizes policies dealing with pollinators and pollinator health to be examples of "honeybee centrism" that lacks more ecologically balanced approaches. In Newfoundland, only about 50 species of native bees have been identified and there is a recognized lack of knowledge about local pollinators (Hicks, 2011). In addition, the importance of pollinator "suites" over single species has been recognized for effective pollination of many crops, including blueberries (Kevan, 1999). In essence, the healthy development of apiculture in Newfoundland and the preservation of endemic pollinator species will require policies and management plans to consider effects of land-use changes and land management practices on honeybee health as well as forage availability and quality. The forestry sector and public land management could be important areas for the development of pollinator-friendly policies and management plans.

CONCLUSION AND RECOMMENDATIONS

The island of Newfoundland occupies a unique place in the beekeeping world. It stands as a bastion of healthy hives, unadulterated by mites and other invasive pests and buffered from many externalities of unsustainable, intensive agriculture. Given these endowments, apiculture in the province holds significant scientific and economic research potential. Market assessment and development of specialty organic hive products within and outside the province could hold great economic possibility. Some honey producers in Newfoundland have already begun capitalizing on the demand from high-end, specialty restaurants. As we can see in the following chapter, important innovations in micropropagation of berry crops could hold promising results in developing the berry industry. Any such development will require adequate pollination services. This necessitates the possibility and the need to maintain co-operation and co-development of apiculture and agriculture with policies that contain mutual goals of long-term sustainability and biosecurity.

The need for co-operation and communication truly translates to all levels if we are to envision apiculture as part of our unique food future in Newfoundland. This means connections between researchers/universities, beekeepers (within and outside the province), farmers/crop-producers, government, and the public. To use the phrase of one interviewee, we need to "take a lesson from the hive." A hive of honeybees is a startlingly beautiful and succinct analogy of community and co-operative sustainability that translates directly to the complexity of ecological interdependence and the food system of which we are a part. If we can envision apiculture as a facet of an ecological reality within which a sustainable food system is embedded, we think we can understand the development of the province's apiculture sector as a social movement for food sovereignty. Indeed, the enthusiasm of beekeepers and the constant demand for nucs in the province can attest to the momentum being built.

Though it was only recently formed in 2014, the NLBKA commodity group has demonstrated rapid advancement in identifying research and commercial needs within the province to grow the honeybee and native bee populations in an ecologically and financially sound manner. As noted by their Communications Manager, Peter Armitage (NLBKA, 2017d), although NLBKA is a young association representing both commercial and hobby beekeepers, their membership is eager to expand beekeeping in the province despite climatic challenges and the limited resources. He notes that they are committed to keeping the island's bee population free of the *Varroa* destructor, small hive beetle, wax moth, American foulbrood, and some other pathogens, pests, and diseases that are the bane of beekeepers in so many other parts of the world, while being committed to advancing apiculture here in an ecologically responsible manner, and recognizing that obviously our *Apis mellifera* colonies are part of a broader ecosystem that includes a large number of *Bombus* species and other native pollinators.

The theme of co-operation holds within it the need for education, for it is in the collaborative networks among parties and stakeholders that lessons are learned and headway is made. This, again, should be considered across the spectrum. From an industry standpoint, the proper educational tools could aid in promoting pollinator-friendly practices and prompt compliance with current and future regulations, thereby aiding a wider understanding of pollinators and best management practices among farmers, fruit producers, municipalities, the general public, and beekeepers themselves. Many of these tools are publicly available through the NLBKA website, although there is a desire to build resources within the provincial government and with agricultural landowners.

Beekeeping is a very knowledge-intensive activity whether it is pursued as a hobby or commercially. Education on best management practices, especially those specific to Newfoundland's climate and biogeography, was raised in both questionnaire responses and interviews. This need for education within the beekeeping industry can be seen in well-established apicultural sectors as well. Nearly three-quarters of the subsidies allocated for education purposes in Austria's Apiculture Programme between 2004 and 2007 were used for training. In addition, those training sessions focused on fundamental knowledge gained the highest attendance numbers (Neuwirth,

Hambrusch, and Wendtner, 2011). Educating beekeepers on best management practices not only could aid in boosting the health and development of the industry, but also could prompt more diligent attention to import regulations and the proper use and transportation of beekeeping equipment.

Educating farmers on pollinator health is recognized as a vital facet of any initiatives to promote pollinator health and diversity (Blackburn, 2012; FAO, 2008; Roche, 2014). Blackburn (2012) suggests the implementation of cost-sharing schemes and incentives-based policies to help farmers and crop producers incorporate pollinator-friendly practices in their operations. Pollinator health is intrinsically linked to ecosystem health. Education on the importance of apicultural development, pollinator health, and best management practices can all help to harmonize the efforts of beekeepers, farmers, and the general public towards viable and sustainable industries and land developments.

Education and co-operation will both facilitate and be propelled by access to necessary information. Such information requires some form of consistent monitoring. Effective and standardized monitoring is a widely recognized need from local to international scales (AMEC, 2010; Byrne and Fitzpatrick, 2009; FAO, 2008; Kozak et al., 2014; OPERA, 2013; Meixner et al., 2010). "Bee monitoring" generally refers to surveillance systems where bee health (either generally or specifically) is observed under practical field conditions (OPERA, 2013).

In order to address a number of challenges related to bee health in Newfoundland, increased effective and standardized monitoring is a necessity. There are many examples of monitoring initiatives striving to increase the efficacy and efficiency of data gathering, including the comprehensive German Bee Monitoring Program (vanEngelsdorp and Meixner, 2010), the ALARM project in Europe (Murray, Kuhlmann, and Potts, 2009), and the US research team on colony collapse disorder (Meixner et al., 2010). We must recognize, however, that even some very basic data can provide valuable information. A recognized lack of capacity (both industry and government) forms a barrier to effective management in Newfoundland. However, in the year since this research began, a voluntary bee registry has been initiated and the NLBKA has shown considerable effort towards galvanizing beekeepers in Newfoundland and co-operating with the provincial government. If executed effectively, a standard registry and yearly updates on bee health, population numbers, distribution, and notable observations can provide a great deal of information.

Finally, effective support programs for apiculture can result from a culmination of proper education, collaboration, and communication. Support programs for apiarists are a vital part of growing the industry, especially given the challenges faced by industry entrants. For example, Romania possesses a vibrant and widely recognized apiculture industry with favourable conditions for bee breeding. However, the cost of production often exceeds the revenue from sales in Romania, thereby necessitating considerable government support in the form of beekeeping production diversification, scientific research, and specialist training (OPERA, 2013). No specific provincial support programs currently exist for apicultural development in Newfoundland. In fact, with funding at approximately 10 per cent of American levels, Canada's Atlantic apiculture industry in general is facing a shortage of support (AMEC, 2010). Nevertheless, a little can go a long way if it is allocated appropriately, whether for facilitating safe importation of new stock, for funding start-up equipment, for training, or for logistical and informational co-ordination. In order for programs to be successful, it will be important to further assess the profile of the industry, target operations with promising development potential, and examine the key barriers currently experienced by hobby and commercial apiaries.

Results from our questionnaire and interviews revealed a great deal of enthusiasm and innovation within the beekeeping community in Newfoundland. However, growth in the small commercial facet of the beekeeping industry is questionable. Therefore, the sustainability of apiculture in Newfoundland hangs ultimately on the precarious nature of the industry profile. The opportunities discussed hold great potential, but this potential cannot be realized unless beekeepers and crop growers on the island can be guaranteed a safe and certain supply of honeybees from season to season. It will be vitally important that growth in apiculture in Newfoundland is strategically developed with consideration of the multiple factors interacting with industry size/profile.

This chapter has summarized some key factors and their basic interactions as they may affect the sustainability of apiculture on the island of Newfoundland. In order to move forward, considerable work and research has yet to be done. Economic feasibility was not considered in this study;

however, economic assessments will be essential for moving forward with developing unique hive products and bringing them to market. There are many gaps in knowledge relating to honeybees within the ecological context of Newfoundland, especially in regard to forage availability, habits, and interactions with native pollinators. Climate interactions and risk assessments on potential effects of climate change on pollinator health in Newfoundland are another large and relevant area of inquiry. Lastly, a thorough and ongoing method of hive monitoring (including hive number, distribution, and health) will be necessary to maintain a database of basic information about the industry. If development and diversification occur, Newfoundland could grow from a hive of possibility into a considerable hive of activity, as well as a sweet and exciting component of a sustainable food future for Newfoundland.

REFERENCES

Allen-Wardell, G., P. Bernhardt, R. Bitner, A. Burquez, S. Buchmann, J. Cane, . . . S. Walker. 1998. "The potential consequences of pollinator declines on the conservation of biodiversity and stability of food crop yields." *Conservation Biology* 12, 1: 8–17. At: http://www.jstor.org/stable/2387457?origin=JSTOR-pdf&seq=1#page_scan_tab_contents.

AMEC. 2010. *Facilitator's Final Report on the Maritime Action Forum on Pollination Research*. Moncton, NB. At: http://www.uoguelph.ca/canpolin/Publications/BNBB_CANPOLIN_Final_Report.pdf.

Aras, P., D. de Oliveira, and L. Savoie. 1996. "Effect of a honey bee (Hymenoptera: *Apidae*) gradient on the pollination and yield of lowbush blueberry." *Journal of Economic Entomology* 89, 5: 1080–83. At: http://jee.oxfordjournals.org/content/89/5/1080.abstract.

Australian Honey Bee Industry Council (AHBIC). 2003. "Biosecurity or disease risk mitigation strategy for the Australian honey bee industry." At: http://www.honeybee.org.au/pdf/BiosecurityPlanFinal.pdf.

Blackburn, T. 2012. "To bee, or not to bee, that is the problem: Managing wild bee decline in Canadian agriculture." Master's thesis, Simon Fraser University. At: http://summit.sfu.ca/item/12243.

Büchler, R., S. Andonov, K. Bienefeld, C. Costa, F. Hatjina, N. Kezic, . . . J. Wilde. 2013. "Standard methods for rearing and selection of *Apis mellifera* queens." *Journal of Apicultural Research* 52, 1: 1–30. doi:10.3896/IBRA.1.52.1.07.

Byrne, A., and Ú. Fitzpatrick. 2009. "Bee conservation policy at the global, regional and national levels." *Apidologie* 40, 3: 194–210. doi:10.1051/apido/2009017.

Canadian Food Inspection Agency. 2013. *Risk Assessment on the Importation of Honey Bee (Apis mellifera) Packages from the United States of America*. Sept. At: http://www.ontariobee.com/sites/ontariobee.com/files/Final%20V13%20Honeybeepackages%20from%20USA_Oct21_2013.pdf.

Canadian Honey Council. 2016a. "Education/resources: Apiary acts and regulations." May. At: http://www.honeycouncil.ca/resources.php.

———. 2016b. "Overview of the Canadian apiculture industry." May. At: http://www.honeycouncil.ca/honey_industry_overview.php.

Cauia, E., A. Siceanu, S. Patruica, M. Bura, A. Sapcaliu, and M. Magdici. 2010. "The standardization of the honeybee colonies evaluation methodology, with application in honeybee breeding programs, in Romanian conditions." *Animal Science and Biotechnologies* 43, 2: 174–80. At: http://www.spasb.ro/index.php/spasb/article/viewFile/861/818.

Chemás, A., and V. Rico-Gray. 1991. "Apiculture and management of associated vegetation by the maya of Tixcacaltuyub, Yucatán, México." *Agroforestry Systems* 13, 1: 13–25. doi:10.1007/BF00129616.

Coby, S. W., W. S. Sheppard, and D. R. Tarpy. 2012. "Status of breeding practices and genetic diversity in domestic U.S. honey bees." In D. Sammataro and J. A. Yoder, eds., *Honey Bee Colony Health: Challenges and Sustainable Solutions*, 25–36. Boca Raton, Fla.: Taylor & Francis Group.

Colla, S. R., M. C. Otterstatter, R. J. Gegear, and J. D. Thomson. 2006. "Plight of the bumblebee: Pathogen spillover from commercial to wild populations." *Biological Conservation* 129, 4: 461–67. doi:10.1016/j.biocon.2005.11.013.

Creswell, J. W. 2009. "Mixed methods procedures." In *Research Design: Qualitative, Quantitative, and Mixed Methods Approaches*, 3rd ed., 203–24. Thousand Oaks, Calif.: Sage Publications.

Currie, R. W., S. F. Pernal, and E. Guzmán-Novoa. 2010. "Honey bee colony losses in Canada." *Journal of Apicultural Research* 49, 1: 104–06. doi:10.3896/IBRA.1.49.1.18.

Delaney, D. A., M. D. Meixner, N. M. Schiff, and W. S. Sheppard. 2009. "Genetic characterization of commercial honey bee (Hymenoptera: *Apidae*) populations in the United States by using mitochondrial and microsatellite markers." *Annals of the Entomological Society of America* 102, 4: 666–73. doi:10.1603/008.102.0411.

Driscoll, D. L., A. Appiah-Yeboah, P. Salib, and D. J. Rupert. 2007. "Merging qualitative and quantitative data in mixed methods research: How to and why not." *Ecological and Environmental Anthropology* (University of Georgia): 18.

Eaton, L. J., and V. O. Nams. 2012. "Honey bee stocking numbers and wild blueberry production in Nova Scotia." *Canadian Journal of Plant Science* 92, 7: 1305–10. doi:10.4141/cjps2012-045.

European Commission. 2013. "Bees and pesticides: Commission goes ahead with plan to better protect bees." 30 May. At: http://ec.europa.eu/food/archive/animal/liveanimals/bees/neonicotinoids_en.htm.

Fletcher, E. 2015. "Honey bee health in Newfoundland and Labrador: The buzz about neonicotinoids, pests and honey bees in Newfoundland and Labrador." Master's research paper, Memorial University of Newfoundland, Grenfell Campus.

Food and Agriculture Organization of the United Nations (FAO). 2008. *Rapid Assessment of Pollinators' Status: A Contribution to the International Initiative for the Conservation and Sustainable Use of Pollinators*. At: http://www.fao.org/uploads/media/raps_2.pdf.

Fries, I. 2012. "Evaluation of *Varroa* mite tolerance in honey bees." In D. Sammataro and J. A. Yoder, eds., *Honey Bee Colony Health: Challenges and Sustainable Solutions*, 21–24. Boca Raton, Fla.: Taylor & Francis Group.

Fürst, M. A., D. P. McMahon, J. L. Osborne, R. J. Paxton, and M. J. F. Brown. 2014. "Disease associations between honeybees and bumblebees as a threat to wild pollinators." *Nature* 506, 7488: 364–66. doi: 10.1038/nature12977.

Graystock, P., K. Yates, S. E. F. Evison, B. Darvill, D. Goulson, and W. O. H. Hughes. 2013. "The Trojan hives: Pollinator pathogens, imported and distributed in bumblebee colonies." *Journal of Applied Ecology* 50, 5: 1207–15. doi:10.1111/1365-2664.12134.

Harpur, B. A., S. Minaei, C. F. Kent, and A. Zayed. 2012. "Management increases genetic diversity of honey bees via admixture." *Molecular Ecology* 21, 18: 4414–21. doi:10.1111/j.1365-294X.2012.05614.x.

Health Canada. 2014. "Update on neonicotinoid pesticides and bee health." 25 Nov. At: http://www.hc-sc.gc.ca/cps-spc/alt_formats/pdf/pubs/pest/_fact-fiche/neonicotinoid/neonicotinoid-eng.pdf.

Hicks, B. 2011. "Pollination of lowbush blueberry (*Vaccinium angustifolium*) in Newfoundland by native and introduced bees." *Journal of the Acadian Entomological Society* 7: 108–18. At: http://www.acadianes.ca/journal/papers/hicks_11-11.pdf.

———. 2014. "The history and present status of honey bee keeping in Newfoundland and Labrador." *The Osprey* 45, 3: 11–14.

Hopwood, J., M. Vaughan, M. Shepherd, D. Biddinger, E. Mader, S. H. Black, and C. Mazzacano. 2012. "Are neonicotinoids killing bees? A review of research into the effects of neonicotinoid insecticides on bees, with recommendations for action." The Xerces Society for Invertebrate Conservation. At: http://www.olyrose.org/articles/are-neonicotinoids-killing-bees_xerces-society1.pdf.

Jaffé, R., V. Dietemann, M. H. Allsopp, C. Costa, R. M. Crewe, R. Dall'olio, and R. F. A. Moritz. 2009. "Estimating the density of honeybee colonies across their natural range to fill the gap in pollinator decline censuses." *Conservation Biology* 24, 2: 583–93. doi:10.1111/j.1523-1739.2009.01331.x.

Javorek, S. K., K. E. Mackenzie, and S. P. Vander Kloet. 2002. "Comparative pollination effectiveness among bees (Hymenoptera: *Apoidea*) on lowbush blueberry (Ericaceae: *Vaccinium angustifolium*)." *Annals of the Entomological Society of America* 95, 3: 345–51. doi:10.1603/0013-8746(2002)095[0345:CPEABH]2.0.CO;2.

Kevan, P. G. 1999. "Pollinators as bioindicators of the state of the environment: Species, activity and diversity." *Agriculture, Ecosystems and Environment* 74, 1: 373–93. doi:10.1016/S0167-8809(99)00044-4.

Keske, C. M. 2008. *Rents, Efficiency, and Incomplete Markets: The Emerging Market for Private Land Preservation and Conservation Easements.* Sarrbrücken, Germany: Aktiengesellschaft.

Kozak, P., S. Pernal, M. Kempers, R. Lafreniere, A. Leboeuf, M. Nasr, . . . D. Ostermann. 2014. *CAPA Statement on Honey Bee Wintering Losses in Canada (2014).* Canadian Association of Professional Apiculturists. At: http://www.capabees.com/content/uploads/2013/07/2014-CAPA-Statement-on-Honey-Bee-Wintering-Losses-in-Canada.pdf.

Lacy, R. C. 1987. "Loss of genetic diversity from managed populations: Interacting effects of drift, mutation, immigration, selection, and population subdivision." *Conservation Biology* 1, 2: 143–58. At: http://www.jstor.org/stable/2385830?origin=JSTOR-pdf&seq=1#page_scan_tab_contents.

Laidlaw, H. H., and R. E. Page. 1997. *Queen Rearing and Bee Breeding*. Cheshire, Conn.: Wicwas Press.

Leboeuf, A., M. Nasr, C. Jordan, M. Kempers, P. Kozak, R. Lafreniere, . . . G. Wilson. 2015. "CAPA Statement on Honey Bee Wintering Losses in Canada (2015)." Canadian Association of Professional Apiculturists. At: http://capabees.org/shared/2015/07/2015-CAPA-Statement-on-Colony-Losses-July-16-Final-16-30.pdf.

Le Conte, Y., J.-L. Brunet, C. McDonnel, C. Dussaubat, and C. Alaux. 2012. "Interactions between risk factors in honey bees." In D. Sammataro and J. A. Yoder, eds., *Honey Bee Colony Health: Challenges and Sustainable Solutions*, 215–22. Boca Raton, Fla.: Taylor & Francis Group.

——— and M. Navajas. 2008. "Climate change: Impact on honey bee populations and diseases." *Revue scientific et technique, International Office of Epizootics* 27, 2: 499–510. At: http://www.researchgate.net/publication/23285587_Climate_change_impact_on_honey_bee_populations_and_diseases.

Meixner, M. D., C. Costa, P. Kryger, F. Hatjina, M. Bouga, E. Ivanova, and R. Büchler. 2010. "Conserving diversity and vitality for honey bee breeding." *Journal of Apicultural Research* 49, 1: 85–92. doi: 10.3896/IBRA.1.49.1.12.

Melhim, A., A. Weersink, Z. Daly, and N. Bennet. 2010. *Beekeeping in Canada: Honey and Pollination Outlook*. Canadian Pollination Initiative. At: http://www.uoguelph.ca/canpolin/Publications/Melhim%20et%20al%20 2010%20Outlook-Beekeeping-in-Canada.pdf.

Murray, T. E., M. Kuhlmann, and S. G. Potts. 2009. "Conservation ecology of bees: Populations, species, and communities." *Apidologie* 40, 3: 211–36. doi:10.1051/apido/2009015.

Neuwirth, J., J. Hambrusch, and S. Wendtner. 2011. "Evaluation of the National Apiculture Programme in Austria, 2004–2007: General lessons learned regarding support programmes." *Rural Areas and Development* 8: 47–58. At: http://purl.umn.edu/138331.

Newfoundland and Labrador, Department of Natural Resources Agrifoods Development Branch. 2009. *Atlantic Canada Agriculture and Agrifood 2009.* At: http://www.nr.gov.nl.ca/nr/agrifoods/marketing/index.html.

Newfoundland and Labrador Beekeeping Association (NLBKA). 2017a. "Newfoundland and Labrador Beekeeping Association research priorities." At: http://nlbeekeeping.ca/data/documents/NLBKA_research_priorities29mar2017.pdf.

———. 2017b. "Newfoundland and Labrador Beekeeping Association mission." At: http://www.nlbeekeeping.ca/about-us/our-mission/.

———. 2017c. "Inventory of honeybee forage on the island of Newfoundland." At: http://www.nlbeekeeping.ca/beekeepers-corner/bee-forage/.

———. 2017d. Personal written communication from Peter Armitage. 3 Apr., regarding Newfoundland and Labrador Beekeeping Association research priorities.

Newfoundland and Labrador Regulation 33/12. 2012. At: http://www.assembly.nl.ca/legislation/sr/annualregs/2012/nr120033.htm.

Nova Scotia Beekeepers. 2016. "Welcome to the Nova Scotia Beekeepers." May. At: http://www.nsbeekeepers.ca/.

Oldroyd, B. P., T. E. Rinderer, J. R. Harbo, and S. M. Buco. 1992. "Effects of intracolonial genetic diversity on honey bee (Hymenoptera: *Apidae*) colony performance." *Annals of the Entomological Society of America* 85, 3: 335–43. At: http://www.usmarc.usda.gov/SP2UserFiles/Place/64133000/PDFFiles/201-300/266-Oldroyd--Effects%20of%20Intracolonial%20Genetic.pdf.

Ontario. 2014. *Pollinator Health: A Proposal for Enhancing Pollinator Health and Reducing the Use of Neonicotinoid Pesticides in Ontario.* At: http://www.omafra.gov.on.ca/english/pollinator/discuss-paper.pdf.

OPERA. 2013. *Bee Health in Europe — Facts and Figures 2013: Compendium of the Latest Information on Bee Health in Europe.* OPERA Research Centre. At: http://operaresearch.eu/files/repository/20130122162456_BEEHEALTHINEUROPE-Facts&Figures2013.pdf.

Potts, S. G., J. C. Biesmeijer, C. Kremer, P. Neumann, O. Schweiger, and W. E. Kunin. 2010. "Global pollinator declines: Trends, impacts and drivers." *Trends in Ecology & Evolution* 25, 6: 345–53. doi:10.1016/j.tree.2010.01.007.

Prince Edward Island Animal Health and Protection Act. 2014. Chapter A-11.1: Animal Health and Protection Act. At: http://www.canlii.org/en/pe/laws/regu/pei-reg-ec271-01/latest/part-1/pei-reg-ec271-01-part-1.pdf.

Rinderer, T. E., L. I. deGuzman, J. Harris, V. Kuznetsov, G. T. Delatte, J. A. Stelzer, and L. Beaman. 2000. "The release of ARS Russian honey bees." *American Bee Journal* 140, 4: 305–07. At: https://afrsweb.usda.gov/SP2UserFiles/Place/64133000/PDFFiles/301-400/381-Rinderer--The%20Release%20of%20ARS.pdf.

Roche, D. 2014. "Ministry for Primary Industries information on bee health for the Primary Production Committee." June. At: http://www.parliament.nz/resource/en-nz/50SCPP_ADV_00DBSCH_INQ_12262_1_A401143/35636202b6ac395976b1010c1df7571639a00ee5.

Rose, T., C. Kremen, A. Thrupp, B. Gemmill-Herren, B. Graub, and N. Azzu. 2015. *Policy Analysis Paper: Mainstreaming of Biodiversity and Ecosystem Services with a Focus on Pollination*. Rome: FAO.

Sammataro, D. 2012. "Global status of honey bee mites." In D. Sammataro and J. A. Yoder, eds., *Honey Bee Colony Health: Challenges and Sustainable Solutions*, 37–54. Boca Raton, Fla.: Taylor & Francis Group.

Shutler, D., K. Head, K. L. Burgher-MacLellan, M. J. Colwell, A. L. Levitt, N. Ostiguy, and G. R. Williams. 2014. "Honey bee *Apis mellifera* parasites in the absence of *Nosema ceranae* fungi and *Varroa destructor* mites." *PLOS ONE* 9, 6: e98599. doi:10.1371/journal.pone.0098599.

State of Hawaii, Plant Industry Division. 2016. "Frequently asked questions about *Varroa* mite." May. At: http://hdoa.hawaii.gov/pi/ppc/varroa-mite-information/frequently-asked-questions-about-varroa-mite/.

Tang, J., J. Wice, V. G. Thomas, and P. G. Kevan. 2007. "Assessment of Canadian federal and provincial legislation to conserve native and managed pollinators." *International Journal of Biodiversity Science & Management* 3, 1: 46–55. doi:10.1080/17451590709618161.

Thrimawithana, A. H., L. Ortiz-Catedral, A. Rodrigo, and M. E. Hauber. 2013. "Reduced total genetic diversity following translocations? A metapopulation approach." *Conservation Genetics* 14, 5: 1043–55. doi:10.1007/s10592-013-0494-7.

University of Guelph. 2016. “Research.” May. At: http://www.uoguelph.ca/honeybee/research.shtml.

vanEngelsdorp, D., J. D. Evans, C. Saegerman, C. Mullin, E. Haubruge, B. Kim, and J. Pettis. 2009. “Colony collapse disorder: A descriptive study.” *PLOS ONE* 4, 8: e6481. doi:10.1371/journal.pone.0006481.

——— and M. D. Meixner. 2010. “A historical review of managed honey bee populations in Europe and the United States and the factors that may affect them.” *Journal of Invertebrate Pathology* 103, 1: S80–S95. doi:10.1016/j.jip.2009.06.011.

Victorian State Department of Environment and Primary Industries. 2013. “Apiculture (beekeeping) on public land policy.” Aug. At: agriculture.vic.gov.au/__data/assets/.../Public-land-apiculture-beekeeping-policy.docx.

Williams, G. 2010. *2009 Newfoundland and Labrador Honey Bee Disease Survey*. At: http://www.faa.gov.nl.ca/publications/pdf/honey_bee_disease_09.pdf.

———, K. Head, K. L. Burgher-MacLellan, R. E. L. Rogers, and D. Shutler. 2010. “Parasitic mites and microsporidians in managed western honey bee colonies on the island of Newfoundland, Canada.” *Canadian Entomologist* 142, 6: 584–88. At: http://www.bioone.org/doi/full/10.4039/n10-029.

13

Technological Advances in the Propagation and Improvement of Newfoundland and Labrador Berries

Samir C. Debnath & Catherine Keske

INTRODUCTION: BERRIES AS A SOURCE OF FOOD SOVEREIGNTY AND FOOD SECURITY

Throughout the food studies genre and across the world, native fruits (including berries) are typically regarded as nutritious food sources that can facilitate food sovereignty and food security. Harvesting and consumption often take place at household and community levels. In addition, both wild and commercially produced berries can be sold or traded at multiple scales, ranging from small markets and co-operatives to nationally certified export products. The nutritional value of berries has been recognized in diverse cultures across the world for millennia. However, relatively speaking, the validation of health benefits through formalized scientific studies such as those presented in this chapter has emerged only recently.

The purpose of this chapter is to provide a scientific account of the health, agronomic, economic, and socio-cultural benefits of four berry crops grown in Newfoundland and Labrador: blueberries (lowbush; *Vaccinium angustifolium* Ait.), partridgeberries (common name: lingonberries; *V. vitis-idaea* ssp. *minus* [Lodd.] Hult.), bakeapples (common name: cloudberries; *Rubus chamaemorus* L.), and cranberries (*V. macrocarpon* Ait.). In doing so, we discuss research programs currently underway at Agriculture and Agri-Food Canada's St. John's Research and Development Centre and at the Provincial Department of Fisheries, Forestry and Agrifoods to advance provincial berry production. First, we present a literature review to provide additional context about the socio-cultural and economic significance of berry crops in the province, and the implications that this has for food sovereignty in Newfoundland and Labrador and in similar regions of the world.

As several authors in this book have noted, for generations Indigenous persons relied on wild berries for nutritional, social, and cultural value. Berries are an example of "country foods," or traditional food sources harvested from local stocks by Indigenous persons across the Canadian North (Van Oostdam et al., 1999). In a study of O-Pipon-Na-Piwin Cree Nation in northern Manitoba, Kamal et al. (2015) reported that the group practice of collecting berries reflects the deeply integrated relationship between food and land embodied by Indigenous persons. Gathering berries provides more than sustenance. It yields *wechihituwin*, a Cree word for "any means of livelihood that is shared and used to help another person, family, or the community." The term emphasizes the fact that "food . . . is not a commodity; it is a set of relationships" (Kamal et al., 2015: 566). Participants in the Kamal et al. study identified berry harvesting as a means for teaching children about sharing and caring for family, as women and their families traditionally engage in berry gathering. They asserted that berry-picking programs could inspire reconnection to the land and strengthen food sovereignty among Indigenous communities in Canada (Kamal et al., 2015).

Several studies discuss the significance of berries in the diet and culture of Newfoundland and Labrador Indigenous communities (Hanrahan, 2008; Schiff and Bernard, Chapter 7). However, a substantial literature also illuminates the social, cultural, and economic benefits of berry picking specifically among Newfoundland and Labrador settler populations across generations. According to Omohundro (1994), Everett, (2007), and Cullum (2008), berries have served as a vital food source for rural and urban households for centuries. Women and their families collected berries from mid-July through October and prepared sauces and jams for the winter months. Berry picking was considered a social pursuit for families, groups of young women, and courtship, and solitary activity for young men. Narváez (1991) noted that households also sold and traded berries in markets beginning in the early days of settlement. Widespread unemployment in the 1930s launched the Newfoundland blueberry industry, during which time men became more actively engaged in berry picking to supplement family income, often receiving store credit in the form of a "berry note." Cullum (2008) presents an excellent account of the development of the Newfoundland blueberry processing and export industry during the twentieth century.

In the folklore literature, several authors have noted that berry grounds facilitated connections between settlers and land, communities, and spiritual realms. Narváez (1991) used ethnographic narratives of Newfoundlanders to document how berry grounds presented ideal economic, social, and environmental conditions for Irish and English settlers to perpetuate Old World superstitions and Celtic harvest celebrations in isolated outport communities. The belief that mythical fairies could abduct lone pickers reinforced the practice of picking berries in a group rather than alone. Narváez (1991) asserted that stories of individuals who were abducted or led "in the fairies" served the purpose for imposing moral lessons about the negative consequences of promiscuity, the dangers of solitude and natural hazards, and the importance of individual contributions towards the collective good.

Studies across the northern boreal ecosystem (one of the world's largest biogeoclimatic zones, comprising 627 million hectares or 29 per cent of North America, including Newfoundland and Labrador and most of Canada) illuminate the significance of berries in facilitating food security and food sovereignty in boreal climates that have scarce agricultural land resources and harsh climates (Keske, Dare, Hancock, and King, 2016; Giuliani, van Oudenhoven, and Mubalieva, 2011). In an ethnobiographic study of the use of wild plants and mushrooms for food and medicinal treatment in rural Sweden, Ukraine, and northwest Russia, Stryamets et al. (2015) conduct 205 semi-structured interviews with rural residents to learn about their use of wild plants for food and medicine. Through photograph identification and translators, the villagers identified *V. oxycoccus* L. (cranberry), *Fragaria vesca* L. (strawberry), *Rubus idaeus* L. (raspberry), *Rubus* spp. (blackberries), *R. chamaemorus* L. (cloudberry or bakeapple) as sources of vitamins and food, and as treatment for high blood pressure, flu and cough, and high temperature. The species identified by these villagers are addressed in the scientific research presented in our chapter.

Giuliani, van Oudenhoven, and Mubalieva (2011) examined the social, economic, and agricultural biodiversity arising from native berry plants in the Tajik Pamir Mountains within the Gorno-Badakhshan province of Tajikistan, a former republic of the Soviet Union. The authors specifically noted how berries serve as a staple for household consumption, food security, and food sovereignty. The native varieties thrive comparatively better than introduced varieties, in part because the native plants naturally ripen earlier, are resistant to cold, drought, and ultraviolet radiation, and can be cultivated on precipitous slopes. Much like Newfoundland and Labrador, in Tajikistan there is a delicate balance between increasing berry production for household consumption and increasing export demand to the extent that households are no longer able to forage for, or consume, the nutritious native food. Green's (2016) ethnographic study of the Sámi food movement in northern Sweden outlines similar paradoxes in the food sovereignty movement. On one hand, heritage foods like lingonberries and cloudberries propel discourse with activist groups about concerns regarding climate change and land access that affect the future production of Sámi food. Berries, reindeer, fish, and other Sámi foods symbolize the Sámi as Indigenous persons who have distinct rights to food, culture, and land because they foster food sovereignty. On the other hand, Green (2016) notes that the increased interest in Sámi ethnic cuisine equally threatens their potential to make a living selling Sámi foods and culinary knowledge, if a food certification process waters down the Sámi's power of collective action or has political ramifications.

In summary, berries have been an important food source for Newfoundland and Labrador Indigenous persons and settlers, just as they have been in other regions of Canada and throughout the northern boreal ecosystem. Berry picking and marketing (including export markets) clearly present complex food sovereignty considerations. The remainder of this chapter focuses on research advancements in the breeding of new cultivars that cross non-native varieties with local native varieties in order to expand commercial production of the products while maintaining otherwise superior traits of the native species.

NEWFOUNDLAND AND LABRADOR ALTERNATIVE CROPS INITIATIVE

As summarized in Simms (2015), the Department of Fisheries, Forestry and Agrifoods partnered with Agriculture and Agri-Food Canada (AAFC) to research and develop new varieties of berries that may be expanded commercially through the province. There is presently a limited retail market for Newfoundland and Labrador berries. Most of the products are processed in value-added secondary products, such as jellies, sauces, and wines. Everett

(2007) also discusses berries as a focal point of cultural tourism, including roadside markets and fairs. Simms (2015) reports that most cultivated berry farms have less than 100 plants and that partridgeberry farms are virtually non-existent.

Most of the major berry crops include the members of the genera: *Vitis* (grapes), *Vaccinium* (blueberry, cranberry, and lingonberry or partridgeberry; Ericaceae), *Fragaria* (strawberry; Rosaceae), *Rubus* (brambles: raspberry and blackberry; Rosaceae), and *Ribes* (currant and gooseberry; Grossulariaceae). Alpine strawberry (*F. vesca* L., Rosaceae), Arctic raspberry (*R. arcticus* L., *R. stellatus* Sm. and their hybrids; Rosaceae), aronia (*Aronia melanocarpa* [Michx.] Elliott, Rosaceae), bilberry (*V. myrtillus* L., Ericaceae), choke cherry (*Prunus virginiana* L.), bakeapple or cloudberry (*R. chamaemorus* L., Rosaceae), edible honeysuckle (*Lonicera caerulea* L., Caprifoliaceae), elderberry (*Sambucus Canadensis* L., Caprifoliaceae), hardy kiwi (*Actinidia arguta* [Siebold & Zucc.] Planch. ex Miq., Actinidiaceae), highbush cranberry (*Viburnum trilobum* Marshall), mora (*R. glaucus* Benth., Rosaceae), Juneberry/saskatoon (*Amelanchier* sp., Rosaceae), muscadine grape (*V. rotundifolia* Mich.,Vitaceae), sea buckthorn (*Hippophae rhamnoides* L., Elaeagnaceae), schisandra (*Schisandra chinensis* [Turcz.] Baill., Schisandraceae), serviceberry (*Amelanchier alnifolia* [Nutt.] Nutt.), and silver buffaloberry (*Shepherdia argentea* [Pursh] Nutt.) are some of the other berry crops (Finn, 1999). Many of these berries are found in Newfoundland and Labrador.

Berry crops are not only a nutritious health-promoting food but are also very attractive for their use as landscape plants. The fruits are highly valued for their varied colour, shape, flavour, and textures. They are highly nutritious and used as snack foods, dessert foods, and in beverages. Berry crops are an excellent source of natural antioxidants that provide protection against harmful free radicals and significantly reduce the probability of the incidence and the mortality caused by cancer, cardiovascular disorders, type II diabetes, and other age- and oxidative stress-related degenerative diseases (Ames, Shigena, and Hegen, 1993; Velioglu, Mazz, Gao, and Oomah, 1998; Rissanen et al., 2003). The berry phytochemicals responsible for antioxidant capacity can largely be attributed to the phenolics, anthocyanins, carotenoids, and other flavonoid compounds. The high contents of anthocyanins, proanthocyanins, flavonols, and catechins in berry crops work as anti-ulcer, antibiotic, anti-diarrheal, and anti-inflammatory agents, and are used in the treatment of allergies, vascular fragility, hypertension, and hypercholesterolemia (Kühnau, 1976; Larson, 1988; Rice-Evans and Miller, 1996).

Lowbush blueberry, partridgeberry (lingonberry), cranberry, and bakeapple (cloudberry) are four of the health-promoting berry crops native to Newfoundland and Labrador. While the former three are commercially important in Canada, bakeapple also has a place as one of the important health-promoting berry crops. *Vaccinium* fruit crops are native to all continents except Australia and Antarctica (Vander Kloet, 1988). These genetically heterozygous dicot angiosperms bear small to moderate-sized fleshy, more-or-less edible fruits on woody perennial vines or shrubs. *Vaccinium* plants are epiphytic or terrestrial and are grown on acidic, sandy, peaty, or organic soils (Vander Kloet, 1988).

Blueberries (lowbush and highbush [*V. corymbosum* L.]) are the most important and fastest-growing fruit crop in the country, with the highest farm-gate value. They account for 28.7 per cent of the total value of all fruits in Canada, with a value of $262 million in 2015. Blueberries were followed by apples, grapes, and cranberries, with farm-gate values of $182 million, $121 million, and $113 million, respectively (Statistics Canada, 2016). In 2015, Canadian production of lowbush and highbush blueberries totaled 103,131 tons and 79,834 tons, respectively, with farm-gate values of $108 million and $154 million (Statistics Canada, 2016). Blueberries are very rich in health-promoting nutrients and are used in a number of foods, including cereals, yogurt, and baked goods. Canadian growers export blueberries to markets across the world, including the US, Europe, China, India, Japan, and Korea (Hein, 2014).

Blueberries are also Canada's most exported fruit crop. Recently, the Canadian federal government has announced a new trade agreement with the European Union that will decrease import taxes and open up markets. This may increase the demand for blueberry exports. In Canada, lowbush blueberries are produced commercially in the provinces of New Brunswick, Nova Scotia, Prince Edward Island, Newfoundland and Labrador, and Quebec (AAFC, 2012).

Partridgeberries in Canada are harvested from the wild, and the demand for partridgeberries outstrips the current supply. This leads to increased market opportunity for Canadian producers. Agriculture and Agri-Food

Canada (AAFC) scientists observed that partridgeberries growing in northern regions of Canada have higher antioxidant content and offer better health benefits compared to European lingonberry (partridgeberry) cultivars. This leads to an opportunity for northern agriculture to produce and expand healthy foods for consumers across Canada and abroad (*Canora Courier*, 2015).

In Newfoundland and Labrador, cranberry farming is relatively new; it started in the late 1990s. In 2014, the Department of Innovation, Business and Rural Development, Department of Natural Resources, and the Atlantic Canada Opportunities Agency announced a $7 million federal–provincial–territorial program to further develop the cranberry industry in the province and to help diversify the economy, reinvigorate communities, and create jobs in rural areas of the province (Forestry and Agrifoods Agency, 2016).

BLUEBERRIES

Blueberries are perennial, woody shrubs bearing fruits in clusters that have a punch of powerful antioxidants. Studies with animals have demonstrated that blueberries can protect the brain from stress and damage caused by neurodegenerative disease, stroke, or aging. Antioxidant and anti-inflammatory effects of blueberry flavonoids might be responsible for these protective effects (Kalt, 2006). Blueberries are commonly used in juice drinks, beverages, baked goods, cereals, yogurt, ice cream, candy, jams, jellies, pies, and many other snacks and delicacies.

Five major blueberry groups are grown commercially: (1) lowbush (*V. angustifolium* Ait.), (2) highbush (*V. corymbosum* L.), (3) half-high (a cross between lowbush and highbush), (4) southern highbush (*V. corymbosum* and hybrids), and (5) rabbiteye (*V. ashei* Reade). Blueberries are native to North America, but they are grown commercially in Asia, Africa, Australia, Europe, New Zealand, and South America (Lehnert, 2008). Natural stands of lowbush blueberries are managed and harvested commercially throughout Atlantic Canada, Quebec, and in Maine. However, with increased consumer awareness and call for nutritious, high-antioxidant berries, demand is now exceeding production.

A program has been started at St. John's Research and Development Centre of Agriculture and Agri-Food Canada (AAFC), where tissue culture techniques and molecular biology are being used to develop highly productive, high antioxidant, superior mid-size blueberry plants (see Figures 13.1 and 13.2). The process involves crossing superior lowbush blueberry with half-high blueberry plants to develop a mid-bush variety. This will be the first time mid-bush blueberries will be produced from crossing half-high with lowbush blueberry plants (Debnath, 2011a). Blueberry hybrids are now under field trial with growers in Newfoundland and Labrador. The Forestry and Agrifoods Agency of the Newfoundland and Labrador government has also partnered with the AAFC St. John's program to look into the commercial production potential of these hybrids. Additional detail about the research methodology and results are available in Simms (2015).

Figure 13.1. Blueberry improvement program.

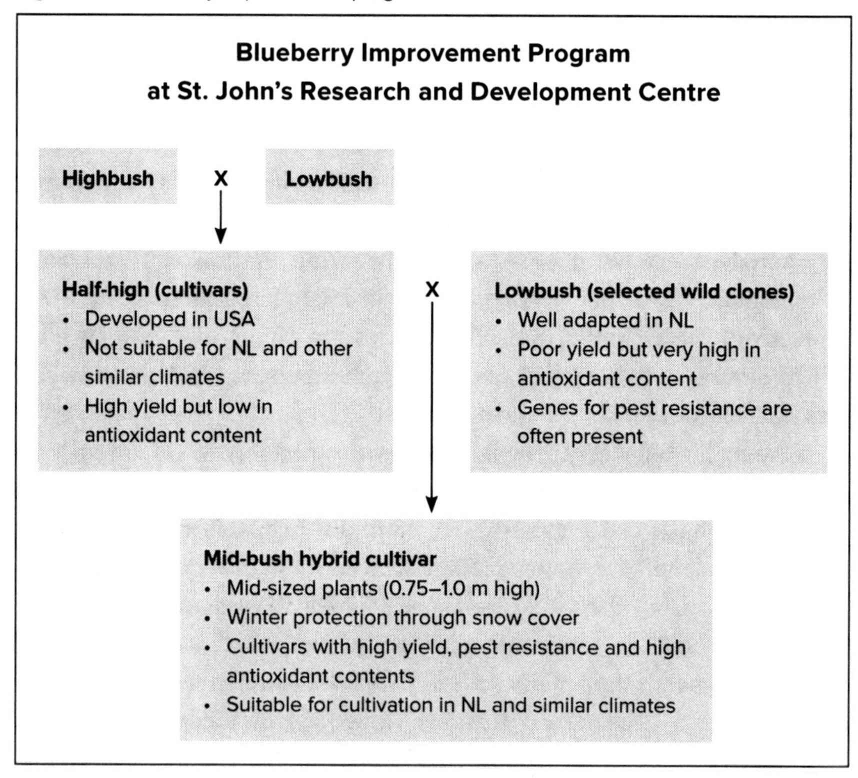

Figure 13.2. Half-high blueberry plants grown in field. (Photo courtesy of Dr. Samir Debnath)

PARTRIDGEBERRY (LINGONBERRY)

The partridgeberry (*V. vitis-idaea* L.; family Ericaceae) is most commonly known in English as lingonberry although it has more than 25 common English names, including bearberry, beaverberry, cougarberry, cowberry, foxberry, lowbush cranberry, mountain bilberry, mountain cranberry, quail-berry, redberry, and red whortleberry (Stang, Gavin, Weis, and Klueh, 1990). While in Newfoundland it is called partridgeberry, redberry is its common name in Labrador. It is a circumboreal woody, dwarf to low-growing, rhizomatous, evergreen shrub (Debnath, 2006). It is an important fruit crop not only for its commercial and medicinal uses, but also for its use as a landscape ornamental ground cover (Dierking and Dierking, 1993). Medicinal uses of partridgeberry fruits and leaves include lowering cholesterol levels, using as bladder and kidney disinfectants, and treating rheumatic disease. While consumption of raw berries stimulates the production of gastric juices (Dierking and Dierking, 1993), partridgeberry leaf extract of arbutin is used for curing stomach disorders (Racz, Fuzi, and Fulop, 1962). Partridgeberry fruits and leaves are a good source of antioxidants, especially from their phenolic compounds, such as anthocyanins, flavonoids, and tannins (Wang et al., 2005; Vyas, Curran, Igamberdiev, and Debnath, 2015), which are believed to reduce the risk of various human degenerative diseases (Prior and Cao, 2000). Extracts from partridgeberry leaves inhibit the expression of hepatitis C virus (Takeshita et al., 2009) and the growth of the human promyelocytic-sensitive leukemia HL60 cell (Skupien, Oszmianski, Kostrzewa-Nowak, and Tarasiuk, 2006). Specifically, proanthocyanidin from blueberry leaves was found to suppress the expression of the subgenomic hepatitis C virus RNA. As lingonberry also contains proanthocyanidin, it is believed to act in a similar way, as this is the effect of proanthocyanidin.

Lingonberries are native to north-temperate regions in Asia, Europe, and North America. They grow wild in diverse habitats, largely in acidic soils ranging to pure peat bogs, in lowland to upland and mountain areas (Gustavsson, 1997). There are two subspecies: *V. vitis-idaea* ssp. *vitis-idaea* (L.) Britton, which is the larger lowland race, and the dwarf arctic-montane race, *V. vitis-idaea* ssp. *minus* (Lodd.) Hult. (see Figures 13.3 and 13.4). The former is distributed in Europe and Asia and the latter is found over Greenland, Iceland, northern Asia, North America, and Scandinavia (Hultén, 1971). While the *vitis-idaea* subspecies has two "flushes" of flowers (Dierking and Dierking, 1993) with two possible annual harvests (summer in August and fall in late October to November), the *minus* type only flowers once (Penney, Gallagher, Hendrickson, Churchill, and Butt, 1997; Heidenreich, 2010).

The partridgeberry is a potentially important new berry crop in Canada (Kuropatwa, 2015). Newfoundland and Labrador is the largest North American partridgeberry producer (Penney et al., 1997). About 140,000 kg of lingonberries are harvested annually from native stands for processing, mostly for export (Jamieson, 2001). An increasing demand for high-quality berries has intensified the need to select superior native plants for horticulture. A program was initiated at the AAFC St. John's Research and Development Centre to develop hybrids between European and Canadian lingonberries. The program also aims to select highly productive individual plants with superior berry qualities and to propagate them vegetatively to maintain these valuable characteristics. In collaboration with the provincial government, AAFC St. John's Research and Development Centre is evaluating these hybrids for potential commercial cultivation (Simms, 2015).

Figure 13.3. Wild (*V. vitis-idaea* ssp. *minus*) and cultivated lingonberries (*V. vitis-idaea* ssp. *vitis-idaea*) grown in a tree cage at St. John's Research and Development Centre, NL, Canada. (Photo courtesy of Dr. Samir Debnath)

CRANBERRY

The cranberry (*V. macrocarpon* Ait.) is a slender, creeping, woody, evergreen perennial vine. It is native to North America and is distributed in moist, acidic soils, peat bogs, and marshes and swamps with a temperate climate (Vander Kloet, 1983). Cranberries are consumed as fresh, whole berries as well as in juices, gelatinized products, and capsules (Guay, 2009; Cimolai and Cimolai, 2007). They contain very high levels of vitamin C, anthocyanidins,

Figure 13.4. Greenhouse-grown partridgeberry plants (*V. vitis-idaea* ssp. *minus*, left; *V. vitis-idaea* ssp. *vitis-idaea*, right). (Photos courtesy of Dr. Samir Debnath)

flavonoids, triterpinoids, catechins, organic acids, and fructose (Guay, 2009), as well as pectin, cellulose, and anthocyanins and proanthocyanidins that help to prevent urinary tract infections. Cranberry consumption can help in reducing the risk for heart disease and can inhibit cancer cell growth (Zdepski et al., 2011). Cranberry tannins (anthocyanidins and proanthocyanidins) work as a natural plant defence system against microbes (Guay, 2009; Cimolai and Cimolai, 2007). Proanthocyanidins can prevent infections in the urinary tract by reducing adhesion of *Escherichia coli* (Leahy, Speroni, and Starr, 2002). They can also prevent bacterial adhesion in the stomach and oral cavity (Zdepski et al., 2011). The stomach ulcer-causing bacterium, *Helicobacter pylori,* was found to be prevented by proanthocyanidins from attaching to isolated stomach cells (Burger et al., 2000). The red colour of cranberry fruit is due to the presence of anthocyanins. Anthocyanins are believed to have important therapeutic values, including antioxidant, anti-tumour, anti-ulcer, and anti-inflammatory activities (Kamei et al., 1995; Koide, Kamei, Hashimoto, Kojima, and Hasegawa, 1996; Cristoni and Magistretti, 1987; Wang et al., 1999).

Although natural stands of cranberries are harvested in Newfoundland and Labrador each year, the Newfoundland and Labrador cranberry industry has continued to grow, with 121 hectares of cultivated area in 2014 (Simms, 2015).

BAKEAPPLE OR CLOUDBERRY

The cloudberry or bakeapple (*R. chamaemorus* L.) belongs to the family Rosaceae. The English common names of *R. chamaemorus* are cloudberry (most commonly used), bakeapple (in Newfoundland and Labrador), knotberry and knoutberry (in England), aqpik or low-bush salmonberry (in Alaska), and averin or evron (in Scotland). It is a boreal circumpolar, rhizomatous perennial dioecious herb, generally found in bogs (Thiem, 2003). The leaves and berries are very rich in vitamin C, tannins, and ellagic acid content (Amakura, Okada, Tsuji, and Tonogai, 2000). The latter is an antioxidant important for anti-carcinogenic, anti-mutagenic, hepatoprotective, and anti-microbial properties (Puupponen-Pimiä et al., 2001). In traditional medicine, it is used to treat scurvy and diarrhea (Thiem, 2003).

MICROPROPAGATION IN BERRY CROPS

As the berry crops are genetically heterozygous, they do not produce plants from seeds that are genetically identical to their respective seed parents. Most berry crops are propagated vegetatively to preserve desired genetic characteristics and to achieve a fruit-bearing condition rapidly. Although berry crops can be propagated vegetatively by cuttings or by division, these methods are slow and labour-intensive and few propagules are produced from a single stock plant. In vitro propagation of a specific genotype can potentially multiply plants more rapidly than conventional vegetative propagation methods. The commercial use of this technology is for mass propagation of a specific genotype and of parental stocks for hybrid seed production in a breeding program, pathogen-free (indexed) germplasm maintenance, year-round plant production, and as an initial step in a nuclear stock crop production system. Propagation by stem cuttings is unproductive, as they do not produce enough rhizomes (Debnath, 2005). Micropropagated plants were found to be superior to those obtained by conventional stem rhizome production cuttings, in terms of plant vigour and berry yield in lingonberries (Gustavsson and Stanys, 2000). Enhanced vegetative growth in tissue culture plants compared to stem cutting plants is also reported by Debnath (2005).

Complete new plants through tissue culture can be derived in three ways: (1) axillary shoot proliferation; (2) adventitious shoot regeneration via organogenesis; and (3) plantlet regeneration via somatic embryogenesis. Various culture conditions, basal media, and growth regulators have been developed for in vitro propagation of berry crops (Debnath, 2003, 2007a, 2007b, 2011b; Graham, 2005; Skirvin, Motoike, Coyner, and Norton, 2005).

Micropropagation via Axillary Shoot Proliferation

Micropropagation via axillary shoot proliferation is the most reliable and applied method for propagating true-to-type berry plants. In this method, apical buds or nodal segments harbouring an axillary bud are cultured to proliferate multiple shoots. The stages include: (1) initiation of aseptic culture; (2) shoot proliferation; (3) rooting of microshoots; and (4) acclimatization. For *Vaccinium* species (blueberry, cranberry, partridgeberry), explants (apical buds or nodal segments) are placed on a low ionic-concentration nutrient medium containing no or low levels of auxins and higher levels of cytokinins to promote axillary budding while preventing excessive callus formation (Debnath, 2003, 2007b; Debnath and McRae, 2001a, 2001b).

Semi-solid or liquid media containing major and minor elements, vitamins, amino acids, and a carbon source are used for in vitro culture. Cultures are maintained at 20–27°C under cool white fluorescent tubes (30 $\mu mol\ m^{-2}\ s^{-1}$) with a 16-hour photoperiod (Debnath, 2007b). Proliferated shoots are subcultured onto a fresh medium. Media with low ionic concentrations are suitable for *Vaccinium* culture (George, 1996). Several basal media have been used for shoot culture of *Vaccinium* species, including woody plant medium (WPM; Lloyd and McCown, 1980); modified Murashige and Skoog (MS; Murashige and Skoog, 1962); Zimmerman's Z-2 (Zimmerman and Broome, 1980); PMN (Eccher, Noè, Piagnani, and Castellis, 1986); and modified cranberry medium (Debnath and McRae, 2001b) supplemented with a cytokinin, such as N6-(2-isopentenyl)adenine (2iP), thidiazuron (TDZ), zeatin or zeatin riboside with or without an auxin (Debnath and McRae, 2001a; Debnath, 2007a). While Zimmerman and Broome (1980) used modified Anderson's medium for highbush blueberry in vitro culture (Anderson, 1975), Lyrene (1980) cultured rabbiteye blueberry on modified Knop's medium for micropropagation (Knop, 1965).

Zeatin was found more effective for axillary shoot proliferation of *Vaccinium* species (Reed and Abdelnour-Esquivel, 1991; Debnath, 2004). However, Gonzalez, Lopez, Valdes, and Ordas (2000) reported that in highbush blueberry, the best shoot multiplication took place on a semi-solid gelled culture medium containing 25 μM 2iP. However, bakeapple responded better in a liquid basal medium containing 6-benzylaminopurine (BA; Debnath, 2007c). An increased in vitro shoot multiplication rate of lowbush blueberry was noticed on a modified cranberry medium with 2–4 μM zeatin and 20 g l^{-1} sucrose (Debnath, 2004). Light plays an essential role in in vitro cultures of berry crops. Cultures exposed to lower irradiance (15 μmol m^{-2} s^{-1}) had better shoot vigour in lowbush blueberries (Debnath, 2004). Noè and Eccher (1994) observed that, as compared to the control treatment (55 μmol m^{-2} s^{-1}), strong light had negative effects on in vitro shoot proliferation in highbush blueberry. Repeated subculture enhances shoot multiplication in *Vaccinium* species (Debnath, 2004).

Adventitious Shoot Regeneration

Adventitious shoot regeneration can be obtained either directly from cultured explants or indirectly from callus tissue developed on the cultured plant material. Regeneration has two morphogenic pathways: (1) organogenesis, which is the formation of unipolar organs (shoots or roots); and (2) somatic embryogenesis, which is the production of somatic embryos with a root and a shoot meristem (Ammirato, 1985). Shoot organogenesis from cranberry leaves and stem segments has been reviewed in the literature (Polashock and Vorsa, 2003; McCown and Zeldin, 2005). Steps for shoot regeneration in berry crops can be divided into: (1) bud formation on the explants; (2) bud elongation and shoot formation; and (3) rooting of the microshoots to form plantlets (Qu, Polashock, and Vorsa, 2000). Factors such as type of culture media, plant growth regulators and their combinations, physical environment, developmental stage of explants, explant types, and genotypes affect adventitious shoot regeneration. Shoots were regenerated from cranberry leaves by culturing on a semi-solid gelled basal medium containing TDZ and 2iP (Qu et al., 2000). Young expanding basal leaf segments of lowbush blueberries, when placed on an agar-gelrite solidified nutrient medium containing 2.3–4.5 μM TDZ and maintained for weeks in darkness with their adaxial side touching the culture medium, produced multiple buds and shoots within six weeks of culture. When transferred to a medium containing 2.3–4.6 μM zeatin, these cultures produced usable shoots after one additional subculture (Debnath, 2009b). Somatic embryogenesis has not been reported in *Vaccinium* species but was found successful in strawberries (Donnoli, Sunseri, Martelli, and Greco, 2001; Biswas, Islam, and Hossain, 2007; Kordestani and Karami, 2008; Zhang, Folta, Thomas, and Davis, 2014). TDZ can be used for the production of somatic embryogenesis in strawberries (Husaini et al., 2008). In strawberries, cold (Husaini, Mercado, Teixeira da Silva, and Schaart, 2011) and dark (Donnoli et al., 2001; Husaini et al., 2011) treatments of the culture were important for the induction of somatic embryogenesis.

Rooting and Acclimatization

Micropropagated shoots in berry crops can be rooted both in vitro and ex vitro (Qu et al., 2000; Debnath and McRae, 2001a, 2001b; Debnath, 2009b). Excised shoots can be placed onto a growth regulator-free medium for rooting in vitro in gelled media (Qu et al., 2000; Debnath and McRae, 2001a, 2005). Rooted plantlets can be transferred in a medium containing 3 parts peat to 2 parts perlite (v/v) and maintained in a mist chamber with very high (95 per cent) relative humidity (RH) followed by transferring into a greenhouse with 85 per cent RH for acclimatization.

Microshoots can also be rooted ex vitro in shredded sphagnum moss (Qu et al., 2000) or in a peat-perlite medium for *Vaccinium* species (Debnath, 2005, 2009b). An auxin-pretreatment was found effective for ex vitro rooting of blueberries where in vitro-derived shoots were treated with 4.9 mM 3-indolebutyric acid (IBA) before planting them on a peat-perlite medium (Debnath, 2009b). However, Debnath and McRae (2005) established an efficient one-step protocol for cranberry micropropagation where shoots are multiplied and then rooted in the same medium containing 2–4 μM zeatin.

Bioreactor Micropropagation

Micropropagation on gelled medium is difficult to automate and the production cost is high compared to a bioreactor system containing a liquid medium. In vitro culture in liquid media allows extended subculture periods and reduces both cost and labour in terms of agar, volume of medium, and subculture periods (Sandal, Bhattacharya, and Ahuja, 2001). Plants grown in a liquid medium can take up more nutrients than those grown on agar, leading to better shoot and root growth (Sandal et al., 2001). However, liquid culture may limit the gas exchange of the plant materials that may cause asphyxia, hyperhydricity, and plant malformation (Detrez, Ndiaye, and Dreyfus, 1994). Some of these problems can be overcome by using growth retardants, which arrest rapid proliferation, and temporary immersion bioreactors (TIB; Ziv, Chen, and Vishnevetsky, 2003). In TIB, the explants are alternately exposed to air and a liquid medium. Other alternatives include the use of a raft support for explants over stationary liquid, adding liquid medium to an established culture on agar, paper bridges, cellulose blocks or sponges, and mist bioreactors (Etienne and Berthouly, 2002).

Bioreactors are self-contained, sterile environments. They capitalize on liquid nutrient or liquid/air inflow and outflow culture systems and are designed for intensive culture with control over microenvironmental conditions (agitation, aeration, dissolved oxygen, etc.; Paek, Chakrabarty, and Hahn, 2005; see Figure 13.5). The manual handling per millilitre suspension is considerably less, since they are able to optimize growth conditions. The application of bioreactor micropropagation in berry crops is still at the infancy stage. Protocols using a bioreactor system combined with a semi-solid gelled medium have been developed in lowbush, highbush, half-high, and hybrid blueberries and bakeapple at the AAFC in St. John's (Debnath, 2007c, 2009a, 2011c, 2017). Lowbush blueberry cultures were established on a gelled, modified cranberry basal medium (BM) containing 5 μM zeatin or 10 μM 2iP. Eight-week-old shoots were then transferred into a bioreactor that contains liquid BM with 1–4 μM zeatin, where multiple shoots were obtained (Debnath, 2009a).

Figure 13.5. Blueberry micropropagation in a bioreactor containing a liquid medium. (Photo courtesy of Dr. Samir Debnath)

Clonal Fidelity of Micropropagated Plants

Clonal fidelity, or true-to-type, of tissue culture plants is a prerequisite for the application of micropropagation. Somaclonal variation, which can be both heritable (genetic) and non-heritable (epigenetic), has been observed in micropropagated berry plants (Swartz, Galletta, and Zimmerman, 1981). Somaclonal variation can be examined in micropropagated plants by their morphological, biochemical, physiological, and genetic characteristics. DNA markers allow direct comparisons of different genetic material, independent of environmental influences (Weising, Nybom, Wolff, and Meyer, 1995). The degree of similarity between banding patterns can provide information about genetic similarity or difference between the samples studied. DNA markers, including random-amplified polymorphic DNA (RAPD), simple (short) sequence repeat (SSR), amplified fragment length polymorphism

(AFLP), inter simple sequence repeat (ISSR), and expressed sequence tag-polymerase chain reaction (EST-PCR), are powerful tools to verify clonal fidelity in tissue culture plants. Debnath (2011a) used EST-PCR markers to monitor clonal fidelity in *Vaccinium* species where the EST-PCR markers showed similar monomorphic amplification profiles in micropropagated blueberry plants. Similar results were also obtained by Goyali, Igamberdiev, and Debnath (2015) using microsatellite markers.

CONCLUSIONS

Lowbush blueberry, cranberry, partridgeberry (lingonberry), and bakeapple (cloudberry) are four health-promoting wild berry crops native to Newfoundland and Labrador. Although they are both grown as crops and harvested from the wild, current demand exceeds their production; improved cultivars need to be developed in order to keep up with their rates of consumption. Hybrid blueberry and partridgeberry cultivars present promising business development opportunities for Newfoundland and Labrador growers. Plant tissue culture, combined with molecular approaches, is an important tool for berry-crop improvement programs. Combined with conventional breeding methods, a cost-effective and efficient in vitro system could significantly accelerate cultivar development programs in berry crops. True-to-type propagules and clonal fidelity are prerequisites for in vitro propagation of berry crops. Molecular markers are powerful tools for monitoring trueness-to-type of micropropagated berry plants. While bioreactor micropropagation via axillary shoot proliferation is a reliable and efficient method for mass propagation of berry plants in minimum time, methods of regenerating adventitious shoots have potential for micropropagation, if the clonal fidelity is retained.

Although the breeding of new berry cultivars presents exciting potential for expanding the province's agricultural production, it is important to keep food sovereignty and food security considerations at the forefront. In order to uphold the basic principles of food security and food sovereignty, households would need to have continued access to wild berries. As discussed earlier in the chapter, it's uncertain whether commercial berry production would complement or detract from current practices of picking berries in the wild. Household consumption could be displaced if there is a surge in demand for those products. The ownership of commercial production and processing, as well as how the profits are redistributed within the province, are also important considerations for maintaining food sovereignty. Ensuring the availability of local ingredients to maintain a consistent product is a constant concern for secondary processors of sauces and jellies. Likewise, any culinary or agricultural tourism that depends on the availability of berries necessarily requires their availability to enhance consumer experience. Furthermore, the introduction of hybrid varieties also presents the potential to affect the taste, nutrition, and production of native crops, although the extent of this is currently unknown and being studied. However, much like in other areas of the world, berries in Newfoundland and Labrador are a nutritious, native food source that has become inextricably linked to local cuisine. Advancing berry production could add momentum to the awareness and scientific validation of the healthful benefits of berries and the importance of nutritious food. This may, in turn, inspire other local agricultural production and result in increased food sovereignty and food security for the province.

REFERENCES

Agriculture and Agri-Food Canada. 2012. "Crop profile for lowbush blueberry in Canada, 2011" (AAFC No. 11751E). At: http://www.agr.gc.ca/eng/?id=1299249991460.

Amakura, Y., M. Okada, S. Tsuji, and Y. Tonogai. 2000. "High-performance liquid chromatographic determination with photodiode array detection of ellagic acid in fresh and processed fruits." *Journal of Chromatography A* 896, 1/2: 87–93. doi:10.1016/S0021-9673(00)00414-3.

Ames, B. N., M. K. Shigena, and T. M. Hegen. 1993. "Oxidants, antioxidants and the degenerative diseases of aging." *Proceedings of the National Academy of Sciences* 90, 17: 7915–22. doi:10.1073/pnas.90.17.7915.

Ammirato, P. V. 1985. "Patterns of development in culture." In R. R. Henke, K. W. Hughes, M. J. Constantin, A. Hollaender, and C. M. Wilson, eds., *Tissue Culture in Forestry and Agriculture*, 9–29. New York: Plenum Press.

Anderson, W. C. 1975. "Propagation of rhododendrons by tissue culture. Part I: Development of a culture medium for multiplication of shoots." *Combined Proceedings International Plant Propagators' Society* 25: 129–35.

Biswas, M. K., R. Islam, and M. Hossain. 2007. "Somatic embryogenesis in strawberry (*Fragaria* sp.) through callus culture." *Plant Cell, Tissue and Organ Culture* 90, 1: 49–54. doi:10.1007/s11240-007-9247-y.

Burger, O., I. Ofek, M. Tabak, E. I. Weiss, N. Sharon, and I. Neeman. 2000. "A high molecular mass constituent of cranberry juice inhibits *Helicobacter pylori* adhesion to human gastric mucus." *FEMS Immunology & Medical Microbiology* 29, 4: 295–301. doi:10.1111/j.1574-695x.2000.tb01537.x.

Canora Courier (Canora, Saskatchewan). 2015. "Lure of the Lingonberry: Win-win for northern agriculture and heart health." Sept. At: http://www.canoracourier.com/.

Cimolai, N., and T. Cimolai. 2007. "The cranberry and the urinary tract." *European Journal of Clinical Microbiology and Infectious Diseases* 26, 11: 767–76. doi:10.1007/s10096-007-0379-0.

Cristoni, A., and M. J. Magistretti. 1987. "Antiulcer and healing activity of *Vaccinium myrtillus* anthocyanosides." *Farmaco Prat* 42, 2: 29–43.

Cullum, L. 2008. "'It was a woman's job, I 'spose, pickin' dirt outa berries': Negotiating gender, work, and wages at Job Brothers, 1940–1950." *Newfoundland and Labrador Studies* 23, 2.

Debnath, S. C. 2003. "Micropropagation of small fruits." In S. M. Jain and K. Ishii, eds., *Micropropagation of Woody Trees and Fruits*, 465–506. Dordrecht, Netherlands: Kluwer Academic Publishers.

———. 2004. "*In vitro* culture of lowbush blueberry (*Vaccinium angustifolium* Ait.)." *Small Fruits Review* 3, 3/4: 393–408. doi:10.1300/j301v03n03_16.

———. 2005. "Morphological development of lingonberry as affected by *in vitro* and *ex vitro* propagation methods and source propagule." *HortScience* 40, 3:760–63. At: http://hortsci.ashspublications.org/.

———. 2006. "Influence of propagation method and indole-3-butyric acid on growth and development of *in vitro*- and *ex vitro*-derived lingonberry plants." *Canadian Journal of Plant Science* 86, 1: 235–43.

———. 2007a. "Propagation of *Vaccinium in vitro*: A review." *International Journal of Fruit Science* 6, 2:47–71. doi:10.1300/j492v06n02_04.

———. 2007b. "Strategies to propagate *Vaccinium* fruit nuclear stocks for Canadian industry." *Canadian Journal of Plant Science* 87, 4: 911–22. doi:10.4141/p06-131.

———. 2007c. "A two-step procedure for *in vitro* multiplication of cloudberry (*Rubus chamaemorus* L.) shoots using bioreactor." *Plant Cell, Tissue and Organ Culture* 88, 2: 185–91. doi:10.1007/s11240-006-9188-x.

———. 2009a. "A scale-up system for lowbush blueberry micropropagation using a bioreactor." *HortScience* 44, 7: 1962–66. At: http://hortsci.ashspublications.org/.

———. 2009b. "A two-step procedure for adventitious shoot regeneration on excised leaves of lowbush blueberry." *In Vitro Cellular & Developmental Biology — Plant* 45, 2: 122–28. doi:10.1007/s11627-008-9186-2.

———. 2011a. *Conventional Methods Combined with Biotechnology in Blueberry Improvement* (AAFC No. 11427E). At: http://publicentrale-ext.agr.gc.ca/pub_view-pub_affichage-eng.cfm?publication_id=11427E.

———. 2011b. "Adventitious shoot regeneration in a bioreactor system and EST-PCR based clonal fidelity in lowbush blueberry (*Vaccinium angustifolium* Ait.)." *Scientia Horticulturae* 128, 2: 124–30. doi:10.1016/j.scienta.2011.01.012.

———. 2011c. "Bioreactors and molecular analysis in berry crop micropropagation: A review." *Canadian Journal of Plant Science* 91, 1: 147–57. doi:10.4141/cjps10131.

———. 2017. "Temporary immersion and stationary bioreactors for mass propagation of true-to-type highbush, half-high, and hybrid blueberries (*Vaccinium* spp.)." *Journal of Horticultural Science & Biotechnology* 92, 1: 72–80. doi:10.1080/14620316.2016.1224606.

——— and K. B. McRae. 2001a. "An efficient *in vitro* shoot propagation of cranberry (*Vaccinium macrocarpon* Ait.) by axillary bud proliferation." *In Vitro Cellular & Developmental Biology — Plant* 37, 2: 243–49. doi:10.1007/s11627-001-0043-9.

——— and ———. 2001b. "*In vitro* culture of lingonberry (*Vaccinium vitis-idaea* L.): The influence of cytokinins and media types on propagation." *Small Fruits Review* 1, 3: 3–19. doi:10.1300/j301v01n03_02.

——— and ———. 2005. "A one-step *in vitro* cloning procedure for cranberry (*Vaccinium macrocarpon* Ait.): The influence of cytokinins on shoot proliferation and rooting." *Small Fruits Review* 4, 3: 57–75. doi:10.1300/j301v04n03_05.

Detrez, C., S. Ndiaye, and B. Dreyfus. 1994. "*In vitro* regeneration of the tropical multipurpose leguminous tree *Sesbania grandiflora* from cotyledon explants." *Plant Cell Reports* 14, 2/3: 87–93. doi:10.1007/bf00233767.

Dierking, S. W. Jr., and W. Beerenobst. 1993. "European *Vaccinium* species." *Acta Horticulturae* 346: 299–304. doi:10.17660/actahortic.1993.346.39.

Donnoli, R., F. Sunseri, G. Martelli, and I. Greco. 2001. "Somatic embryogenesis, plant regeneration and genetic transformation in *Fragaria* spp." *Acta Horticulturae* 560: 235–40. doi:10.17660/actahortic.2001.560.45.

Eccher, T., N. Noè, C. Piagnani, and S. Castellis. 1986. "Effects of increasing concentrations of BAP and 2iP on *in vitro* culture of *Vaccinium corymbosum.*" *Acta Horticulturae* 179: 879–81. doi:10.17660/ActaHortic.1986.179.155.

Etienne, H., and M. Berthouly. 2002. "Temporary immersion systems in plant micropropagation." *Plant Cell, Tissue and Organ Culture* 69, 3: 215–31. doi:10.1023/a:1015668610465.

Everett, H. 2007. "A welcoming wilderness: The role of wild berries in the construction of Newfoundland and Labrador as a tourist destination." *Ethnologies* 291, 2: 49–80.

Finn, C. 1999. "Temperate berry crops." In J. Janick, ed., *Perspectives on New Crops and New Uses*, 324–34. Alexandria, Virginia: ASHS Press.

Forestry and Agrifoods Agency. 2016. "Cranberries." At: http://www.faa.gov.nl.ca/agrifoods/plants/berries/cranberry.html.

George, E. F. 1996. *Plant Propagation by Tissue Culture. Part 2: In Practice.* Edington, UK: Exegetics.

Giuliani, A., F. van Oudenhoven, and S. Mubalieva. 2011. "Agricultural biodiversity in the Tajik Pamirs." *Mountain Research and Development* 31, 1: 16–26. doi:http://dx.doi.org/10.1659/MRD-JOURNAL-D-10-00109.1.

Gonzalez, M. V., M. Lopez, A. E. Valdes, and R. J. Ordas. 2000. "Micropropagation of three berry fruit species using nodal segments from field-grown plants." *Annals of Applied Biology* 137, 1: 73–78. doi:10.1111/j.1744-7348.2000.tb00059.x.

Goyali, J. C., A. U. Igamberdiev, and S. C. Debnath. 2015. "Propagation methods affect fruit morphology and antioxidant properties but maintain clonal fidelity in lowbush blueberry." *Hortscience* 50, 6: 888–96. At: http://hortsci.ashspublications.org/.

Graham, J. 2005. "*Fragaria* strawberry." In R. Litz, ed., *Biotechnology of Fruit and Nut Crops*, 456–74. Wallingford, UK: CAB International.

Green, A. S. 2016. "Tastes of sovereignty: An ethnography of Sámi food movements in Arctic Sweden." Ph.D. dissertation, Department of Applied Anthropology, Oregon State University.

Guay, D. R. 2009. "Cranberry and urinary tract infections." *Drugs* 69, 7: 775–807. doi:10.2165/00003495-200969070-00002.

Gustavsson, B. A. 1997. "Breeding strategies in lingonberry culture (*Vaccinium vitis-idaea*)." *Acta Horticulturae* 446: 129–37. doi:10.17660/actahortic.1997.446.18.

——— and V. Stanys. 2000. "Field performance of 'Sanna' lingonberry derived by micropropagation vs. stem cuttings." *HortScience* 35, 4: 742–44. At: http://hortsci.ashspublications.org/.

Hanrahan, M. 2008. "Tracking social change among the Labrador Inuit and Inuit-Metis: What does the nutrition literature tell us?" *Food Society Culture* 11, 3: 315–33. doi:10.2752/17517-4408X347883.

Heidenreich, C. 2010. "The lowdown on lingonberries." *New York Berry News* 9, 6 (June). At: http://www.fruit.cornell.edu/berry/production/pdfs/Lingonberries.pdf.

Hein, T. 2014. "Blueberry breeding update: New mid-bush blueberry varieties in development, suited for Canadian climates." *Ag Annex* (Feb.). At: http://www.agannex.com/production/blueberry-breeding-update.

Hultén, E. 1971. *The Circumpolar Plants. Vol. 2. Dicotyledons*. Stockholm: Almqvist & Wiksell.

Husaini, A. M., S. Aquil, M. Bhat, T. Qadri, M. Z. Kamaluddin, and M. Z. Abdin. 2008. "A high-efficiency direct somatic embryogenesis system for strawberry (*Fragaria ×ananassa* Duch.) cultivar Chandler." *Journal of Crop Science and Biotechnology* 11: 107–10.

———, J. A. Mercado, J. A. Teixeira da Silva, and J. G. Schaart. 2011. "Review of factors affecting organogenesis, somatic embryogenesis and *Agrobacterium tumefaciens*-mediated transformation of strawberry." *Genes, Genomes and Genomics* 5: 1–11. At: http://www.globalsciencebooks.info/.

Jamieson, A. R. 2001. "Horticulture in Canada — Spotlight on the Atlantic provinces." *Chronica Horticulturae* 41, 3: 8–11. At: http://www.ishs.org/chronica-horticulturae.

Kalt, W. 2006. "*Vaccinium* berry crops and human health." *Acta Horticulturae* 715: 533–38. doi:10.17660/actahortic.2006.715.82.

Kamal, A. G., R. Linklater, S. Thompson, J. Dipple, and Ithinto Mechisowin Committee. 2015. "A Recipe for Change: Reclamation of Indigenous Food Sovereignty in O-Pipon-Na-Piwin Cree Nation for Decolonization, Resource Sharing, and Cultural Restoration." *Globalizations* 12, 4: 559–75. doi:http://dx.doi.org/10.1080/14747731.2015.1039761.

Kamei, H., T. Kojima, M. Hasegawa, T. Koide, T. Umeda, T. Yukawa, and K. Terabe. 1995. "Suppression of tumor cell growth by anthocyanins *in vitro*." *Cancer Investigation* 13, 6: 590–94. doi:10.3109/07357909509024927.

Keske, C. M. H., J. B. Dare, T. Hancock, and M. King. 2016. "The connectivity of food security, food sovereignty, and food justice in boreal ecosystems: The case of Saint-Pierre and Miquelon." *Spatial Justice*, Special Issue on Food Justice and Agriculture 9, 1. At: http://www.jssj.org/article/la-connexion-entre-la-securite-alimentaire-la-souverainete-alimentaire-et-la-justice-alimentaire-dans-les-ecosystemes-boreals-le-cas-de-saint-pierre-et-miquelon/.

Knop, W. 1965. "Quantitative Untersuchungen über die Ernahrungsprozesse der Pflanzen." *Landwirtsch Vers Stn* 7: 93–107.

Koide, T., H. Kamei, Y. Hashimoto, T. Kojima, and M. Hasegawa. 1996. "Antitumor effect of hydrolyzed anthocyanin from grape rinds and red rice." *Cancer Biotherapy & Radiopharmaceuticals* 11, 4: 273–77. doi:10.1089/cbr.1996.11.273.

Kordestani, G. K., and O. Karami. 2008. "Picloram-induced somatic embryogenesis in leaves of strawberry (*Fragaria ananassa* L.)." *Acta Biologica Cracoviensia Series Botanica* 50, 1: 69–72. At: http://www2.ib.uj.edu.pl/abc/.

Kühnau, J. 1976. "The flavonoids. A class of semi-essential food components: Their role in human nutrition." In G. H. Bourne, ed., *World Review of Nutrition and Dietetics*, vol. 24, 117–91. New York: Karger.

Kuropatwa, R. 2015. "Growers wanted to test lingonberry hybrids." *Fruit Research, The Western Producer* 93, 20 (14 May): 62. At: http://www.producer.com/.

Larson, R. A. 1988. "The antioxidants of higher plants." *Phytochemistry* 27, 4: 969–78. doi:10.1016/0031-9422(88)80254-1.

Leahy, M., J. Speroni, and M. Starr. 2002. "Latest development in cranberry health research." *Pharmaceutical Biology* 40, 1: 50–54. doi:10.1076/phbi.40.7.50.9172.

Lehnert, D. 2008. "Blueberry production is skyrocketing worldwide." *Fruit Growers News*, 30 May. At: http://fruitgrowersnews.com/.

Lloyd, G., and B. McCown. 1980. "Commercially feasible micropropagation of mountain laurel, Kalmia latifolia, by use of shoot-tip culture." *Combined Proceedings International Plant Propagators' Society* 30: 421–27.

Lyrene, P. M. 1980. "Micropropagation of rabbiteye blueberries." *HortScience* 15: 80–81. At: http://hortsci.ashspublications.org/.

McCown, B. H., and E. L. Zeldin. 2005. "*Vaccinium* spp. cranberry." In R. Litz, ed., *Biotechnology of Fruit and Nut Crops*, 247–61. Wallingford, UK: CAB International.

Murashige, T., and F. Skoog. 1962. "A revised medium for rapid growth and bio assays with tobacco tissue cultures." *Physiologia Plantarum* 15, 3: 473–79. doi:10.1111/j.1399-3054.1962.tb08052.x.

Narváez, P. 1991. "Newfoundland berry pickers in the Fairies: Maintaining spatial, temporal and moral boundaries through legendry." In P. Narváez, ed., *The Good People: New Fairylore Essays*, 336–68. New York: Garland.

Noè, N., and T. Eccher. 1994. "Influence of irradiance on *in vitro* growth and proliferation of *Vaccinium corymbosum* (highbush blueberry) and subsequent rooting in vivo." *Physiologia Plantarum* 91, 2: 273–75. doi:10.1034/j.1399-3054.1994.910221.x.

Omohundro, J. 1994. *Rough Food: The Seasons of Subsistence in Northern Newfoundland*. St. John's: ISER Books.

Paek, K. Y., D. Chakrabarty, and E. J. Hahn. 2005. "Application of bioreactor systems for large scale production of horticultural and medicinal plants." *Plant Cell, Tissue and Organ Culture* 81, 3: 287–300. doi:10.1007/s11240-004-6648-z.

Penney, B. G., C. A. Gallagher, P. A. Hendrickson, R. A. Churchill, and E. Butt. 1997. "The wild partridgeberry (*Vaccinium vitis-idaea* L. var. minus Lodd) industry in Newfoundland and Labrador and the potential for expansion utilizing European cultivars." *Acta Horticulturae* 446: 139–42. doi:10.17660/actahortic.1997.446.19.

Polashock, J. J., and N. Vorsa. 2003. "Cranberry (*Vaccinium macrocarpon* Ait.)." In G. Khachatourians, A. McHughen, R. Scorza, W.-K. Nip, and Y. H. Hui, eds., *Transgenic Plants and Crops*, 383–96. New York: Marcel Dekker.

Prior, R. L., and G. Cao. 2000. "Antioxidant phytochemicals in fruits and vegetables: Diet and health implications." *HortScience* 35: 588–92. At: http://hortsci.ashspublications.org/.

Puupponen-Pimiä, R., L. Nohynek, C. Meier, M. Kähkönen, M. Heinonen, and K.-M. Oksman-Caldentey. 2001. "Antimicrobial properties of phenolic compounds from berry extracts." *Journal of Applied Microbiology* 90, 4: 494–507. doi:10.1046/j.1365-2672.2001.01271.x.

Qu, L., J. Polashock, and N. Vorsa. 2000. "A high efficient *in vitro* cranberry regeneration system using leaf explants." *HortScience* 35: 948–52. At: http://hortsci.ashspublications.org/.

Racz, G., I. Fuzi, and L. Fulop. 1962. "A method for determination of the arbutin content of cowberry leaves (Folium *vitis idaea*)." *Rumanian Medical Review* 6: 88–90.

Reed, B. M., and A. Abdelnour-Esquivel. 1991. "The use of zeatin to initiate *in vitro* cultures of *Vaccinium* species and cultivars." *HortScience* 26: 1320–22. At: http://hortsci.ashspublications.org/.

Rice-Evans, C. A., and N. J. Miller. 1996. "Antioxidant activities of flavonoids as bioactive components of food." *Biochemical Society Transactions* 24, 3: 790–95. doi:10.1042/bst0240790.

Rissanen, T. H., S. Voutilainen, J. Virtanen, B. Venho, M. Vanharante, J. Mursu, and J. Salonen. 2003. "Low intake of fruits, berries and vegetables is associated with excess mortality in men: The Kuopio Ischaemic Heart Disease Risk factor (KIHD) study." *Journal of Nutrition* 133, 1: 199–204. At: http://jn.nutrition.org/.

Sandal, I., A. Bhattacharya, and P. S. Ahuja. 2001. "An efficient liquid culture system for tea shoot proliferation." *Plant Cell, Tissue and Organ Culture* 65, 1: 75–80. doi:10.1023/a:1010662306067.

Simms, D. 2015. "Berry crops — Commercialization potential." *AgriView Newsletter* 9, 3: 8. At: http://www.nlfa.ca/#!newsletter/ckn8.

Skirvin, R. M., S. Motoike, M. Coyner, and M. A. Norton. 2005. "*Rubus* spp. cane fruit." In R. Litz, ed., *Biotechnology of Fruit and Nut Crops*, 566–82. Wallingford, UK: CAB International.

Skupień, K., J. Oszmiański, D. Kostrzewa-Nowak, and J. Tarasiuk. 2006. "*In vitro* antileukaemic activity of extracts from berry plant leaves against sensitive and multidrug resistant HL60 cells." *Cancer Letters* 236, 2: 282–91. doi:10.1016/j.canlet.2005.05.018.

Stang, E. J., G. Gavin, G. G. Weis, and J. Klueh. 1990. "Lingonberry: Potential new fruit for the northern United States." In J. Janick and J. E. Simon, eds., *Advances in New Crops*, 321–23. Portland, Oregon: Timber Press.

Statistics Canada. 2016. "Fruit and vegetable production, 2015." Statistics Canada Catalogue no. 11-001-X. At: http://www.statcan.gc.ca/daily-quotidien/160203/dq160203a-eng.htm.

Stryamets, N., M. Elbakidze, M. Ceuterick, P. Angelstam, and R. Axelsson. 2015. "From economic survival to recreation: Contemporary uses of wild food and medicine in rural Sweden, Ukraine and NW Russia." *Journal of Ethnobiology and Ethnomedicine* 11, 53: 1–18. doi:10.1186/s13002-015-0036-0.

Swartz, H. J., G. J. Galletta, and R. H. Zimmerman. 1981. "Field performance and phenotypic stability of tissue culture-propagated strawberries." *Journal of the American Society for Horticultural Science* 106: 667–73.

Takeshita, M., Y. Ishida, E. Akamatsu, Y. Ohmori, M. Sudoh, H. Uto, H. Tsubouchi, and H. Kataoka. 2009. "Proanthocyanidin from blueberry leaves suppresses expression of subgenomic hepatitis C virus RNA." *Journal of Biological Chemistry* 284, 32: 21165–76. doi:10.1074/jbc.m109.004945.

Thiem, B. 2003. "*Rubus chamaemorus* L. — A boreal plant rich in biologically active metabolites: A review." *Biological Letters* 40, 1: 3–13. At: http://www.biollett.amu.edu.pl/.

Vander Kloet, S. P. 1983. "The taxonomy of *Vaccinium* section Oxycoccus." *Rhodora* 85: 1–43.

———. 1988. *The Genus Vaccinium in North America*. Ottawa: Canadian Government Publishing Centre.

Van Oostdam, J., A. Gilman, E. Dewailly, P. Usher, B. Wheatley, H. Kuhnlein, . . . V. Jerome. 1999. "Human health implications of environmental contaminants in Arctic Canada: A review." *Science of the Total Environment* 230, 1 (July): 1–82. doi:10.1016/S0048-9697(99)00036-4.

Velioglu, Y. S., G. Mazza, L. Gao, and B. D. Oomah. 1998. "Antioxidant activity and total phenolics in selected fruits, vegetables, and grain products." *Journal of Agricultural and Food Chemistry* 46, 10: 4113–17. doi:10.1021/jf9801973.

Vyas, P., N. H. Curran, A. U. Igamberdiev, and S. C. Debnath. 2015. "Antioxidant properties of lingonberry (*Vaccinium vitis-idaea* L.) leaves within a set of wild clones and cultivars." *Canadian Journal of Plant Science* 95, 4: 663–69. doi:10.4141/cjps-2014-400.

Wang, H., M. G. Nair, G. M. Strasburg, Y.-C. Chang, A. M. Booren, J. I. Gray, and D. L. DeWitt. 1999. "Antioxidant and antiinflammatory activities of anthocyanins and their aglycon, cyanidin, from tart cherries." *Journal of Natural Products* 62, 2: 294–96. doi:10.1021/np980501m.

Wang, S. Y., R. Feng, L. Bowman, R. Penhallegon, M. Ding, and Y. Lu. 2005. "Antioxidant activity in lingonberries (*Vaccinium vitis-idaea* L.) and its inhibitory effect on activator protein-1, nuclear factor-kB and mitogen-activated protein kinases activation." *Journal of Agricultural and Food Chemistry* 53, 8: 3156–66. doi:10.1021/jf048379m.

Weising, K., H. Nybom, K. Wolff, and W. Meyer. 1995. *DNA Fingerprinting in Plants and Fungi*. Boca Raton, Fla.: CRC Press.

Zdepski, A., S. C. Debnath, A. B. Howell, J. Polashock, P. V. Oudemans, N. Vorsa, and T. P. Michael. 2011. "Cranberry." In K. M. Folta and C. Kole, eds., *Genetics, Genomics and Breeding of Berries*, 41–63. Enfield, NH: Science Publishers.

Zhang, Q., K. M. Folta, and T. M. Davis. 2014. "Somatic embryogenesis, tetraploidy, and variant leaf morphology in transgenic diploid strawberry (*Fragaria vesca* subspecies *vesca* 'Hawaii 4')." *BMC Plant Biology* 14, 1: 23. doi:10.1186/1471-2229-14-23.

Zimmerman, R. H., and O. C. Broome. 1980. "Blueberry micropropagation." In *Proceedings of the Conference on Nursery Production of Fruit Plants through Tissue Culture: Applications and Feasibility*, 44–47. Beltsville, Md: United States Department of Agriculture.

Ziv, M., J. Chen, and J. Vishnevetsky. 2003. "Propagation of plants in bioreactors: Prospects and limitations." *Acta Horticulturae* 616: 85–93. doi:10.17660/actahortic.2003.616.6.

Epilogue

The Newfoundland and Labrador Food System Feedback Loop, and Growing a Sustainable Food System

Catherine Keske

I've just encountered the Newfoundland and Labrador food feedback loop, and its ironies, first-hand while I'm away from the province. The winter academic term has ended, and I've returned to my home state of Colorado to enjoy the remaining days of the ski season in the southern Rockies. On my way home I stop in Georgetown, a historic mining town established in 1859 during the Pikes Peak gold rush. Though it's a short drive to several major ski resorts and most of the homes are covered with a cheery colour of paint, Georgetown clearly hasn't experienced infusions of wealth and income on par with nearby Vail and Aspen.

I step into Kneisel & Anderson, a quaint grocery in a Georgetown row house owned by the same family since 1883. It's renowned for its interesting selection of international products, and I've been coming here for years. However, this visit is different from years past because I'm more enlightened about foods from Atlantic Canada. I spot a small 100 ml jar of cloudberry jam priced at US$16, more than double the price back in Newfoundland. They're asking slightly less for lingonberry spread. The proprietor notices my interests, and as she finishes up my hand-cut steaks she proudly proclaims that this is probably the only place in the state that carries cloudberry jelly or jam. Even better, she says, there are lingonberries from Newfoundland for sale in the freezer.

"Partridgberries!" I exclaim. "How did you come by that?"

She explains that she works with a distributor from Minnesota and she elaborates on the arduous importation process. All of this results in high prices for me, as a consumer in Colorado who has developed an affinity for traditional Newfoundland and Labrador delicacies. Exportation can drive prices higher for these same products back in the province, as cloudberry jam is somewhat expensive back in Newfoundland and Labrador. Yet without international trade, I wouldn't have the option of buying partridgeberries in Colorado, or Colorado-based Smart Wool brand socks when I need them in Newfoundland and Labrador.

I offer the grocer and a few interested customers some of my "Come From Away" knowledge that cloudberries are known as "bakeapples" (anglicized from the French phrase, *Baie qu'appelle?* or What is this berry called?), and I provide an update about the recovery of the northern cod stock. The proprietor jots down information about Fogo Island Ltd. Perhaps she'll place future orders with them, she says.

I ruminate upon this experience during my drive home, marvelling at how economic forces unite remote corners of the globe that otherwise seem disconnected. I recall enjoying Atlantic salmon for dinner the previous night at the ski resort. I feel fortunate to be able to link the region where the partridgeberries were grown to the venerable store where they were sold, thousands of kilometres away. I conducted cost-comparison shopping and delayed the purchase of a relatively obscure product grown in Newfoundland and Labrador because I could find a less expensive and fresher version of it in a few weeks. Yet I was able to purchase a few staples for my return trip that I otherwise wouldn't be able to buy easily or inexpensively in Newfoundland and Labrador. Many people in the province aren't in the fortunate situation to make these decisions, because they lack the means, opportunities, and information.

I further contemplate how the sale of characteristically Newfoundland foods like fish and berries in far-away places affects the availability of food back in Newfoundland and Labrador. As this book discusses in depth, people in households and communities constantly grapple with food security and the availability of fresh, nutritious, and culturally appropriate food within the province. However, I also feel proud that I was able to share some insights into the sustainable Fogo Island Co-operative Society with a grocer who, along with her customer base, might be able to appreciate the community's story, as well as its products. An important take-away, I conclude, is that we all need to exercise mindfulness and to recognize the need for policies to ensure that food exports don't displace the availability of local food. Furthermore, communities need to be actively engaged in their food system, including remaining involved in how food is produced and distributed. This includes governance over exports, as well as encouraging the consumption of nutritious and healthy food. At the heart of this, I think to myself, Newfoundland and Labrador policies need to uphold both food sovereignty and food security.

The ironies I encountered that day reflect, to a certain extent, the schism between widespread interest in "food from somewhere," as discussed by Foley and Mather in Chapter 9, and the uneven distribution and availability of food, addressed by several contributing authors including Traverso-Yepez, Sarkar, Gadag, and Hunter (Chapter 5); Schiff and Bernard (Chapter 7); and Lowitt and Neis (Chapter 8). Modern-day lifestyles impose competing time commitments. Pre-packaged meals and imported produce are cost-competitive, particularly when time constraints are considered, which crowds out local food production. Unlike previous generations, most modern-day North American households do not plan their lifestyles around agricultural seasons. The same can be said of fishing, which was at one time the sacred custom around which Newfoundland and Labrador communities and commerce were built. As shown in many places throughout this book, there is a need for engaged community conversations, grassroots efforts, and restructured societal food policies at multiple scales in order to ensure that all have access to nutritional and high-quality food.

On a positive note, this is a tremendous opportunity to ask the question, "What would a sustainable Newfoundland and Labrador food system look like?" Interestingly, conditions are ripe so that a "sustainable scenario" could unfold on either end of the spectrum, from self-provisioning to export-reliance, with several variations in between.

SELF-PROVISIONING

At one end of the spectrum of a potentially sustainable food system are household and community self-provisioning approaches that don't involve money or markets. As Adrian Tanner discusses in Chapter 6, at least at one point in time self-provisioning was a key sustainable practice. His research has shown that in previous generations, Indigenous peoples maintained a nutritional diet with community-oriented lifestyles that placed value on production processes facilitating medium- and long-range food planning and predictable access to food sources.

Several other authors indicate that household self-provisioning continues to be practised within many Newfoundland and Labrador households today by using wild game, home gardening, and berry gathering to supplement imported foods (Omohundro, 1994; Roseman and Royal, Chapter 2; Lowitt and Neis, Chapter 8). Others have discussed the modern-day presence of household self-provisioning throughout Canada and in other regions of the world, most notably in low-income areas (Murton, Bavington, and Dokis, 2016).

However, as chronicled by several authors, including Omohundro (1994), subsistence and self-provisioning lifestyles require hard work. This is compounded by rapid advancements in the communications industry that make rural residents increasingly aware of missed opportunities. At the time of this writing, popular reality television productions such as *Life Below Zero*, *Yukon Men*, and *Alaskan Bush People* stream images of northern families and aloof mavericks navigating harsh elements to achieve a simple (but seemingly content) existence. Unfortunately, stories of those who have not deliberately chosen their situation are rarely featured, nor is the legacy of tragic consequences to Indigenous persons from sustained disruption to their culture and food-provisioning practices, as further described in Chapter 7 by Schiff and Bernard.

The reality is that if self-provisioning is a "default position" rather than a lifestyle choice, then it is difficult to conclude that these self-provisioning households are truly empowered. Also, food supplies, most notably wild caribou herds, are less available throughout the North than once was the case, which makes traditional food sources and subsistence livelihoods more difficult to achieve. As described by Foley and Mather (Chapter 9), historically, the province has used seafood for international trade, not just for consumption. Ultimately, this led to overexploitation and the collapse of the northern cod-fishing industry, as well as to a more recent decline for other species such as northern shrimp and snow crab. All of this equates to a lack of availability of seafood as local food. With the northern cod fisheries now just beginning to recover, perhaps there is no better time than the present to ensure local control over seafood and other food resources so that some household self-provisioning could be sustained over time.

Newfoundlanders and Labradorians have a strong tradition of self-provisioning. As noted by Myron King in Chapter 10, the traditional ecological knowledge (TEK) passed down over the generations is remarkable, and validated (perhaps even enhanced) by advances in technology. Several local projects initiated by the non-profit Food First NL organization authenticate the importance of local knowledge to encourage continued food provisioning and self-reliance skills among youth and rural residents.[1]

In summary, there is considerable substance to the legendary Newfoundlander colloquialism, "Gotta get me moose, b'y." If necessary, many Newfoundlanders and Labradorians can return to traditional knowledge of the land and sea to survive. However, as described by Lynne Phillips in Chapter 1, this is not without cost. While there was some degree of self-sufficiency in previous generations, there was also considerable suffering. Policies to advance food security and food sovereignty within the province should respect (and to some degree encourage) self-provisioning, while recognizing that that is likely not a feasible, sufficient, or even desired situation for many households over time. Self-provisioning households and communities need conditions to create resilience and empowerment, rather than to perpetuate marginalization.

AGRICULTURAL PRODUCTION BUILT AROUND SELF-SUFFICIENCY

For some, food security and a sustainable agricultural system invoke the notion of a self-contained agricultural system that relies on limited trade outside of the province or outside of Canada. Publications such as *Cows Don't Know It's Sunday* (Murray, 2002) chronicle the vibrant community of farmers in the St. John's vicinity producing food for local markets through the first half of the twentieth century. Potatoes, cabbage, carrots, beets, turnips, beans, and rhubarb complemented limited animal husbandry, including dairy production. On the Northern Peninsula, John Omohundro's *Rough Food* (1994) describes in great detail the lifestyles and infrastructure (including the famed roadside gardens) built around the vicissitudes of weather and transhumance for generations. Works by these authors show that self-sufficiency was possible, at least at one time. However, does attaining food security through self-sufficiency mean that Newfoundlanders and Labradorians should return to producing and distributing their food almost entirely within the province?

There is a fair degree of agreement among residents, academics, and policy-makers that the province needs to strive to become more self-sufficient in its agricultural production, but only time will tell how much locally grown food will be available at markets or grocery stores and how consumers will respond. There has been a recent uptick in provincial government initiatives to expand the province's commodity production. In early 2017, Newfoundland and Labrador Premier Dwight Ball announced that 64,000 new hectares of land would soon be made available from the government specifically for agricultural production, in order for the province to "increase self-sufficiency by 20 percent by 2022" (CBC News, 2017).

According to scientists and policy-makers at Newfoundland and Labrador's Agrifoods Development Branch (Kavanagh and Ellsworth, 2016), the provincial government has aligned its research and policy initiatives around increasing the availability of a stable and steady supply of fresh, healthy, locally available food. Kavanagh and Ellsworth note that Newfoundland and Labrador already provides 10 per cent of the root vegetables consumed during peak season, and 100 per cent of the fluid milk, chicken, and eggs to supply the province's needs year-round, as well as many other agricultural products consumed within the province. With the Premier's self-sufficiency targets, the current production rate is targeted to double within five years.

As is outlined in the paragraphs that follow, Kavanagh and Ellsworth detail several ongoing projects sponsored by the Agrifoods Development Branch, including experimental field trials involving wheat, corn, and legumes (Forestry and Agrifoods, 2016). The provincial Crop Rotation Project is in its third season of examining the benefits of rotating high-value annual crops with the goals of decreasing inputs (fertilizer and pesticides), improving nutrient and water-use efficiency, and increasing crop productivity above levels that would be achieved through mono-cropping.

The provincial Seed Potato Program also has been in progress for several years. The Agrifoods Branch believes there is considerable room to grow the market for Newfoundland and Labrador potatoes, in part because they have been a traditional dietary staple for centuries and store well over long winters. Recently, the government has taken over a seed propagation facility to develop and propagate disease- and pest-resistant seed potatoes. In this program, climatically suitable varieties are developed at the Nuclear Seed Potato

Propagation Facility and then sent to the provincial Glenwood Seed Potato Farm to increase the availability of certified, disease-free traditional and new varieties of Newfoundland potato seed stock. Certified seeds are known to be from stock free of potato wart and golden nematode (two federally regulated pests found in the province and common to home gardens), and they are not exposed to pathogens through ferry transport (Fitzpatrick, 2015).

We share a similar optimism that the future holds promise for producing a sustainable provincial food system. Growing more of the food being consumed within the province would advance food security and food sovereignty goals. However, the sobering reality is that land policies and weather present very real limitations for industry growth potential. According to Kavanagh and Ellsworth, less than 2 per cent of the land is arable, and Crown lands comprise 88 per cent of the land in the province. Permission to expand production agriculture on Crown lands is often a protracted and complex process, taking six months to three years, though recent progress has been made to expand agricultural production on Crown lands and to facilitate an online application process (Newfoundland and Labrador, 2016). Midsummer frost warnings are not uncommon, which makes agricultural production tricky. In fact, the 2015 growing season was cheekily described as "July-uary" (Bird, 2015). The unpredictable weather and microclimates make large-scale provincial agricultural production a particularly risky venture, even for an industry already known for risk.

In the presence of an increasingly globalized market, there is a high cost for self-contained agricultural production. Some countries engage in protectionism and self-reliance, but most nations are like Canada, which has actively engaged in and expanded international trade for generations. Nations import products when the opportunity costs of production are high, and they supply products for trade when the opportunity cost of production is relatively low. Hence, the most likely scenario for a fully self-contained Canadian agricultural system would arise if the costs of international or regional trade are simply too great, perhaps from national safety or security (including bio-security) concerns. This type of self-contained agricultural production system would likely require considerable government financial support to ensure that farms remain economically viable, given the nation's (and each province's) relatively small and diffusely spread rural populations.

Realistically, it is more a question of what percentage of the province's food will be grown, processed, and distributed within Newfoundland and Labrador in the future, and this could be influenced by a number of factors. Provincial Agrifoods programs that target the production of key crops, while expanding rotational systems to address environmental targets (such as soil health and water quality), could expand if costs are contained and if residents notice an improvement in shelf life and quality of food. Greenhouse growing options are also being revisited. Review of international trade agreements with the US and Mexico (Canada's biggest trade partners) and the impending departure of the UK from the European Union present a great deal of uncertainty. Renegotiating trade agreements with these countries could affect the quantity and types of food available in Canada. In summary, it is clear that increased food production in the province is a priority for producers, consumers, and the government to meet food security and food sovereignty goals, but only time will tell how this unfolds.

FOOD DESERT VERSUS "SMALL IS BEAUTIFUL": THE FEASIBILITY OF LOCALLY SOURCED FOOD

It is important to reflect on the benefits and limitations of the previously described scenarios in the context of global food production. Although it is not true for every situation, in general — and more so than at any other time in history — consumers are exposed to food from distant corners of the world. With the swipe of a smart phone, consumers can gain information directly from the manufacturer about their favourite snacks or view advertisements for products not otherwise locally available. Global trade agreements and transport efficiencies make it possible to purchase out-of-season produce in large supermarkets. It can be said that it's always harvest time somewhere.

Without exercising mindfulness, there is very real potential for the province to become almost entirely dependent on imported food, effectively evolving into what has been termed a "food desert" (Larson and Guilliland, 2008; Raja, Ma, and Yadef, 2008). Food deserts apply to situations where food is prohibitively expensive and local populations don't have access to resources for food production (Halweil, 2002). Some might argue that Newfoundland and Labrador is already a food desert (particularly in the northern regions of Labrador). It is often stated that 90 per cent of the food consumed within the province has been produced elsewhere, and that only a small supply (3–10 days) of food is within the province at any given time, adding concern about prolonged disruptions in ferry services (Food First NL, 2015). No matter how accurate this projection, it is indisputable that the province already is highly reliant on imported food from mainland Canada and elsewhere. However, as has been noted throughout *Food Futures*, to some extent Newfoundlanders and Labradorians have relied on imports for centuries, though in recent times there has been renewed concern about the consequences of being in this state for a sustained period of time.

Elsewhere across North America and the rest of the world, locally sourced food has been promoted as an important contribution to a sustainable food system (Feenstra, 1997; Halweil, 2002; Feagan, 2007). Ostensibly, there are lower environmental impacts and costs from reduced transport requirements. Fresh, nutritionally healthy food may be supplied directly to consumers in a way that involves fewer preservatives and improved shelf life, and allows profits to be retained by local businesses. In Chapter 9, Foley and Mather note that smaller, environmentally conscious initiatives reflect a new food regime that has formed to overcome the limitations of large-scale, corporate-driven food production, and that consumers increasingly desire "food from somewhere." Foley and Mather make the case that this model also applies to fisheries and seafood, and that this food regime is presently unfolding in Newfoundland and Labrador.

Other chapters illustrate the considerable potential for locally based food production initiatives to take root and gain momentum over time. The multi-dimensional, holistic benefits are highlighted in Emily Doyle and Martha Traverso-Yepez's example of the Harbour Grace school greenhouse (Chapter 3) and by the Centreville-Wareham-Trinity community gardens, as discussed by Vodden, Keske, and Islam (Chapter 4).

Several initiatives reflect propulsion towards local food production. As previously discussed, the provincial government's efforts are an encouraging indicator of recent momentum to boost the availability of locally produced food supplies. Holistic farming and permaculture efforts that embrace regional environmental conditions also have attracted media attention (Bird, 2015). Memorial University, the Canadian Forest Service, and other funding agencies have made recent investments in laboratories and infrastructure to advance agricultural and environmental research that could theoretically be expanded at greater scale across the province (Kavanagh and Ellsworth, 2016; Keske, 2015).

Several successful research programs featured in this book provide hope that local, sustainable food production is possible, particularly for niche markets. Couturier and Rideout, in Chapter 11, chronicle decades of successful aquaculture at MUN's Marine Institute and the environmentally and nutritionally desirable outcomes of this work. The aquaculture program is poised to continue adapting to the province's food needs and to continue expanding as a global leader in aquaculture research. Debnath and Keske (Chapter 13) highlight research behind the province's berry production, which has expanded considerably during the past decade. These foods can be incorporated into household self-provisioning, and they may resonate with markets outside of Atlantic Canada. As shown by Walke and Wu (Chapter 12), honeybees and apiculture reflect a niche agricultural sector with potential to generate worldwide interest. Furthermore, the bees exemplify how some of the province's unique natural advantages could be sustainably cultivated.

As discussed in several chapters (specifically, Lowitt and Neis, Chapter 8, as well as Traverso-Yepez, Sarkar, Gadag, and Hunter, Chapter 5), the success of these initiatives greatly depends on consumers' willingness to recognize and support local production. If Newfoundland and Labrador residents exercise food sovereignty and take control of defining their own food system, local production will align with more expansive provincial and national initiatives. Ostensibly, increased production means that more food may be readily available, and there is potential to operationalize at a local level to ensure an equitable distribution of food to vulnerable populations. The right elements exist for grassroots, small-scale production to build momentum across the province.

Of course, incentives matter. Consumers respond to food prices as a function of their available income. At this writing, after several years of rising income and wealth, the province has fallen on difficult financial times in large part due to a precipitous decline in oil prices, tied to one of the province's (and Canada's) most valuable commodities. With an abrupt decline in wealth and income and the path to a rebound unclear, most households have an increased awareness of household expenditures, including food. In light of the challenges with the economy, it could be argued that there is no better time than the present for Newfoundlanders and Labradorians to redefine their food system and to create what they envision as a sustainable food system from the bottom up.

International market forces are also an important consideration. Under the renowned Hecksher-Ohlin economic model, goods like agricultural staples that require relatively more intensive resources are imported, and relatively abundant goods are exported (O'Rourke and Williamson, 1999). To ensure the availability of fresh products produced in Newfoundland and Labrador, local prices must remain competitive with international prices, and residents must have sufficient income to acquire the locally produced food. Several of the agricultural products discussed in this book (honey, berries, fish, and seafood) reflect locally available, distinctly provincial products. However, there is perpetual market tension between keeping the products local to ensure local supply and exporting the products at a higher price. If more revenue can be created through exporting products at an international price, all the while ensuring that financial benefits generated throughout the supply chain are distributed locally, does this embody the spirit of the local food movement? As discussed by Foley and Mather (Chapter 9), numerous authors have chronicled the irony of the export-based cod industry, which formed a pivotal base for the Newfoundland and Labrador food system. Have times changed so that history won't repeat itself?

GROWING A SUSTAINABLE FOOD SYSTEM IN NEWFOUNDLAND AND LABRADOR

Now that greater emphasis is being placed on sustainable food production, locally or provincially based food system initiatives may gain traction. However, to fully appreciate the potential for expanding local food production, some attention should be given to the definitions of "sustainability" and "sustainable development."

The most common definition of "sustainable development," set forth by the 1987 World Commission on Environment and Development (Brundtland Commission), is that the needs of today's generation are met without compromising the ability of future generations to meet their needs. Food and water are quintessential examples of resources consumed on a daily basis. The management of these resources requires consideration today, as well as for future generations. So, how does a society sustainably grow its food sector for the future? Daly, Cobb, and Cobb (1994) and others draw attention to what is known as "three pillars of sustainability," reflecting human, natural/environmental, and financial/physical capital. Thus, a sustainable food system involves ensuring that human, environmental, and physical capital is maintained in the present as well in the future.

Several studies offer guidance on how to measure sustainable food production and agricultural systems (Figge and Hahn, 2004, 2005; Van Passell et al., 2009; Hou et al., 2014), although in reality the practice remains relatively academic. Furthermore, for basic necessities like food, ensuring its availability in the future may be compromised by the ongoing need for immediate consumption. Despite difficulties, the message of ensuring a balance among human, natural/environmental, and financial/physical capital is also an appropriate one to consider. So, too, are the specific targets within each of the pillars. For example, should efforts to ensure that food is available to all ("human capital") prioritize the most vulnerable populations? Should subsets within these populations, such as children who require protein for long-term brain development, be prioritized? What effect does this weighting system have on consumers or the food production system as a whole and what imbalance may result?

The environmental trade-offs incurred by food production also merit reflection. Several chapters address the environmental trade-offs arising from increased agrifood production, with Couturier and Rideout (Chapter 11) presenting trade-offs in the context of life-cycle assessment (LCA) studies and economic considerations. Though greenhouse gas emissions are relatively low for aquaculture compared to terrestrial agriculture, there are downside ecosystem risks from introduced species and intensified production. Many projects are "environmentally sustainable" in some ways, but miss other environmental targets. In other specific examples presented in the book, pesticides associated with increased agricultural production may damage the natural advantages of Newfoundland bees. Hydroelectric power generation delivers low greenhouse gas emissions, but potentially imposes other environmental consequences from its infrastructure development, including methylmercury accumulation in fish. As Foley and Mather (Chapter 9) and Myron King (Chapter 10) discuss, a sustainable approach to food production must also address environmental sustainability and the well-being of farmers and fishers. The trade-offs between transferring environmental capital to human capital in an unsustainable manner have been addressed by many authors, though it is worth emphasizing here that the long-term ecological impacts and consequences from overharvesting will not be known for years to come. The transfer between environmental and human capital also applies to provisioning ecosystem services such as agricultural production, where the long-term human impacts on the ecosystem may be felt (or otherwise not be known) for generations.

This brings us to our concluding, overarching point about how to successfully create the conditions for a sustainable food system in Newfoundland and Labrador. As has been discussed throughout the book, it is critical for local communities to define and advance what they envision as a sustainable food system. We have shown that momentum and interest are building through non-profit organizations such as Food First NL, university-based research and engagement efforts such as FARM, and initiatives funded by the provincial government.

However, a successful food movement requires local engagement throughout the remote areas of the province as well as within urban centres. Newfoundlanders and Labradorians must, quite literally, grow their own food culture. Growing our own food culture hints at increased empowerment over our food lives, and provides opportunity to cultivate and innovate local food production strategies to ensure that healthy, nutritious, culturally appropriate food is distributed to all. In other words, food sovereignty will help Newfoundland and Labrador attain food security.

In conclusion, much like generations that have come before ours, navigating the elements will likely continue in the years to come. So, too, will the important role that communities play in ensuring that everyone eats and that there is a sustainable food system for "The Rock."

NOTES

1. For more information about several stellar Food First NL projects, see the Food First NL website at: https://foodfirstnl.squarespace.com/about-us/.

REFERENCES

Agriculture and Agri-Food Canada. "Growing Forward 2." At: http://www.agr.gc.ca/eng/about-us/key-departmental-initiatives/growing-forward-2/?id=1294780620963.

Bird, L. 2015. "An island orchard: How one couple is working towards Newfoundland food security." CBC News, 7 Nov. At: http://www.cbc.ca/news/canada/newfoundland-labrador/crow-brook-orchard-increasing-food-security-1.3303953.

CBC News. 2017. "Crops on Crown countryside: N.L. farmers get access to more land: Premier Dwight Ball announces additional 64,000 hectares." At: http://www.cbc.ca/news/canada/newfoundland-labrador/more-crown-land-farmers-premier-dwight-ball-1.3985567.

Daly, H. E., J. B. Cobb, and C. W. Cobb. 1994. *For the Common Good: Redirecting the Economy toward Community, the Environment, and a Sustainable Future.* Boston: Beacon Press.

Davey, W. J., and R. MacKinnon. 2016. *Dictionary of Cape Breton English.* Toronto: University of Toronto Press.

Discovery Television. *Alaskan Bush People.* At: http://www.discovery.com/tv-shows/alaskan-bush-people/.

———. *Yukon Men.* At: http://www.discovery.com/tv-shows/yukon-men/.

Feagan, R. 2007. "The place of food: Mapping out the 'local' in local food systems." *Progress in Human Geography* 31, 1: 23–42. doi: 10.1177/0309132507073527.

Feenstra, G. W. 1997. "Local food systems and sustainable communities." *American Journal of Alternative Agriculture* 12, 1: 28–36. doi:https://doi.org/10.1017/S0889189300007165 .

Figge, F., and T. Hahn. 2004. "Sustainable value added — measuring corporate contributions to sustainability beyond eco-efficiency." *Ecological Economics* 48, 2: 173–87. doi: 10.1016/j.ecolecon.2003.08.005.

——— and ———. 2005. "The cost of sustainability capital and the creation of sustainable value by companies." *Journal of Industrial Ecology* 9, 4: 47–58. doi:10.1162/108819805775247936.

Fitzpatrick, A. 2015. *Nuclear Seed Potato Propagation Facility 2014–2015 Final Report.* At: http://www.faa.gov.nl.ca/publications/pdf/NSPPF_2014_15.pdf.

Food First NL. 2015. "Everybody eats: A discussion paper on food security in Newfoundland and Labrador." Nov. At: http://www.foodfirstnl.ca/our-projects/everybody-eats.

Forestry and Agrifoods, NL Government. 2016. At: http://economics.gov.nl.ca/E2016/ForestryAndAgrifoods.pdf.

Halweil, B. 2002. *Home Grown: The Case for Local Food in a Global Market.* Washington: Worldwatch Institute.

Hou, L., D. Hoag, C. M. Keske, and C. Lu. 2014. "Sustainable value of degraded soils in China's Loess Plateau: An updated approach." *Ecological Economics* 97: 20–27. doi:10.1016/j.ecolecon.2013.10.013.

Kavanagh, V., and S. Ellsworth. 2016. Personal written communication, Apr. Agrifood Development Branch, Government of NL.

Keske, C. M. H. 2015. "Food security, food sovereignty, and the agricultural supply chain in Newfoundland and Labrador." Department of Economics Visiting Speaker Series, Support for Scholarship in the Arts, by Vice President Academic, St. John's, 30 Jan. doi:10.13140/RG.2.1.3909.9764. At: https://www.researchgate.net/publication/303518068_Food_security_food_sovereignty_and_the_agricultural_supply_chain_in_Newfoundland_and_Labrador.

Larson, K., and J. Guilliland. 2008. "Mapping the evolution of 'food deserts' in a Canadian city: Supermarket accessibility in London, Ontario, 1961–2005." *International Journal of Health Geographics* 7, 16: 1–16. doi:10.1186/1476-072X-7-16.

Murton, J., D. Bavington, and C. Dokis, eds. 2016. *Subsistence under Capitalism*. Montreal and Kingston: McGill-Queen's University Press.

National Geographic. *Life Below Zero*. At: http://channel.nationalgeographic.com/life-below-zero/.

Newfoundland and Labrador. 2016. *The Way Forward: A Vision for Sustainability and Growth in Newfoundland and Labrador*. At: https://www.gov.nl.ca/pdf/the_way_forward.pdf.

O'Rourke, K. H., and J. G. Williamson. 1999. *Globalization and History: The Evolution of a Nineteenth Century Atlantic Economy*. Cambridge, Mass.: MIT Press.

Raja, S., C. Ma, and P. Yadef. 2008. "Beyond food deserts: Measuring and mapping racial disparities in neighborhood food environments." *Journal of Planning Education and Research* 27, 4: 469–82. doi:10.1177/0739456X08317461.

Van Passel, S., G. Van Huylenbroeck, L. Lauwers, and E. Mathijs. 2009. "Sustainable value assessment of farms using frontier efficiency benchmarks." *Journal of Environmental Management* 90, 10: 3057–69. doi:10.106/j.jenvman.2009.04.009.

World Commission on Environment and Development (Brundtland Commission). 1987. *Our Common Future: Report of the World Commission on Environment and Development*. Oxford: Oxford University Press.

Contributors

Karine Bernard is a Professional Dietician and Master of Science candidate in Community Health and Humanities at Memorial University of Newfoundland, St. John's.

Cyr Couturier is Research Scientist and Chair of Aquaculture Programs at the Fisheries and Marine Institute, Memorial University of Newfoundland, St. John's.

Samir C. Debnath is a Research Scientist for Agriculture and Agri-Food Canada at St. John's Research and Development Centre, St. John's.

Emily Doyle is a Ph.D. candidate in the Division of Community Health and Humanities, Faculty of Medicine, Memorial University of Newfoundland.

Paul Foley is Assistant Professor in theSchool of Science and the Environment, Memorial University, Grenfell Campus.

Veeresh Gadag is Professor of Biostatistics, Division of Community Health and Humanities, Faculty of Medicine, Memorial University of Newfoundland.

Kelly Hunter received her M.Sc. in Community Health from the Faculty of Medicine, Memorial University of Newfoundland.

Jannatul Islam received his MA in Environmental Policy at Memorial University of Newfoundland, Grenfell Campus.

Catherine Keske is Associate Professor of Management at the University of California-Merced, and Adjunct Professor in the School of Science and Environment at Memorial University of Newfoundland, Grenfell Campus.

Myron King is a Research Associate at the Environmental Policy Institute, Memorial University, Grenfell Campus and a Ph.D. student at the University of Hull, Hull International Fisheries Institute.

Kristen Lowitt is Assistant Professor in the Department of Geography and Environment at Mount Allison University.

Charles Mather is Associate Professor in the Department of Geography, Memorial University of Newfoundland, St. John's.

Barbara Neis is a University Research Fellow in the Department of Sociology at Memorial University, St. John's.

Lynne Phillips is Professor of Anthropology and former Dean of Faculty of Humanities and Social Sciences, Memorial University of Newfoundland, St. John's.

Keith Rideout is Coordinator of Programs, School of Fisheries, Fisheries and Marine Institute, Memorial University of Newfoundland, St. John's.

Sharon R. Roseman is Professor of Anthropology and Associate Dean (Research and Graduate Programs) in the Faculty of Humanities and Social Sciences, Memorial University of Newfoundland, St. John's.

Diane Royal is a Master's candidate in Anthropology at Memorial University of Newfoundland, St. John's.

Atanu Sarkar is Associate Professor of Environmental and Occupational Health, Division of Community Health and Humanities, Faculty of Medicine, Memorial University of Newfoundland.

Rebecca Schiff is Associate Professor in the Department of Health Sciences, Lakehead University.

Adrian Tanner is Honorary Research Professor in Anthropology at Memorial University of Newfoundland.

Martha Traverso-Yepez is Associate Professor of Health Promotion and Social Determinants of Health, Division of Community Health and Humanities, Faculty of Medicine, Memorial University of Newfoundland.

Kelly Vodden is Vice-President (Research) at Memorial University of Newfoundland, Grenfell Campus.

Stephan Walke earned his Bachelor of Sustainable Resource Management degree at Memorial University, Grenfell Campus.

Jianghua Wu is Associate Professor, Sustainable Resource Management, in the School of Science and the Environment, Memorial University, Grenfell Campus.

Yellow pepper heart. (Photo by Kimberley Devlin)